Hot-melt Extrusion: Pharmaceutical Applications

Hot-melt Extrusion: Pharmaceutical Applications

Edited by

DENNIS DOUROUMIS
School of Science, University of Greenwich, UK

WILEY
A John Wiley & Sons, Ltd., Publication

This edition first published 2012

Registered office
John Wiley & Sons Ltd, The Atrium, Southern Gate, Chichester, West Sussex, PO19 8SQ, United Kingdom

For details of our global editorial offices, for customer services and for information about how to apply for permission to reuse the copyright material in this book please see our website at www.wiley.com.

Library of Congress Cataloging-in-Publication Data

A catalogue record for this book is available from the British Library.

ISBN: 9780470711187

Set in 10/12pt Times by Aptara Inc., New Delhi, India.

Dedication

As you set out for Ithaca
hope your road is a long one,
full of adventure, full of discovery.
Laistrygonians, Cyclops,
angry Poseidon – don't be afraid of them:
you'll never find things like that on your way
as long as you keep your thoughts raised high,
as long as a rare excitement
stirs your spirit and your body.

ITHACA (Konstantinos Kavafis, 1911)

This book is dedicated to my wonderful wife Eleni-Angeliki and my lovely son George-Alexander and daughter Eugenia-Erene. I thank them for their continuous support and patience.

Contents

List of Contributors

Jessica Albers, Evonik Industries AG, Kirschenallee, 64293 Darmstadt

Ana Almeida, Laboratory of Pharmaceutical Technology, Ghent University, Harelbekestraat 72, B-9000 Gent, Belgium

Joshua Boateng, University of Greenwich, School of Science, Medway Campus, Central Avenue, ME4 4TB, Kent, UK

Bart Claeys, Laboratory of Pharmaceutical Technology, Ghent University, Harelbekestraat 72, B-9000 Gent, Belgium

Dennis Douroumis, University of Greenwich, School of Science, Medway Campus, Central Avenue, ME4 4TB, Kent, UK

Adam Dreiblatt, Century Extrusion, 2412 W. Aero Park Ct., Traverse City, MI 49686 USA

Tom Geilen, Thermo Fisher Scientific, Dieselstrasse 4, 76227 Karlsruhe, Germany

Thobias Geissler, Thermo Fisher Scientific, Dieselstrasse 4, 76227 Karlsruhe, Germany

Costas G. Gogos, Department of Chemical, Biological, and Pharmaceutical Engineering New Jersey Institute of Technology, Newark, NJ, USA

Andreas Gryczke, Ernst-Ludwig-Straße 19a, 64560 Riedstadt, Germany

Sandra Guns, Laboratory of Pharmacotechnology and Biopharmacy, Catholic University of Leuven Campus Gasthuisberg O & N2, Herestraat 49, 3000 Leuven, Belgium

Abhay Gupta, FDA-CDER, Division of Product Quality Research White Oak Life Sciences Building 64, 10903 New Hampshire Ave, Silver Spring, MD 2099, USA

Mark Hall, The Dow Chemical Company, Midland Michigan, US

Chris Heil, Thermo Fisher Scientific, 5225 Verona Rd, Madison, WI 53711 USA

Pernille Høyrup Hemmingsen, Egalet Ltd, DK-3500 Værløse, Denmark

Jeffrey Hirsch, Thermo Fisher Scientific, 5225 Verona Rd, Madison, WI 53711 USA

Masoor A. Khan, FDA-CDER, Division of Product Quality Research White Oak Life Sciences Building 64, 10903 New Hampshire Ave, Silver Spring, MD 2099, USA

Dirk Leister, Thermo Fisher Scientific, Dieselstrasse 4, 76227 Karlsruhe, Germany

Huiju Liu, Department of Chemical, Biological, and Pharmaceutical Engineering New Jersey Institute of Technology, Newark, NJ, USA

Andrew Loxley, Particle Sciences Inc., 3894 Courtney St #180, Bethlehem PA 18017, USA

Keith Luker, Randcastle Extrusion Systems, Inc., 220 Little Falls Rd. Unit 6 Cedar Grove, NJ 07009

Guy Van den Mooter, Laboratory of Pharmacotechnology and Biopharmacy, Catholic University of Leuven Campus Gasthuisberg O & N2, Herestraat 49, 3000 Leuven, Belgium

Kathrin Nollenberger, Evonik Industries AG, Kirschenallee, 64293 Darmstadt

Michael Read, The Dow Chemical Company, Midland Michigan, US

Martin Rex Olsen, Egalet Ltd., DK-3500 Værløse, Denmark

Jean Paul Remon, Laboratory of Pharmaceutical Technology, Ghent University, Harelbekestraat 72, B-9000 Gent, Belgium

Geert Verreck, Janssen Research & Development, Turnhoutseweg 30, 2340 Beerse, Belgium

Chris Vervaet, Laboratory of Pharmaceutical Technology, Ghent University, Harelbekestraat 72, B-9000 Gent, Belgium

Peng Wang, Department of Chemical Engineering University of Rhode Island, Kingston, RI, USA

Preface

Hot-melt Extrusion (HME) is an emerging continuous processing technology for the development of various solid dosage forms and drug delivery systems. In the last 20 years HME has attracted increased attention from both the pharmaceutical industry and academia. The enormous need for new dynamic manufacturing processes to produce robust finished products makes HME an excellent technology. Although there are several publications on HME applications, this is the first attempt to provide a concrete overview of HME pharmaceutical applications.

The aim of this book is to present a comprehensive review of the theory, instrumentation and wide spectrum of applications. The book is targeted at scientists in academia and industry and graduate students in various research-intensive programs in pharmaceutical sciences and medicine who are dealing with many aspects of drug formulation and delivery, pharmaceutical engineering and processing and polymers and materials science.

Chapters 1 and 2 discuss single- and twin-screw extrusion operational principles, design and critical processing parameters. Chapter 3 is an overview of HME developments in pharmaceutics, and discusses a number of drug delivery systems and physicochemical characterization techniques of HME extrudates. Chapters 4 and 5 deal with theoretical approaches of drug–polymer miscibility estimation and discuss the role, influence and selection of plasticizers in the HME process. Chapters 6 and 7 provide in-depth knowledge of drug products extruded by a wide range of polymers and their applications. More detail is provided in Chapter 8 where the application of HME for the manufacture of thin films is discussed. Chapter 9 is dedicated to the employment of HME for the taste-masking of bitter APIs, and discusses the selection of various excipients for these purposes.

Chapter 10 includes a comprehensive discussion of clinical studies performed by various groups, bioavailability and pharmacokinetics of oral immediate release, oral controlled release and implants. The relatively new manufacturing process of injection molding is introduced in Chapter 11, and aspects such as critical process parameters, excipients, new products and their properties are critically analyzed. A comprehensive discussion of dispersive and distributive mixing is included in Chapter 12 and case studies are presented.

The reader can find important information in Chapter 13 about the scale-up of the hot-melt extrusion process from a lab-scale extruder to a commercial-scale extruder, as well as different scale-up scenarios. Novel applications of HME for the manufacturing of devices and implant systems can be found in Chapter 14, including examples of marketed products.

Chapter 15 is an FDA perspective on HME product and process understanding with special attention given to Quality by Design (QbD) as a tool to understanding HME processing. Finally, Chapter 16 introduces a process analytical technology (PAT) approach by using near-infrared spectroscopy for understanding and controlling the hot-melt extrusion process in the pharmaceutical industry.

I would like to acknowledge the valuable support and cooperation of all the contributing authors throughout this process, to whom I offer a most sincere thank you. Without their dedication and timely submission of material, this book would not have been published.

Dennis Douroumis

1

Single-screw Extrusion: Principles

Keith Luker
Randcastle Extrusion Systems, Inc.

1.1 Introduction

Until recently, single-screw extruders (SSE) have little changed in principle since their invention around 1897. They are mechanically simple devices. A one-piece screw, continuously rotated within a barrel, develops a good quality melt and generates high stable pressures for consistent output. These inherent characteristics, combined with low cost and low maintenance, make it the machine of choice for the production of virtually all extruded products.

Historically, the polymers and particulate they carry (including active pharmaceutical ingredients or API) are subjected to compressive shear-dominated deformation. Compression of particulates, such as API, forces the particulate together into agglomerations under very high pressure *before and during melting*. When this happens, shear deformation is insufficient to break the agglomerations into their constituent parts. Agglomerations within a polymer matrix define a poorly mixed product.

Many ingenious schemes are known to improve the basic screw. Since the 1950s, a variety of mixers have been available. Some of these force material into small spaces for additional shearing. Some divide the flow into many streams so that smaller masses are sheared more effectively. Some make use of pins embedded in the root of the screw and some cut the screw flights. They have one thing in common that limits their effectiveness, however: they are placed after the screw melts the material, and most of a screw is necessarily dedicated to producing a melted polymer. Typically, these mixers are less than four screw diameters long.

Hot-melt Extrusion: Pharmaceutical Applications, First Edition. Edited by Dennis Douroumis.

Since around the 1970s, various barrier or melt separation screws became widely available. These force material over a barrier flight of reduced dimension (compared to the main flight), preventing unmelted material from moving downstream. As the material moves over the barrier flight, it receives additional shearing and is therefore mixed a little bit better. Some screws force material back and forth across barriers which also slightly improves the SSE mixing.

To some degree, all of these inventions are incrementally successful. However, they do not change the fundamentals of compression and shear dominance in the SSE. Until recently, the SSE was therefore an agglomerating machine.

Meanwhile, the twin-screw extruder (TSE), and in particular the parallel intermeshing co-rotating TSE[1], became the dominant continuous compounding mixer for polymers and particulate. This is because it works on a fundamentally different and better principle: *It melts prior to the final compression of the melt.* This means that it prevents agglomeration of the ingredients and has no need to then break up agglomerates formed by compression. Fundamentally, it is not shear dominated. Instead, material moving through the intersection of the screws is extended. Such deformation is elongational. Elongation, instead of pushing API particles together, *pulls them apart*. Unlike the SSEs discussed above, the TSE mixers do not start mixing near the end of the screw. They do not dedicate just a few length-over-diameter or *L*/*D* ratios to mixing; instead, they combine elongational melting and mixing early in the extruder in a first set of kneaders and then repeat the elongational melt-mixing process with additional kneaders. In this way, a substantial part of the TSE length is dedicated to elongational melt-mixing.

However, the TSE has flaws. Not all the material moves through the intermesh region; some material escapes down the channels without moving through the extensional fields. In addition, some material will see the intermesh many times. The key elongational history of the polymer and API will therefore be uneven. Compared to single screws, the TSE is less pressure stable; compared to singles, the TSE does not generate high pressures. (When a gear pump is used to generate high stable pressures they require a sophisticated algorithm that is sensitive to small changes, especially in the starve feeding system.)

Very recently, significant advances in fundamental SSE technology have changed the landscape. Costeux et al. proved in 2011 [1] that the SSE could have dominant elongational flow where melting occurred before compression. There is therefore no need to break up agglomerates. Unlike the TSE, all the material can consistently pass through the elongational mixers. Melting and mixing are started very near the hopper so that a significant part of the total length of the SSE becomes a mixer. These new SSEs retain their advantages of simplicity and low cost. They can still generate high and stable pressures most suitable for hot-melt extrusion (HME) production, even when starve fed without a complex control system.

1.2 Ideal Compounding

In order to understand the SSE for HME, we must understand compounding as we will necessarily have at least an API and a polymer. It is undesirable to have local concentrations

[1] Since this particular TSE dominates the market, the use of TSE throughout this chapter should be understood to mean the parallel intermeshing type.

of API or polymer in the product. Compounding is defined as combining two or more ingredients, but really good compounding has additional requirements. The melt-mixing process should treat the material equally. It should not be overly mixed in one region and under-mixed in another. Mixing should apply the least amount of energy to limit degradation of the components.

Compounding is accomplished by taking local concentrations and reducing them to a satisfactory size where satisfaction depends on the use. This is accomplished by dispersion (breaking solids or globules into smaller concentrations) or distribution (rearrangement of solids or melt).

Local concentrations will occur when polymer pellets are dry-mixed with API. Each pellet is a local concentration that must be distributed to incorporate the API. The API can also be thought of as a local concentration that must be distributed within the polymer pellets. Local concentrations are immediately reduced when working with a powder/powder blend (compared to pellet/powder). The better the mixture, the easier it is for an extruder to further reduce the local concentrations. Nevertheless, no matter how well mixed two powders are, there will be local concentrations at some scale. The job of the extruder is to further reduce these concentrations. This cannot be accomplished through a purely compressive screw since that takes the mixture and, at best, maintains the dry-mix quality[2]. Instead, elongation is required to draw the concentrated regions apart.

An ideal HME mixer would maintain ingredient quality during the compounding process. Both plastics and API degrade due to thermal and mechanical stress. To mix well, there should be an orderly progression through the mixing process that maintains the quality of the ingredients.

Thermally, a single heat history of the shortest possible duration at the lowest temperature is preferred.

Mechanically, an elongationally dominated system, *where all the material has the same elongational history*, is preferred. This will minimize unnecessary mechanical degradation and decrease the thermal processing time to achieve the same result. Since the shear component of the mechanical system builds excessive heat (compared to the elongational component), it should be minimized.

1.3 Basics of the Single-screw Extruder

Low bulk density polymer solids, often mixed with various forms of particulate (such as API), most commonly fall from a hopper into a long, continuously rotated extruder screw within a temperature controlled barrel, as depicted in Figure 1.1.

The screw forces the solid material into a decreasing space along the screw at higher temperatures. There the compressed material is pushed up against the heated container (the barrel). The compression both forces the air out of the hopper and melts the material by pushing the material against the hot metal barrel. The dense/molten material is continuously pumped forward through a shaping die. The material exits the die where it is drawn down in a free molten state through a cooling medium until solid while continually pulled.

[2]Compressive screws, in some circumstances, can take an orderly mix and agglomerate the ingredients. This will occur when the act of squeezing the mixture separates the ingredients. This is more common than is generally realized.

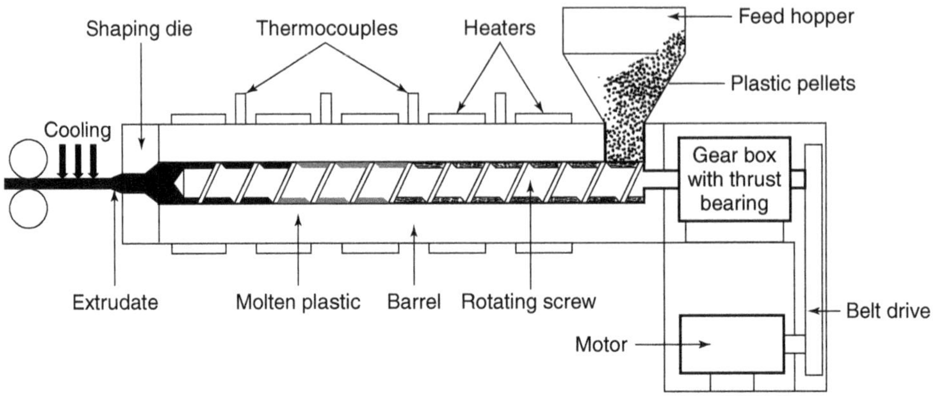

Figure 1.1 Horizontal SSE driven at feed end.

The key to the process is the extruder screw. While many variations can be considered, the classic screw has a constant diameter. The modern screw length is usually 24–50 times its diameter. This is expressed as the length-over-diameter ratio or *L/D* ratio. Screws are, most commonly, made from a solid piece of steel leaving a screw root that is polished. The flights are ground and fit closely within the barrel. Figure 1.2 depicts a general-purpose polymer 24/1 *L/D* screw.

Typically, the one-piece screw is driven from the right through a simple key on a shaft that fits into the gear reducer of the extruder. The general-purpose screw has a flight pitch equal to the screw diameter. A classic general-purpose screw has three parts (the feed, the compression and the metering section), all of equal length.

Most HME extruders are small and many are used for research and development. Sometimes, the very high API cost prohibits use by any but the smallest SSEs.

Plastic pellets for HME extruders are made in bulk. They are the same size for all extruders, often in the range 0.13–0.19 inch. This means that the channel depth must be sufficiently large for pellets to fit. Otherwise, the pellets will jam when entering the screw and such jamming can break the screw. Screws with a 0.18-inch feed channel depth are

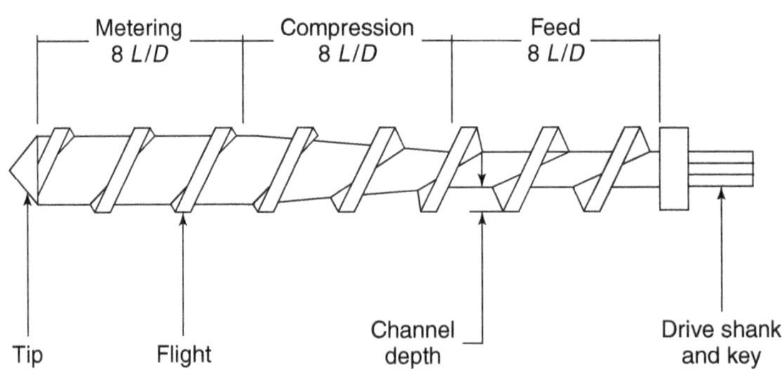

Figure 1.2 A 24/1 L/D general-purpose screw.

recommended. Extruders for the drive system above are available in sizes as small as 1 inch diameter. Smaller screws with a feed-driven drive system become so weak that even a small upset (such as a single oversized pellet) can cause the screw to break.

For many HME applications, 1 inch extruders are too large because of the cost of formulation in the research and development phase. Smaller screws are available with the recommended 0.18-inch feed channel depth in combination with a second type of drive system.

In Figure 1.2, note that the root diameter of the metering section is much larger than the root of the feed section. The strength of the root increases with the cube of diameter. If the screw is driven through the metering section, the screw itself becomes much stronger since the highest torque is transmitted through the biggest root diameter (as depicted in Figure 1.3). This allows screws as small as 0.625 inch diameter. Since the output of an extruder screw decreases with the square of the diameter, a 0.625 inch extruder will only have one-quarter of the output of a 1-inch diameter extruder, a good thing when ingredients are scarce or expensive.

To prevent material from leaking into the gearbox, a special seal is used (see Figure 1.3). This seal has a reverse flight compared to the main part of the screw. By making the pitch and channel depth small, it becomes much more powerful than the main screw and the HME material itself becomes the seal. The seal is best with two channels, one deep channel and one shallow. Material moves from the deep channel (the fill length is longer for a deep channel) and into the shallow channel and back into the main flow, preventing stagnation.

Such seals are particularly useful for pressure stability. Since they act as an accumulator, they suppress small surges or variations in pressure and output. Once the small surges are dampened, it is then possible to use an automatic pressure controller to maintain the pressure as fine as ± 10 psi in the barrel [2].

The discharge-driven design is commonly made as a vertical machine. This exposes the screw to the feedstock for easy entry into the screw.

Below the 0.625-inch diameter, screws usually become too weak for the most common pellet sizes. However, smaller extruders are made including 0.5, 0.375 and even 0.25 inch screws. These are built for free-flowing powders or ground pellets. This puts the general lower limits for HME at about 10 grams per hour.

1.3.1 Screw Feed Section

Referring to Figure 1.2, the 'feed' channel depth (and so the root diameter) is constant and is associated with solids conveying. Temperatures in the feed section are therefore usually set below the melting temperature of the plastic[3]. If the temperatures are set too high, then the material will melt and conveyance (feeding) will stop as material will stick to the screw rather than be conveyed by it.

Pressure in the feed is usually very low and often zero as the screw is acting as a conveyor. Solids conveying needs little torque so accounts for a small percentage of the total motor load.

[3]It is important to realize that the temperature is not directly controlled along the entire length of the screw; only the barrel surrounding the screw is directly temperature controlled.

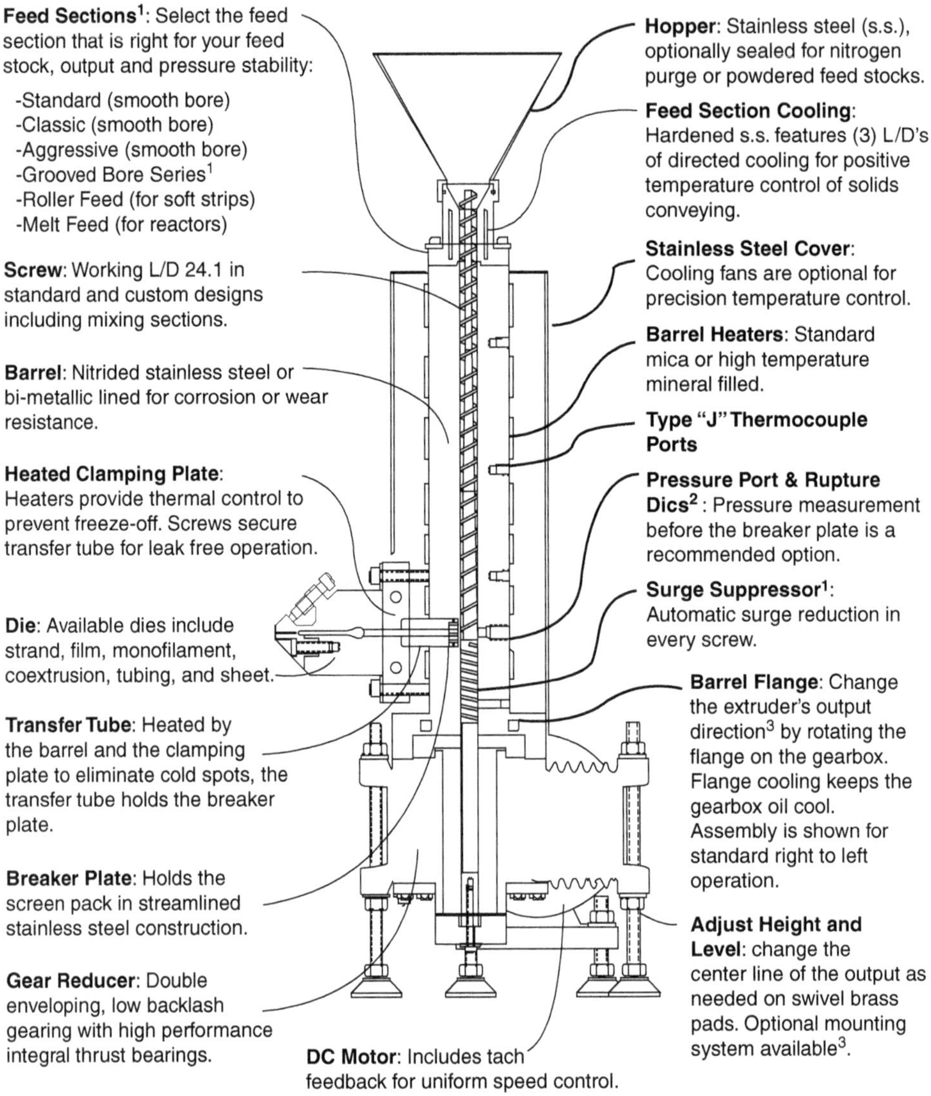

Figure 1.3 Discharge-driven vertical extruder.

Temperature changes to the feed can cause complex changes in extrusion behavior; they can change the barrel friction, which is the driving force for material transport along the screw which is resistive to flow. Feed temperature is therefore usually optimized to control the solids conveying. This provides the most uniform pressure stability, typically measured at the barrel discharge. When the pressure is unstable, the extruder is said to 'surge' because changes in barrel pressure cause changes in output.

Changes in the feed section temperature of the screw change the amount of preheating of the solids. Higher preheat temperatures (especially in smaller extruders) can mean easier

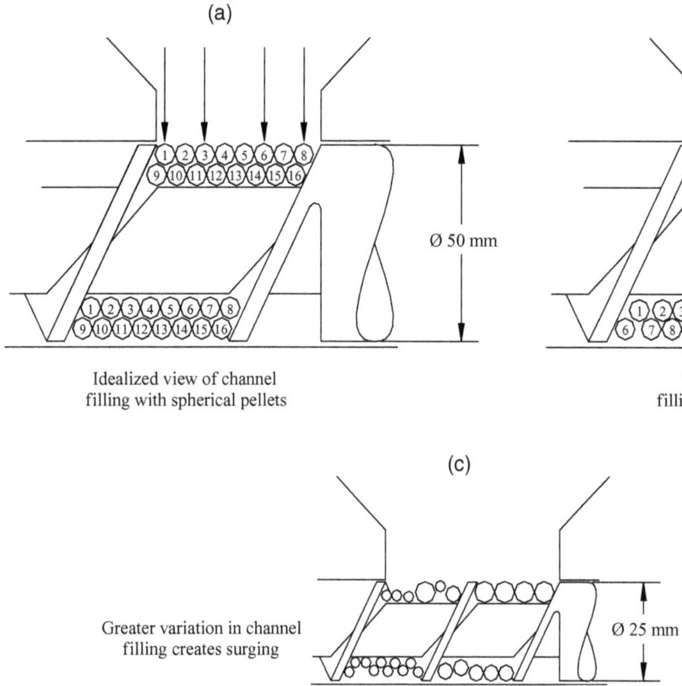

Figure 1.4 *Idealized feeding in small screws.*

material deformation within the decreasing channel of the melting zone, and lower torque which we see as a reduction in motor load. Uniform feeding should not be assumed.

In Figure 1.4a, an idealized (but completely unrealistic) view is shown with each of the perfectly spherical pellets falling perfectly into the screw channel. Figure 1.4b takes a step in an more likely direction, showing gaps between pellets and a small difference in the number of pellets top and bottom of 12–13 (which is not insignificant). Whatever variation exists in the first part of the screw is transferred downstream to some degree.

Figure 1.4c shows a more realistic variation in pellet size. One need only look at a handful of typical pellets to see significant variations in diameter; often the largest is twice the diameter of the smallest. If the pellets were spheres, the mass of the largest pellets would be eight times greater than the smallest pellets. Figure 1.4c shows a large difference (not atypical) in mass in the two flights, as will occur over time. In a large extruder, the average mass is not greatly affected by relative size of the pellets. In smaller extruders, it will cause wide variations in mass transfer and make surging more likely.

Other problems include arching or bridging over the opening, as depicted in Figure 1.5a, and where feeding will cease. This problem can be lessened by adding a stirrer as in Figure 1.5b. It is not only pellets that can arch over the opening but also powders. Ideally, HME materials should be free flowing so that they can easily move into the screw channel. When poured onto a flat plate, they should have an angle of repose of less than 45° [3]. If

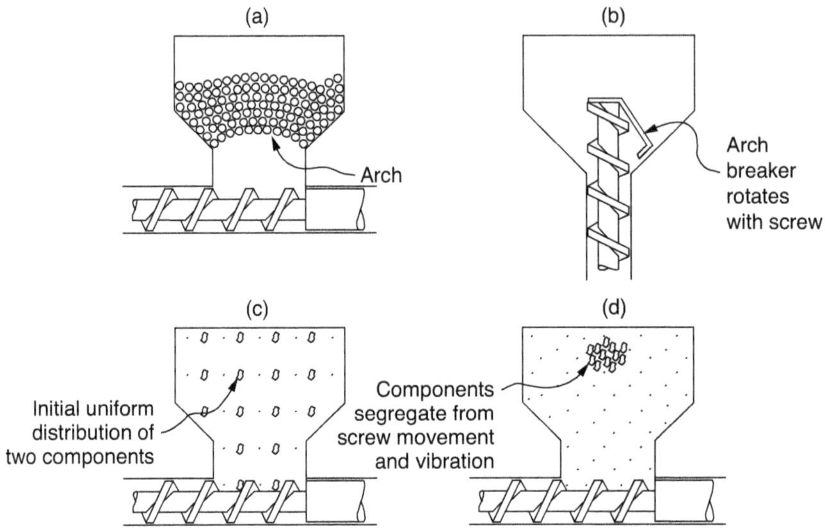

Figure 1.5 *Feedstock arching and material separation.*

the angle is much greater, there will be a tendency to compact within the feed section of the barrel or the hopper.

It should not be assumed that a uniform mixture in the hopper will reach the screw channel in the same uniform mixture. Preferential feeding (because of ingredient size or geometry), the motion of the screw or vibration can change a perfectly uniform mixture (Figure 1.5c) and de-mix it before it can reach the screw channel. This can be solved by starve feeding, that is, reducing the input to the screw to less than the maximum it can take by gravity, with multiple feeders. This will reduce the quality of input mixture, which puts greater demands on the extruder compounding.

Once the material reaches the channel, we can consider solids transportation (Figure 1.6). Once the material has moved into the screw channel, it is propelled down the screw by friction. However, it is not the friction of the screw that drives the material forward but the friction of the barrel. Consider the purely theoretical situation depicted in Figure 1.6, where the screw is stationary and the barrel revolves around the screw. It is clear that the friction of the barrel contacts the mixture and would drag it in a circle but, since the screw flights are pitched, the material is dragged forward against the screw.

This is not a very positive system when compared to positive displacement pumps, and is easily defeated. Since higher barrel than screw friction is required, anything that reduces barrel friction will change or even defeat transport. For example, the addition of a small amount of a slippery liquid to a pellet/API mixture will lubricate the barrel. The resistance of the screw can then be higher than the barrel friction and forward motion will stop.

Another important consideration for HME extrusion is the density of the feedstock. Pellets are the most common feedstock for extrusion. For HME extrusion however, there is a natural desire to work with ground or powdered polymers so that the input mixture is better mixed. This almost always reduces the bulk density of the feedstock as ground

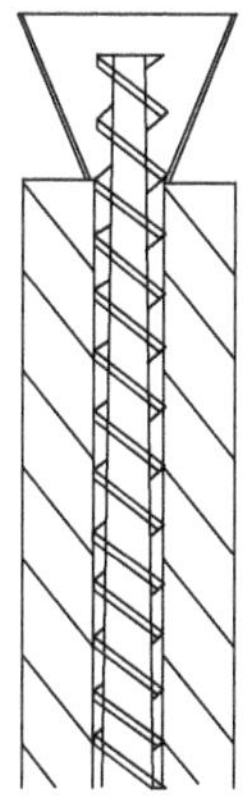

Conveyance occurs when barrel friction on the material mixture is higher than the screw's resistance.

Figure 1.6 *Consider friction when visualizing a stationary screw and rotating barrel.*

material is irregular and does not stack neatly into a dense form like a pellet. If we imagine a lowered bulk density mixture entering the screw channel as in Figure 1.6, the barrel friction is again reduced until it becomes so low that there will be insufficient friction to overcome the resistance of the screw. Flow will stop when barrel slippage is complete.

While the feed section is particularly sensitive to changes in friction, material is driven forward by friction in all parts of the extruder. It should be recognized that friction is poorly understood and that within the HME extruder friction is very complex, changing with temperature and local conditions.

1.3.2 Screw Compressor Section

The second third of the general-purpose screw, also called the transition section, has a channel depth that diminishes over its length to one-third of the feed depth. It is then said to have a 3:1 apparent compression ratio (ACR). This part of the screw is associated with melting and removal of air which is pushed out the hopper. Temperatures are typically set to allow ready deformation of the material. This is necessary because the space along the channel is decreasing and, if temperatures are too low, material will jam in the screw. Most such jams are temporary but this stopping and starting is largely responsible for surging. The jamming is reflected in higher screw torque, which we see as higher motor load.

The ACR must be high enough to squeeze out the air from the feedstock, but not so high that it prevents material from flowing to the next section.

The lower bulk density of some HME mixtures also requires a change to the screw. Since there is a lower bulk density in the feed flights, there is also less mass. A 3:1 ACR is general-purpose only because pellets are the most common feedstock. These dense pellets will generally fill the screw once the air is removed during compression and create a stable flow. However, once the bulk density is reduced as described, the 3:1 screw will no longer deliver sufficient material to fill the screw in a stable fashion and the output is likely to surge. For reduced bulk density powder/powder mixtures, a 4:1 ACR is typical.

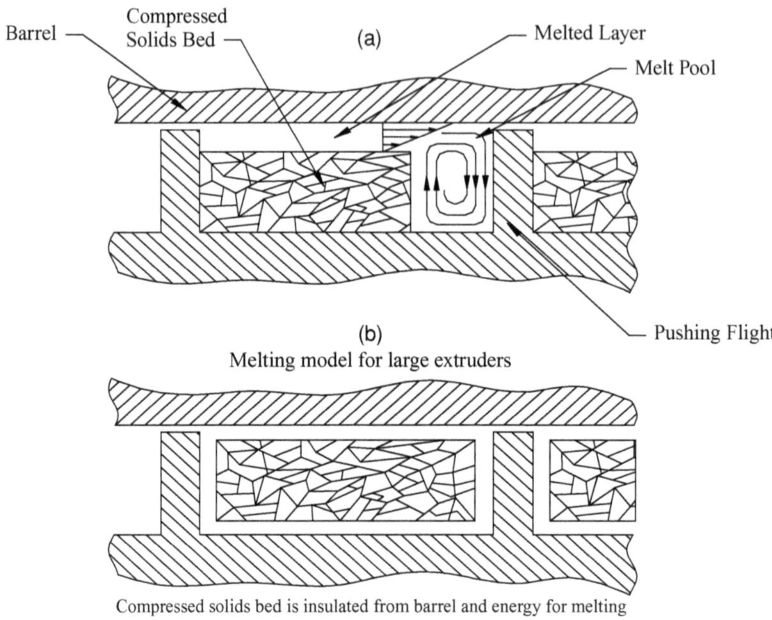

Figure 1.7 *Melting model in SSE.*

As material is compressed, it is pushed up against the metal barrel and forms a solids bed. It is easy to see that API particulate would be compressed into agglomerates between parts of the solids bed. The portion of the solids bed that is pushed up against the heated metal quickly transfers its energy to the polymer–API mixture and a melted layer forms. Since the polymer beneath this layer is insulated by the melted material itself, heat transfer to the rest of the mass in the channel is slowed. Before it can melt, it is scraped by the advancing flight as depicted in Figure 1.7a.

As screws become larger, the heat transfer from the screw root and flights also transfers energy to the outside of the solids bed (Figure 1.7b), resulting in a cold compressed solids bed that ‘floats’ within melted material. The solids bed is now isolated from metal contact; developing a good-quality melt is now unlikely if this solids bed simply proceeds down-channel.

The solution to the problem is to drain the melt as quickly as it occurs, as shown in Figure 1.8. These screws are called melt-separation screws.[4] The oldest design (the Uniroyal) is depicted in the figure, but a variety of designs are available.

A common problem with barrier screws is compaction of the solids bed such that the solids bed stops moving. This prevents the continual flow of material movement through the screw. This compaction can occur anywhere along the barrier, but is most likely at the beginning and end of the barrier. This can cause surging, gel showers, carbon specs and loss

[4]It is common to refer to these as ‘barrier screws’. However, barrier screws can have other purposes besides separating the melt from the solids bed and for mixing in particular.

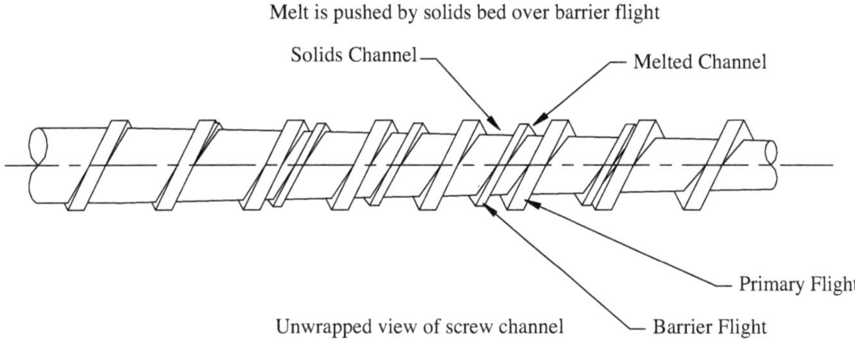

Figure 1.8 *The Uniroyal screw design.*

of output. Han et al. [4] showed that the removal of part of the first barrier could prevent these problems in some circumstances. Nevertheless, it should be noted that barrier screws increase the compressive force on API with a greater tendency towards agglomerations.

1.3.3 Screw Metering Section

Once the materials are melted, they flow through the metering section in molten form. Flow is a combination of the barrel drag flow and pressure flow. The flow is visualized by the straightening of the meter's spiral channel and the barrel is considered as a plane moving across the stationary channel, as in Figure 1.9.

The barrel drags material towards the pushing side of the flight. Since there is only a very small gap between the barrel and screw, material is then driven against the pushing side of the flight, across the screw root and back up the trailing side of the screw. When this is combined with pressure flow in the channel, spiraling occurs. Just as in any tube (in this case a spiraling tube), the centermost material moves forwards more quickly than material at the outside of the tube. The combination of drag flow and pressure flow therefore create, in a very limited way, axial mixing in the metering section.

1.3.4 Mixers

A wide variety of mixers can be added before, during or after the metering section either in general-purpose screws or melt-separation screws. Most of these are distributive mixers with little claim to dispersive capability, and are described in the following sections.

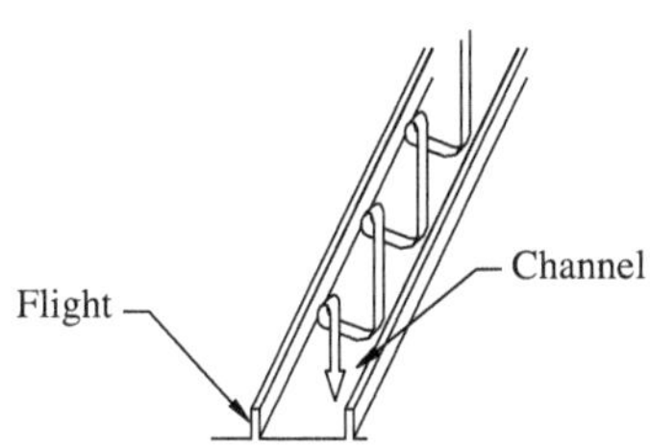

Figure 1.9 *Spiral flow in straightened metering section.*

Figure 1.10 *Pin mixers break up spiral flow.*

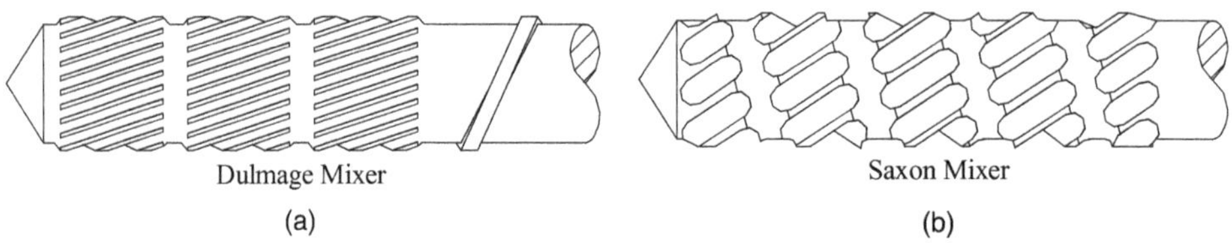

Figure 1.11 *Slotted mixers for energy exchange.*

1.3.4.1 Mixing Pins

Pins can be placed between flights (most commonly round or diamond-shaped) or in lieu of flights, as shown in Figure 1.10. Pins are often made small in diameter because the greater the number of pins, the more the flow is combined and recombined. However, small pins will bend and can fatigue over time and eventually break. Round pins tend to have an area behind them where materials can stagnate, hence other shapes such as diamonds are used. It should be noted that, unless the diamond-shaped pins are properly oriented in the flow and very close together, they too will have stagnant zones.

1.3.4.2 Slotted Mixers

Figures 1.11a and 1.11b show common slotted mixers. The flowing materials are separated into many streams to reduce their large mass into smaller portions. Each small portion rotates in its channel and allows for some heat exchange at the barrel for better thermal mixing before the flows recombine.

1.3.4.3 Variable-depth Mixers

Figures 1.12a and 1.12b depict variable-channel-depth mixers. Presumably, these could influence axial mixing by changing the drag flow and pressure flow components referred to in Figure 1.9.

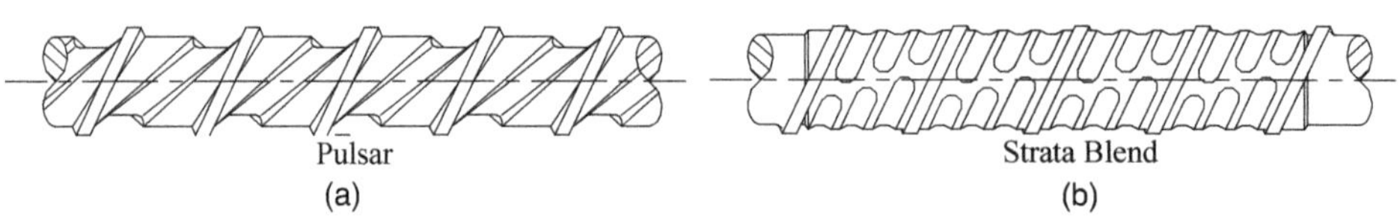

Figure 1.12 *Shifting material by changing root diameter.*

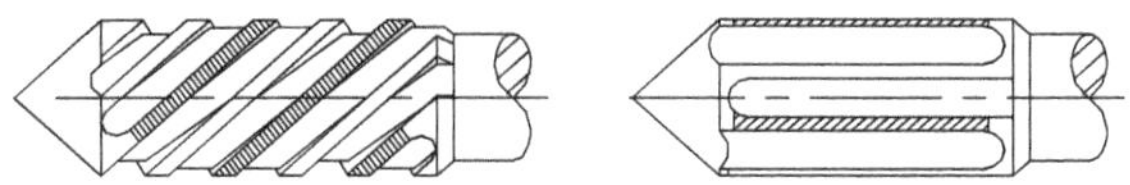

Figure 1.13 Fluted shearing mixers.

1.3.4.4 Shear Mixers

Figures 1.13a and 1.13b depict two long-established mixers where materials are forced through narrow slots for additional shearing, and thought to have mild distributive capability.

1.3.5 Limitations of Conventional Single-screw Mixers

Invariably, the question of which of these mixers is the best mixer and under what circumstances arises. The most consistent testing has been by Dow Chemical using the same procedure to compare various screws and mixers. ABS (acrylonitrile, butadiene, styrene) resin is compounded with about 12% white pigment to form an opaque background and pelletized. Then, black color concentrate pellets are added at 0.5–3.0% to the compounded pellets and a strand is made. When the strand is sectioned, black spirals (from the rotation of the screw) appear. Screws with the least distinct spirals are judged best.

At Antec 2010, the author reported [5]

> "Very consistent mixer studies include the Maddock [...], Stratablend (trademark New Castle) [...], Energy Transfer [...], Variable Barrier Energy Transfer [...], and the DM2 with Eagle Mixing Tip [...]—generally referred to as high performance screws. One study even includes the more complex Twente mixing ring, the Barr sleeve mixer and a Barr ring mixer [...]. These papers describe a spiraling pattern in the extrudate."

All the screws and mixers lessened the distinct spiral pattern of the control Screw somewhat. However, spiraling patterns were still easy to see in all screws. The conventional SSE, while good at melting and pumping, has therefore not approached the mixing performance of the TSE.

1.4 SSE Elongational Mixers

It was generally believed that the SSE could not create substantial elongational flow. By means of its two screws, the TSE readily draws material apart (just as two hands readily draw on the ends of a rubber band to stretch it). The SSE was thought permanently handicapped and therefore incapable of significant stretching flow.

Imagine two counter-rotating metal rolls with a small gap between them. Material (such as rubber, molten polymer or a man's tie) will be drawn into the inlet and expelled on the other. If only one roll is driven, material is still pulled in and expelled; only one surface therefore needs to move to form a pump. A single screw has a moving surface so it is possible to create a pump.

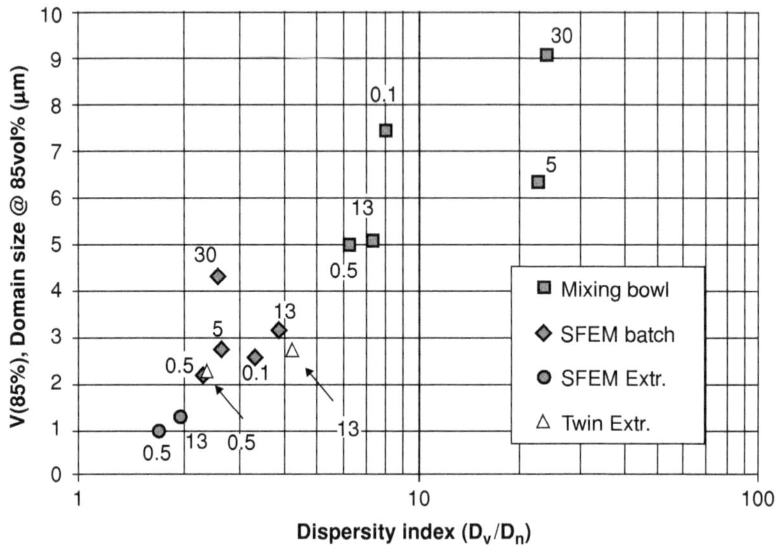

Figure 1.14 *Quantified mixing (courtesy of Dow Chemical, Antec 2011).*

Such pumps can be defeated by pushing more material into the roll gap above than it can remove. The same moving surfaces become resistive to flow. The greater the force applied to push material through the ex-pump, the greater the resistance and the higher the temperature rise. Similarly, the most popular screw designs (the barrier screws, Egan and UC mixers) are intended to work by pushing material through a small gap.

A member of the newer class of elongationally dominant screws was tested using a similar procedure of testing black color concentrates against a pigmented background. This screw has a series of three spiral flow elongational mixers (SFEM) on the screw and is called an Elongator. Even at 200 times magnification, no spiral patterns were found.

This is not surprising. Work had been presented showing a much finer scale of mixing than spirals in rod seen by eye. For example, in 2007 [6] polystyrene was mixed with polyethylene and shown to disperse to 1 µm scale, carbon nano-tubes and ceramic particulate were distributed to the 500 nm scale.

While pictures at very high magnification are helpful, quantitative mixing comparisons have remained elusive. However, at Antec 2011, Costeux et al. [1] presented a paper that quantified mixtures from four different processing tools. It shows that the SFEM Elongator is elongationally dominant because it breaks down blends of high viscosity ratios that cannot be dispersed by shear alone.

Figure 1.14 is a plot of mixtures processed on four different mixers: two batch and two continuous. The numbers indicate the MI (melt index) of the PE component in the 70% PP (polypropylene) 30% PE (polyethylene) blend.

The vertical axis shows the distribution of the minor phase domains. Domains are actually like spheres: the lower the number, the smaller the diameter and the better the mixing. The horizontal axis shows the domain volume divided by the number of domains. The lower the dispersity index, the more uniform the domains.

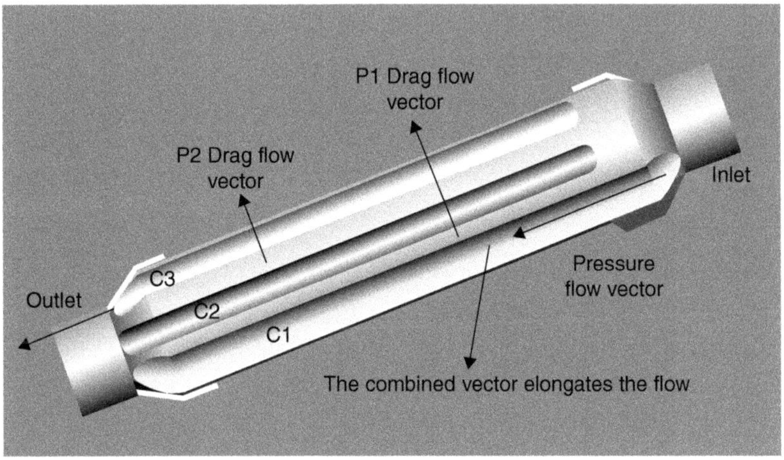

Figure 1.15 *Flow vectors in AFEM Recirculator.*

An atomic force microscope (AFM) was used to examine the samples and the minor phase domains were then digitized. Mixing was quantified by digitizing the domains of the minor phase and applying image analysis to extract 3D domain distributions.

The plot shows that the SFEM batch mixer mixed better than the mixing bowl batch mixer. For the continuous devices, the SFEM single screw was superior to the twin.

It is important to remember that these domains are spherical. The volume of a sphere decreases rapidly with diameter. If the diameter of a sphere is reduced by half, its volume is reduced by an eighth. If the mixing bowl makes spheres with a diameter of 5 μm (as in the 0.5 MI material), they will have volume of 65.5 μm^3. Both the twin-screw and the SFEM batch mixer made 2 μm diameter spheres of the same material with a volume of 4.2 μm^3, making 15 times as many smaller spheres than the bowl mixer. The twin-screw and SFEM batch mixer therefore disperse 15 times better than the mixing bowl.

Likewise, since the SFEM Elongator screw mixer made 1 μm domains (a volume of 0.5 μm^3) for the same material, the SFEM Elongator mixed 8 times better than the twin-screw. This new class of mixers can therefore mix particulate and polymers in ways that have not been possible previously.

There are two types of elongational mixers: the SFEM Elongator and the AFEM (axial fluted elongation mixer) Recirculator (Figure 1.15). The AFEM Elongator has a first axial channel (C1) next to a first pump (P1). Although P1 has the same geometry as a barrier, it has a different behavior than a barrier. A barrier screw forces material over the barrier in an attempt to shear the material, which requires a great deal of pressure. However, if the pressure is low in the channel, then the same barrier geometry becomes a pump which pulls on viscous materials.

Elongational flow is created as the material moves down the first channel (pushed down the channel by pressure flow) and is pulled by the pump by drag flow at an angle. The pressure flow moves in the direction of the inlet vector down the C1 channel (parallel to the screw axis) while the drag flow moves the material at 90° (a right angle to the screw

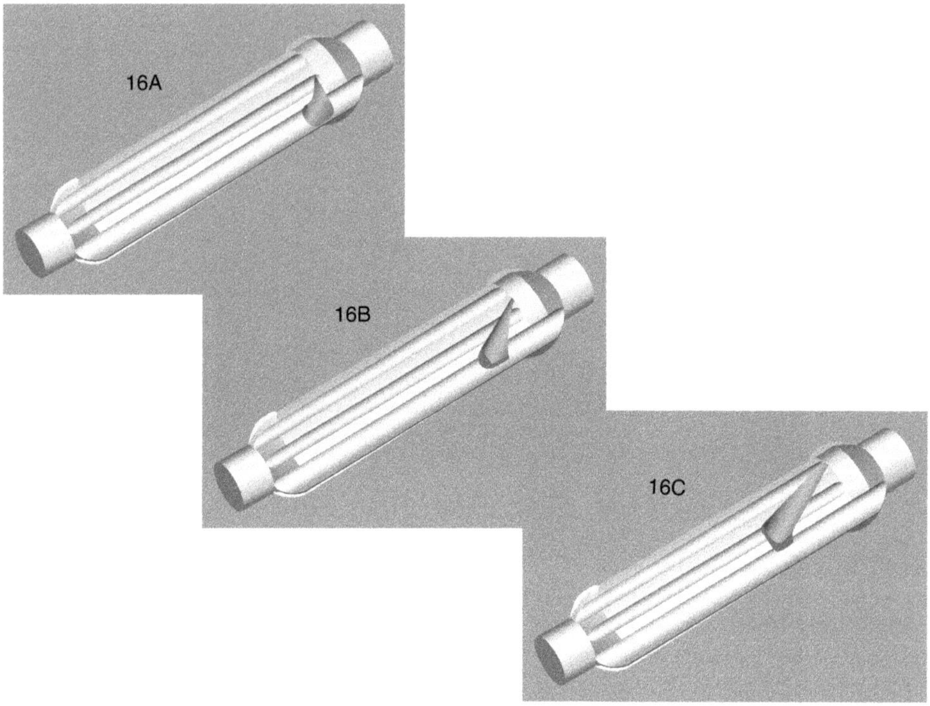

Figure 1.16 The length of the plane (compared to its width) increases as the C1 channel is increasingly starved. The width can increase to the limit of the P1 pumping capacity.

axis) as shown by the drag flow vector. The combined vector is 45° when the vectors are of equal length; this becomes stretched.

When the P1 pumping capacity is higher than the input to the channel, 100% of the material is stretched as it moves to the pump. The drag flow pumping capacity is easily kept higher than the input to the channel by starve feeding. The fill length of C1 depends on the amount of starve feeding. If the pump's capacity is exceeded, some material can flow out the end of C1 and not be stretched.

A colored section of material (representing a local concentration of API), surrounded by clear material entering C1, will be drawn by P1 at the entry as shown in Figure 1.16. This section will start to deform into a plane as show in Figure 1.16a. As the local concentration moves down the channel, the mass in the channel is reduced and the plane becomes larger as in Figure 1.16a. Eventually, the entire local concentration of API is converted into a plane. This can be thought of as the short local cylinder becoming a wider and longer flat surface.

While Figure 1.16 is representative, the actual plane created in the transverse direction is much more elongated than shown.

The process begins with shear as shown in Figure 1.17. Because the pressure is near zero in C1, the flow over P1 is nearly pure shear. This is very unlike the flow over a barrier flight which is primarily pressure flow, necessary to force material over the barrier. Pressure flow

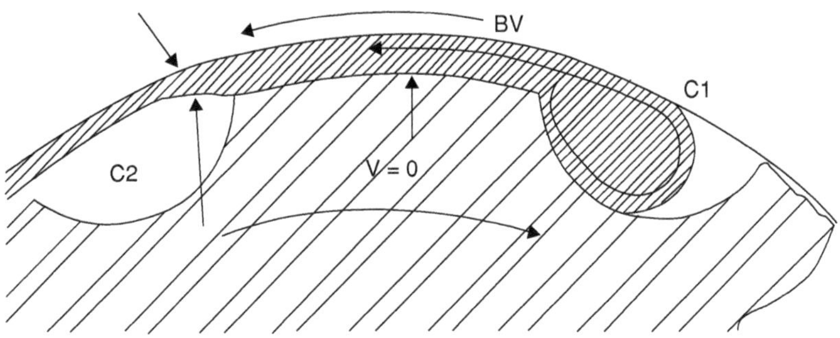

Figure 1.17 Film stretches again over C2.

does not contribute much to mixing. Instead this 'pure' shear flow reorients the material as the material sticks to the barrel and the screw. This reorganizes the flow and further distributes the API.

After the shear flow, material reaches a second channel C2 which lowers the resistance of the screw drastically on the material near the screw. However, it does not change the adhesion to the barrel. This creates a powerful 2D stretching and converts the material into a thin film at the two arrows in Figure 1.17.

The thin film at zero pressure is useful for degassing when a volatile is present. A hole can be placed in the barrel and the gas extracted from the thin film by pressure flow towards the downstream flight. The pressure may also push material into the empty end (the end away from the hopper) of the C2 and C3 channels. In this case, some material can move upstream in the C2 and C3 channels and recirculate. The amount of recirculation depends on the distance from the AFEM Recirculator to the downstream flight and the amount of C1 fill. If the C1 channel is very starved, there can be more recirculatory flow than input flow. This is usually only advantageous for extremely difficult mixing problems (such as nano-compounding) rather than HME applications. If the C1 fill is greater than the capacity of P1, then there will be no recirculation and material can exit C1.

There is another useful feature of the AFEM Recirculator. During an upward surge in pressure, the fill length of C1 will lengthen and will shorten during the trough of the surge. The mixer will therefore act as an accumulator and will dampen pressure and flow instabilities. Each additional mixer increases the dampening to make the final flow surprisingly stable, even when the screw is starve fed.

The second type of screw in this class is the SFEM Elongator, as shown in Figure 1.18. The design is very similar to the AFEM Recirculator, but has a pitched design and flights connected to the mixer. The pitch increases the forwarding ability of the screw, which minimizes the residence time within the screw. The connected flights mean that the output of any channel flows immediately onto a flight and then moves downstream. Recirculation therefore does not occur, which makes the SFEM Elongator better for most HME applications. Figure 1.18 has two sets of C1, P1, C2, P2, C3 surrounded by a flight clearance.

Most of the same principles outlined for the AFEM Recirculator apply to the SFEM Elongator. That is, the first channel has 3D elongation, the first pump has shear undiminished by pressure flow to reorganize the material top to bottom and it has 2D stretching as the

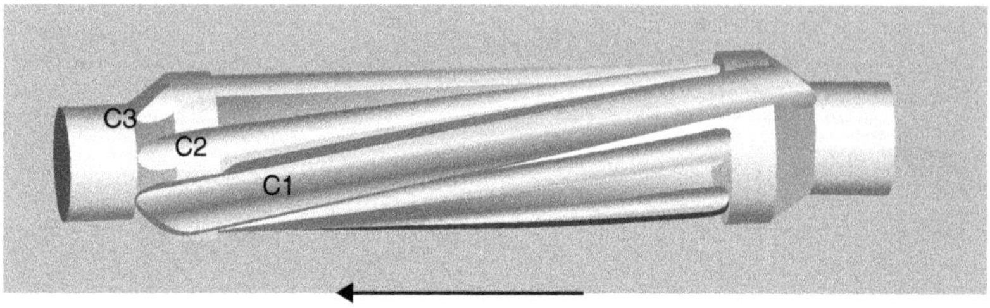

Figure 1.18 *The SFEM Elongator stretches but does not recirculate.*

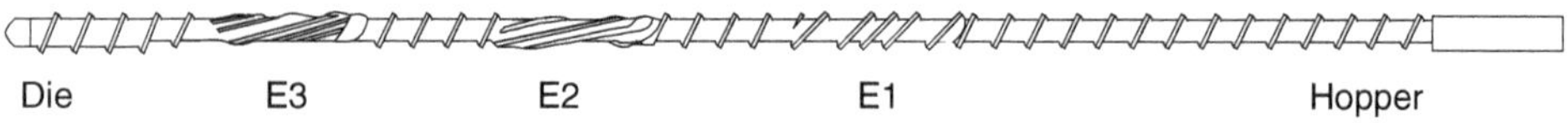

Figure 1.19

material moves from P1 to P2 over the second channel. When the material is pumped from C3, the material moves immediately onto the connected downstream flight and is forwarded so that recirculation does not occur, keeping the residence time short.

The stretching flows in the channels have been observed. This is done by mixing color concentrate into the polymer, stopping the screw and cooling the material as quickly as possible on the screw. The stretching flows are preserved by the color within the cooled pullouts. This is more easily observed for flood-fed conditions where the C2 channel is filled. This makes extraction of the cold screw and subsequent sectioning easier. Figure 1.19 depicts the schematic of the SFEM Elongator screw.

The thick frozen section shown in Figure 1.20 is from the second of three Elongators (E2) where 2% blue color concentrate was mixed with polypropylene [1]. The arrows point to the material's movement downstream within the channels. Traces of color on the root of the screw show the angle of rotation within the C2 and C3 channel.

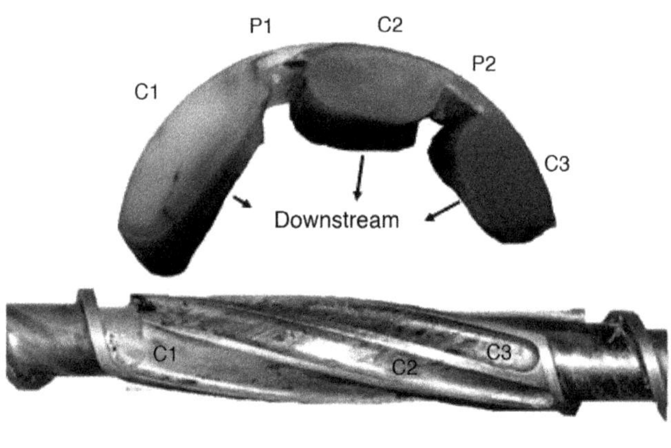

Figure 1.20 *A thick frozen section that was cooled and then removed from the screw. For a better understanding of the figure, please refer to the color section.*

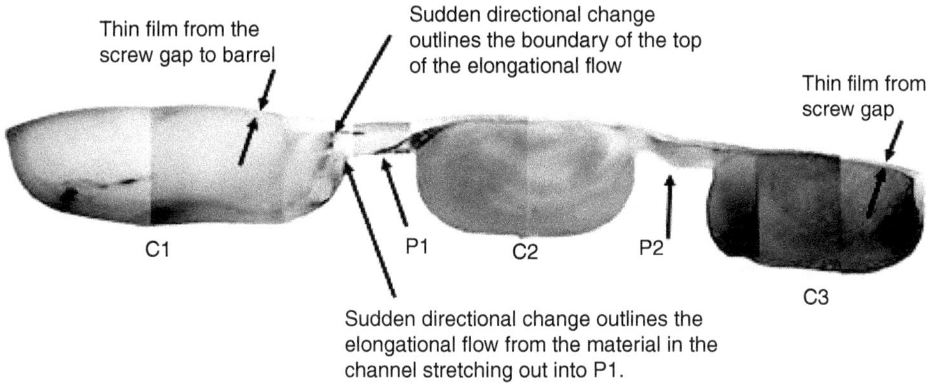

Figure 1.21 *A composite of nine pictures shows the elongational mixing flows. For a better understanding of the figure, please refer to the color section.*

Figure 1.21 is the same cross-section as shown in Figure 1.20, but thin-sectioned and magnified for additional detail by stitching together 8 separate pictures. A dramatic difference in the quality of the mixture is seen as the material moves from C1 to C2 and from C2 to C3.

Flights of the screw are located to the far left and far right of the cold pull-out in Figure 1.21. These flights have a clearance to the barrel of about 0.004 inch. Material 'leaks' through the gap and is seen as a thin film. This film is dragged along by the barrel through the mixer until finally disappearing in C3.

In C1, the individual pellets are not visible as they are in a typical compressed solids bed of material. Instead, having flowed through E1 and the following channels, the material arrives in C1 as a highly viscous melt.

Material in C1 is then dragged to the right towards P1. The two arrows within CI outline the boundaries of the elongational flow as the material approaches P1. Within P1, shear flow occurs. Immediately after P1, the exiting material is dragged into a very thin layer just beneath the leakage flow material. This is the stretching flow that is primarily 2D and greatly extends the material because of the high barrel velocity. The thickness of this stretched material appears much less than the leakage flow thickness.

A similar process occurs as material moves from C2 and approaches P2, as depicted in Figure 1.22. Material over P2 must now contain the material in C3 so it is not sheared in the same way as P1, as can be seen by the flow lines in P2 and the change in the leakage flow dimensions. The material near the output of P2 elongates and converges, as shown by the dotted lines. This is similar to the flow immediately after P1.

In an HME extruder, starve feeding is preferred to flood feeding. Flood feeding fills the screw to its maximum capacity from the hopper. Flood feeding can allow compression of the API before the first SFEM Elongator and can cause agglomeration.

Starve feeding uses various volumetric feeders or a gravimetric feeder to limit the flow into the screw. Because the flow in the solids channel is less than the capacity of the screw, the partially empty channel has no pressure on the API. This prevents compression and

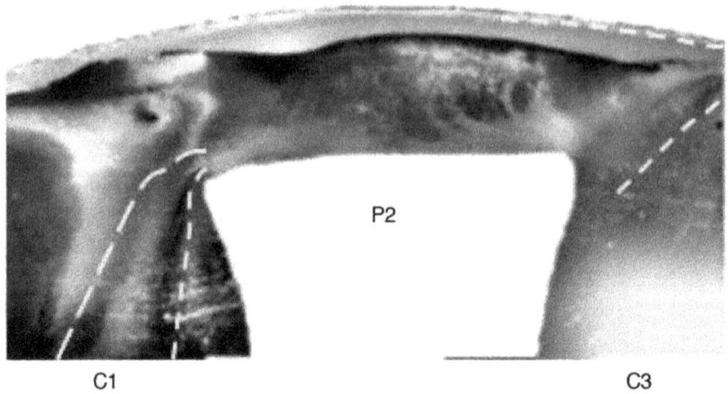

Figure 1.22 Dotted lines show the converging (extensional) mixing flows at P2. For a better understanding of the figure, please refer to the color section.

agglomeration before the material enters C1. By feeding at a rate less than the P1 capacity, C1 will empty, C2 will not fill as in the flood-fed example above and all the processed material will move into C3. It will then have the same thermal and mechanical processing history. Since the system is elongationally dominant (rather than shear dominant) it will have the lowest energy input to accomplish the mixing.

1.5 Summary

To compound well means to mix all the ingredients with the same thermal and mechanical heat history using the least amount of energy. A bell curve showing the number of mixing cycles does not describe an orderly mixing process. The TSE by adding more mixers to increase the amount of material moving through a brief extensional field, creates a the greater the disparity between the over-mixed and the non-mixed ingredients. A good compounder requires a mixer that treats all the input in a predictable, orderly fashion.

The AFEM Recirculator is best used for mixtures where the API is extremely small or of low concentration and where the ingredients are thermally stable and can withstand a longer residence time.

The SFEM Elongator SSE is a most orderly compounder and is best for thermally sensitive materials and where venting is necessary. It accepts a disorderly mixture with local concentrations and methodically organizes it by elongating all of the input in the same way. The modern elongational SSE limits the amount of mixing to what is needed to achieve a certain goal. Because of its ability to exert elongational forces from the beginning of the cycle, the proper amount of mixing can be applied which limits the thermal and mechanical processing history.

The SSE is a low-cost processor that can compound and develop sufficiently high and stable pressures to make an HME product from a single heat history. When fitted with elongational screw technology, the SSE mixes as well as and better than any of the other technologies.

References

(1) Costeux, S. et al. (2011) Facile TPO dispersion using extensional mixing. Antec 2011.
(2) Luker, K. (1995) Surge suppression: a new means to limit surging. Antec 1995.
(3) Rauwendaal, C. (1986) *Polymer Extrusion*. Hanser Publishers, New York.
(4) Han, K.S. et al. (1996) Elimination of a restriction at the entrance of barrier flighted extruder screw sections. *Plastic Engineering Magazine*, April, xx–xx.
(5) Luker, K. (2010) Comparison of flow striations of various SSE mixers to the recirculator and elongator mixers. Antec 2010.
(6) Luker, K. (2007) Summary results of a novel single screw compounder. Antec 2007.
(7) Luker, K. and Cunningham, T.M. (2009) Investigation into a high output polypropylene screw and its mixing mechanism. Antec 2009.

2

Twin-screw Extruders for Pharmaceutical Hot-melt Extrusion: Technology, Techniques and Practices

Dirk Leister, Tom Geilen, and Thobias Geissler
Thermo Fisher Scientific, Karlsruhe, Germany

2.1 Introduction

A large number of today's materials used are plastics and undergo hot-melt extrusion (HME) processes in their course of manufacturing. HME and its related downstream processes (e.g. injection molding) are widely used and accepted in the modern plastics industry.

As the process is very robust and fairly easy to scale up, it is used from laboratory scale where only a few hundred grams of materials are being used up to production scale of >60 tons per hour output.

In the 1980s, the company BASF was among the first to apply the technology of HME for pharmaceutical applications [1]. The process has became more and more popular in pharmaceutical research and production since then, as it offers some features which cannot be accessed by other means. HME helps to overcome poor bioavailability of active pharmaceutical ingredients (API), as well as creating new modified-release drug systems and can serve as a unit operation to taste mask the bitterness of a tablet. Many more fields of application for HME have been discovered since then.

By its nature, the HME process is continuous where API are dispersed into a carrier matrix; this is most often a polymer with suitable properties for HME (e.g. glass transition temperature, thermal stability or hygroscopicity). To achieve optimal process conditions, additional excipients such as plasticizers and solubilizers can be added into the material

Hot-melt Extrusion: Pharmaceutical Applications, First Edition. Edited by Dennis Douroumis.

mixture. As by definition the extrusion step itself is the moment were the molten material is pressed through a die to give shape, the common understanding of the process also involves dosing of the ingredients into the extruder and the compounding/mixing step prior to this event.

The purpose of this chapter is to describe all relevant instruments and operations involved in the HME process and give an understanding of its functionality, list relevant equipment requirements for pharmaceutical usage as well as describe a practical approach to successfully set up an HME process.

2.2 Extruder Types and Working Principle

An extruder consists of a barrel that can be heated and cooled, enclosing one or multiple screws which convey, compound and subsequently force a polymer melt through a die mounted at the barrel end. While the material is transported inside the barrel, the originally solid polymer/API/excipient mixture is plasticized due to induced shear force and/or applied heating which enables the terminal extrusion step. In order to expel any volatiles, air or moisture that appears during the melting process, a venting step to the atmosphere or with applied vacuum can be necessary prior to the final extrusion step.

As mentioned, the origin of extrusion is the plastic industry; to accommodate for the optimal processing of the different thermoplastic polymers, a variety of extruder types have therefore been developed. They can be classified by the number of the screws being used [2], as listed in Table 2.1.

To obtain a well-dispersed mixture of API and excipients, the mixing capabilities of an extruder are crucial. As a matter of fact, the co-rotating twin-screw extruder (TSE) with intermeshing and thus self-cleaning screw elements is superior in this respect over the single-screw extruder. The screws of a TSE can be normally built up from individual screw elements with different properties, giving a very flexible option to set up transport zones, mixing zones and degassing zones along the barrel as the individual application requires. The screw shaft flexibility is the key success factor of co-rotating parallel TSE; a counter-rotating design is less flexible in that respect.

With its horizontally split barrel and easy to handle and to clean design, the TSE also meets the requirement of pharmaceutical research and development where lack of these properties can reduce development time considerably.

The individual parts of a TSE and the related process parameters and values are depicted in Figure 2.1 and described in more detail in the following sections.

Table 2.1 *Classification of different extruder types.*

Single-screw extruder	Twin-screw extruder	Multiple-screw extruder
Smooth barrel	Co-rotating screws	Rotating center shaft
Barrel with grooves	Counter rotating screws	Static center shaft
	Intermeshing screws	
	Non-intermeshing screws	

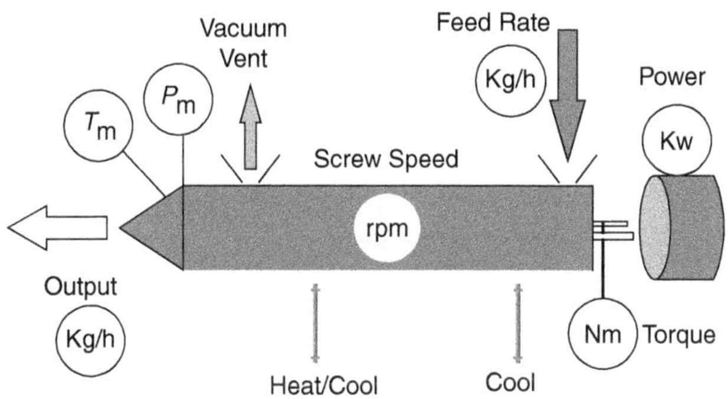

Figure 2.1 *Schematic of a TSE.*

2.3 Individual Parts of a TSE

2.3.1 Drive Unit

The motor to drive the twin screws together with a gearbox and a safety clutch makes up the drive unit of a TSE. The main energy that is required for the melt process inside the extruder is generated by the drive unit and transferred via the rotating screws into the polymer mixture.

To allow for a smooth start-up of the extrusion process, the external capability to heat and/or cool the barrel and the respective processing zones within is used. Later on in the process, when the required energy is mainly delivered from the drive unit, the heating and cooling can be used to fine-tune the conditions.

The mechanical power P_m [W] that is introduced into the system is defined by the equation:

$$P_m = \frac{2\pi n}{60} \times M$$

where n is the screw speed in revolutions per minute or rpm (min^{-1}) and M (Nm) is the torque on the screw shafts, derived from the servo motor drive. As a process value, the power consumption over time of the drive unit is normally monitored and recorded during the process.

2.3.2 Screws

The screws used widely in current TSEs consist of a screw shaft onto which individual screw elements are arranged. The flexibility to use elements with different geometries allows an optimal screw configuration to be set up for a given process.

The length of the screws operating in the extruder is normally given in terms of *L/D* ratio (the length of the screw divided by the outer screw diameter). Typical screw lengths are in the range of 25–40 *L/D*. The screw speed is a process parameter that can be altered by the user and is monitored and recorded in rpm (min^{-1}).

It is mandatory that product contact parts in pharmaceutical applications are inert to the material that later on becomes the final drug dosage form. Leaching of contaminations

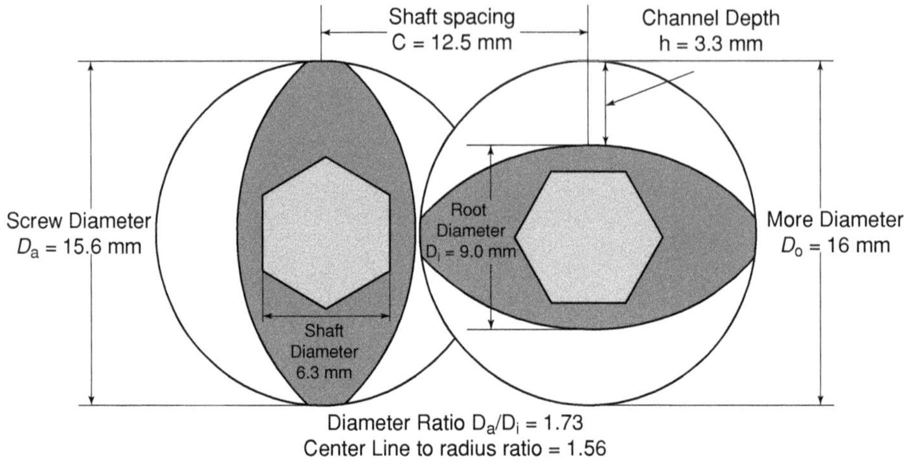

Figure 2.2 *Example screw dimensions of a Thermo Fisher Scientific Pharma 16 HME.*

into the material stream or reactions with the same must therefore be prevented. For pharmaceutical or food applications, screw elements manufactured from surgical steel grades (1.42123/X15TN or similar) are used.

The maximum torque that can be delivered into the polymer melt and the free working volume inside the barrel depend on the shaft design as well as on the ratio of inner to outer diameter of the twin lead screw elements; see Figure 2.2 for example screw dimensions.

The shown geometry ($D_a/D_i = 1.73$) for a 16 mm extruder provides a large free volume and ensures a channel depth of at least 3 mm to accept standard polymer granules. In contrast to polymer applications, where sometimes stiff and hard fillers (e.g. glass fibers, ceramic powders) require a higher torque, the large free volume design is preferred for pharmaceutical applications as it allows higher throughput and better mixing at milder conditions to protect sensitive API.

The torque is a measure of the energy absorbed by a mass unit (specific energy) which is necessary to run a distinct process. The applied torque is therefore monitored as an important process parameter. The specific energy is a characteristic of an extrusion process and widely used for scale-up purposes. Together with the Residence Time Distribution (RTD), the specific energy should be kept constant when transferring a process from lab scale (e.g. 16 mm) to production (e.g. 48 mm or even higher). This approach is applicable when the two TSEs in question increase in scale to the same geometrical proportions (e.g. same D_a/D_i ratio).

During the start-up of an extrusion process, the screw shafts can become blocked because of solid materials or product residues from previous runs resulting in a large torque load. This should be avoided by starting up the process in mild conditions (low rpm of the screws and low feed rate). When operating the extruder at nominal speed, it is also necessary to stay below maximum torque so that the safety clutch or motor protection does not trigger.

2.3.3 Screw Elements

The flexibility of the segmented screw is a key success factor of co-rotating parallel twin-screw extruders. It allows the screw configuration to be changed to meet various application requirements.

The number and geometry of screw elements are increased over time, mainly driven by the requirements for specific applications. A comprehensive overview of available elements was prepared by Kohlgrüber [2].

In this text we will concentrate on the most common elements used to build up the screw configurations which are required in HME processes: transporting/conveying, melting, mixing and shaping (pressure build up).

2.3.3.1 Conveying Elements

Conveying elements have a self-wiping twin lead geometry and are used in feeding, conveying and venting sections along the screw configuration. The helix of a standard element has a pitch of 1 *L/D*. The helix pitch may vary to increase or decrease the free volume and conveying speed. All feed screws shown in Figure 2.3 have a length of 1 *L/D* to visualize the different helix pitch.

2.3.3.2 Mixing Elements

Mixing sections are created by combining multiple single mixing elements (Figure 2.4). The offset (30°, 60° or 90°) between adjacent elements determines the conveying and mixing properties. The conveying properties decrease with increasing offset angles while the mixing properties increase. In extreme 90° offset, sections have pure mixing and no conveying capabilities.

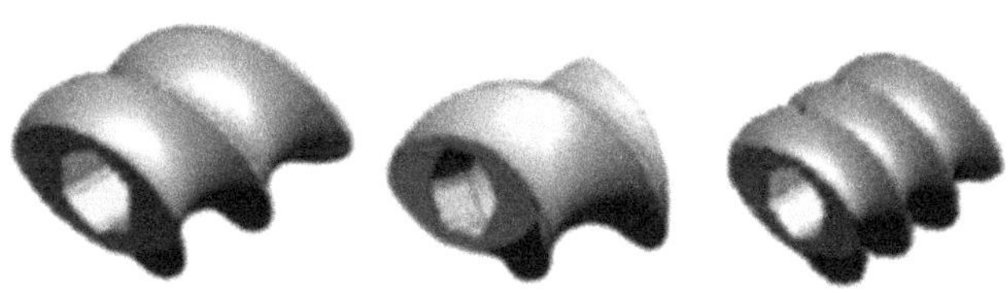

Figure 2.3 *Conveying screw elements with pitches 1* L/D, *3/2* L/D *and 2/3* L/D *(left to right).*

Figure 2.4 *Single mixing element (1/4* L/D*).*

Figure 2.5 Mixing block consists of 5 time 1/4 L/D mixing elements with 30° offset.

Mixing sections with 30° or 60° offset (Figure 2.5) can convey the material forward or backward depending on the direction of their pitch. Most common mixing elements have a length of 1/4 *L*/*D*. Longer elements (1/2 *L*/*D*) introduce higher shear while shorter elements (1/8 *L*/*D*) improve dispersive mixing.

2.3.4 Distributive Flow Elements

Distributive flow elements generate a low-energy mixing. The distributive flow dominates over shear flow. These types of elements may be used to incorporate liquids into a melt further downstream of the extruder barrel. Each element combines an outer grooved and inner plain diameter disk. The orientation of the elements alters on both shafts. Due to the notched outer disk, distributive flow elements have a non-self-wiping geometry. Distributive flow elements are available with different notch depth (full (1), 1/2 and 1/4 depth; Figure 2.6) and also with different disk thicknesses, which leads to different element lengths.

Figure 2.6 Distributive flow element full-depth notch, 1/4 L/D length.

2.3.5 Discharge Feed Screw

Discharge feed screws (Figure 2.7) have a single lead geometry to generate the required extrusion pressure to shape the final product at the end of the extrusion process.

Sophisticated modeling approaches can be used to determine the conveying and mixing behavior for a given screw geometry. This is however beyond the scope of this chapter, and is discussed elsewhere [2]. Since a specific polymer and HME formulation shows certain adherence to the barrel wall, the calculated values and models do not always reflect the real life behavior perfectly and therefore have to be refined iteratively after comparison to experimental runs. Table 2.2 gives a qualitative description of conveying and mixing

Figure 2.7 *Discharge feed screw.*

Table 2.2 *Conveying and mixing properties of screw elements, where the number of +/− indicates strength of property.*

	Conveying	Mixing
Feed screws	++++++++	+
30° forward	+++++	++
60° forward	++++	+++++
90° alternate	zero	++++++++
60° reverse	−−−−	++++++
Reverse feed screws	−−−−−−−−−	++

properties of the above described screw elements. This is a good estimation from which is assemble the desired screw configuration.

2.3.6 Barrel

Being also a product contact part, the barrel material used in pharmaceutical HME has the same constraints regarding contamination and reaction with the melt stream as the screw elements.

For the housing for the screws to operate, the barrel construction should also offer some degree of flexibility to support the user in setting up the specific layout very easily. It should be possible for ports for material in-feed and venting to be placed anywhere along the barrel according to process requirements. A segmented barrel approach is therefore widely used. Different extruder manufacturers have taken different approaches to realize a segmented barrel; we describe the horizontal split barrel in more depth here.

The horizontal split barrel consists of a lower section, the liner and the top barrel assembly. To give best access to both parts and enable easy and thorough cleaning (required for compliance with current good manufacturing practice (cGMP) cleaning validation approaches), these parts can be removed and disassembled easily as shown in Figure 2.8.

The upper and lower barrel parts are held together with a clamp mechanism that ensures tight closing along the whole barrel length. The upper barrel is divided into 4 pieces, each of length 10 *L*/*D*. Each element has 2 ports that can be open or closed individually, which results in 8 individual process zones each of length 5 *L*/*D*.

The ports can be closed, which is normally the case for transport, mixing and extrusion zones, or opened for feeding (solids or liquids) or venting.

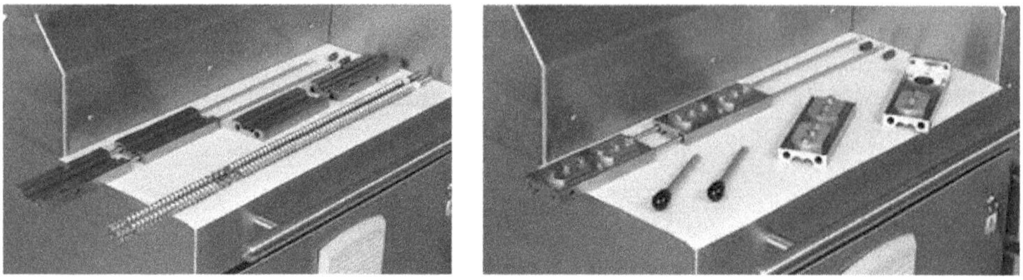

Figure 2.8 Split barrel assembly completely disassembled. Left: bottom liner and screw and right: top barrel assembly.

Over the length, the 40 *L*/*D* barrel is therefore divided into 8 individual temperature-controlled zones (5 *L*/*D* each) that can be set according to the process requirements. The temperature at the extruder die can also be set individually, whereas the feeding zone is normally only cooled.

2.4 Downstreaming

Before the actual extrusion step occurs, the molten and compounded polymer melt is pressed through the die that is mounted onto the very end of the extruder barrel (Figure 2.9). This step determines the shape of the extrudate and helps to facilitate the further downstream processing into the desired product.

Widely used is the strand pelletization line where one or more strands are extruded, collected on an air-cooled conveyor belt and then cut into small cylinders in a strand pelletizer (Figure 2.10). The speed of the rotating knife inside the strand pelletizer can be controlled to achieve different cylinder lengths. Another common device for downstream processing is the chill roll: the melt is squeezed between two chilled rolls where it is cooled and calibrated into a belt of defined thickness. The resulting belt will be broken into

Figure 2.9 Strand die with one bore and 2 mm diameter.

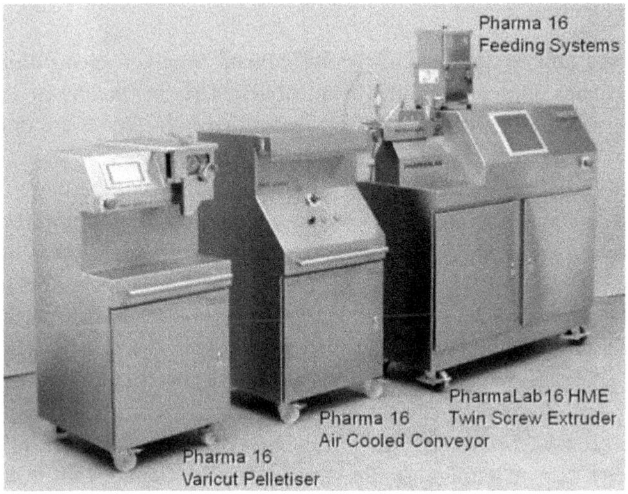

Figure 2.10 *16mm TSE for pharmaceutical usage complete with strand pelletizing unit. For a better understanding of the figure, please refer to the color section.*

smaller flakes at the end of the unit. These are only two examples from a high number of possibilities. In general, the melt can be shaped into any kind of design. The melt can also be casted on foils to obtain patches for transdermal therapeutic systems (TTS). In general, it can be said that the melt needs to be processed in a cooling and in a shaping step. Using calendaring or injection molding can result in final tablets which do not need to be treated any further.

2.5 Individual Processing Sections of the TSE

The pharmaceutical HME process allows the number of individual unit operations to be reduced in order to obtain a final dosage form in which the process steps are arranged in a series of adjacent zones linked to each other. An overview of the HME is depicted in Figure 2.11 and individual sections are described in the following.

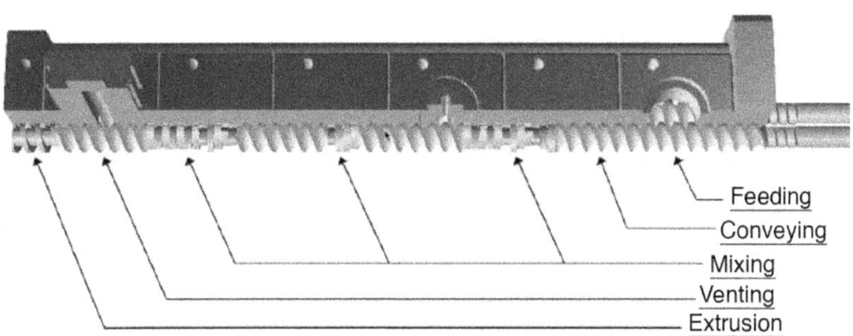

Figure 2.11 *Overview of processing zones of the TSE (cross-section).*

2.5.1 Feeding Section

The polymer, API and excipients can be introduced together as a pre-blend or as individual feed streams into the feeding zone of the TSE by means of a volumetric or gravimetric feeder. Details of the gravimetric loss-in-weight feeder are described later in Section 2.6.

It is important to realize that the screw speed of the TSE and the output of the feeding system are independent of each other. As well as the mixing capabilities, this is the distinct difference between single- and twin-screw extruders. With single-screw extruders, there are normally no additional feeding devices so that the throughput of the system depends only on the screw speed of the extruder. The polymer material sits in a feed hopper and enters via gravity through the feed throat (an opening near the rear of the barrel) into the extruder. There, it comes into contact with the rotating screw that forces the polymer beads further down the barrel.

A co-rotating twin-screw extruder has excellent conveying capacity. For this reason it is necessary to meter raw materials into the barrel. This has the advantage that several feed streams can be dosed in a controlled way. More importantly, the screw speed of the TSE can be changed to achieve different mixing effects; only the feed rate of the feeding system affects the throughput of the whole system.

The powder falls onto the rotating screws and is transported towards the die. In this section, conveying screw elements are used. Depending on the granularity of the used material, the width of the pitch can vary.

As described in Section 2.3.6, the split barrel provides the possibility of opening additional feed ports along the screw. This approach is most commonly used for introducing heat and shear-sensitive API at a later stage of the process, as the material is only exposed to the appropriate energy to achieve the desired mixing without degrading.

The barrel is usually cooled at the feeding section to prevent components with a low melting point adhering to the feed hopper wall and thus blocking the inlet.

2.5.2 Conveying/Melting Section

By conveying the powders through the melting zone, the induced energy and shear force starts softening and melting the polymer and other components. This process is heavily dependent upon:

- screw speed and filling level;
- melting point of the individual components;
- particle size;
- residence time/throughput; and
- screw configuration.

The use of plasticizers in the formulation (e.g. poloxamers such as Lutrol F68) can help to lower the melting point and thus achieve processing conditions that do not harm or degrade the used API.

2.5.3 Mixing Section

HME aims to produce stable solid dispersions. This can include embedding the drug molecule in either a crystalline or amorphous state, or dispersing it at a molecular level into the polymer carrier and stabilizing it there.

For nearly all mixing applications, a well-dispersed and well-distributed mixture is required. This can be achieved by the different arrangements of the mixing elements described in Section 2.3.3 on the screw shaft. The distributive mixing aims to homogenize the melt temperature and distributes all solid and liquid particles to obtain an excellent content uniformity. The goal of dispersive mixing, however, is to break down any solid particles, such as crystalline drug molecules or not-yet-molten polymers, resulting in the above-mentioned solid dispersion.

The quality of mixing is dependent upon the screw speed, throughput, viscosity of the melt and the screw geometry. The greater the screw speed and the smaller the throughput, the better is the mixing performance.

The design of the mixing elements also affects the mixing quality. Narrow disc elements provide a better mixing performance. If back mixing is applied, e.g. by incorporating 90° elements, the mixing performance can be further improved. To obtain a good dispersive mixing, it is important to induce a certain shear stress. Wide-disc mixing elements (length > 1/4 L/D) provide effective dispersion with the tradeoff that the melt temperature will increase. These factors have to be evened out in order to avoid overheating the melt in a certain area and causing degradation of the components or API. This is a commonly used approach to introduce multiple small mixing sections along the screw length to achieve the overall desired mixing performance. Between the mixing sections, the melt temperature has time to reduce slightly.

2.5.4 Venting Section

Powders introduced into the extruder can have entrained air either in pores or adherent to their surface. When the powder becomes compressed, this air needs to be disposed of.

Residual moisture or solvent can also lead to gas formation when the melt temperature increases. It is therefore necessary to allow venting of these gases. If venting prior to the extrusion step is not efficient, it will result in bubbles that occur in the melt stream after the die and prior to downstreaming. This has to be prevented, as an uneven melt stream cannot be processed into a homogeneous final product.

Venting can be achieved by opening the top barrel section to the atmosphere over a conveying section after mixing has occurred. A specially designed vent insert is normally used in the top section of the barrel to allow a large opening with low gas speeds, so no product is drawn into the vent stream. If necessary, a vacuum can be applied to support the degassing.

2.5.5 Extrusion Section

Just before the die, pressure is built up by using discharge feed screws. This allows a 100% fill level and ensures a constant melt flow through the die to provide an even shaping.

Normal transport elements would result in melt pumping and an uneven melt stream at the die outlet.

2.6 Feeding of Solids

To achieve a constant material flow into the extruder a gravimetrical loss-in-weight feeder (Figure 2.12) is usually used in pharmaceutical processes. The ability to constantly monitor and control the feed rate results in a much more precise feed constancy over time.

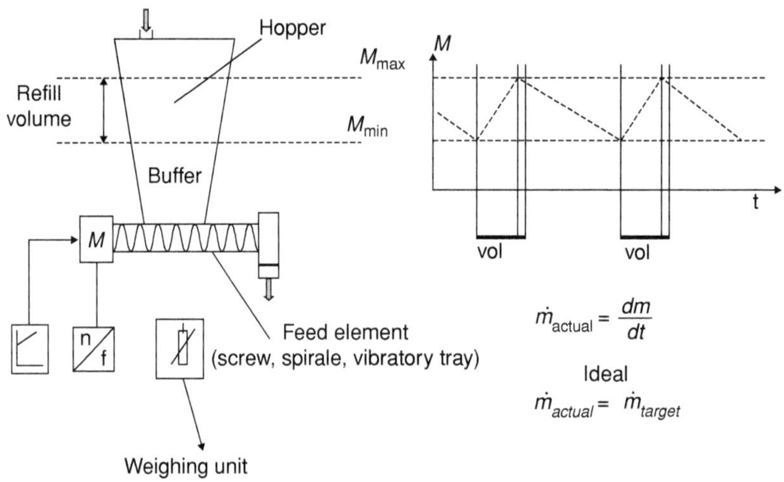

Figure 2.12 *Operating principle of loss-in-weight feeders.*

Driven from a motor, a feeding device (e.g. helix or screw) meters powder material through the horizontal discharge into the extruder. The design of the feeding device is highly dependent upon the flow properties of the material. A weighing unit constantly monitors the loss-of-weight of the material in the feeder hopper. According to the set throughput (A_{target} in kg/h), the feeder control determines appropriate motor speed to achieve $\dot{m}_{actual} = \dot{m}_{target}$.

To ensure a continuous operation and a stable filling level of the feeding device, the hopper is refilled within a volume of material within certain boundaries (M_{max} and M_{min}). As the weighing unit is disturbed during refill the control switches off and the rotation speed of the feeding device is kept constant during this period, resulting in a volumetric feeding. In order to minimize the time of volumetric operation, the refill should occur at high frequencies with a low refill volume.

2.7 TSE Operating Parameters

An important feature of the TSE is the fact that the throughput of the material stream fed into the extruder using a gravimetric feeder is independent of the screw speed of the extruder itself. The material metered by the gravimetric feeder will determine the throughput of the

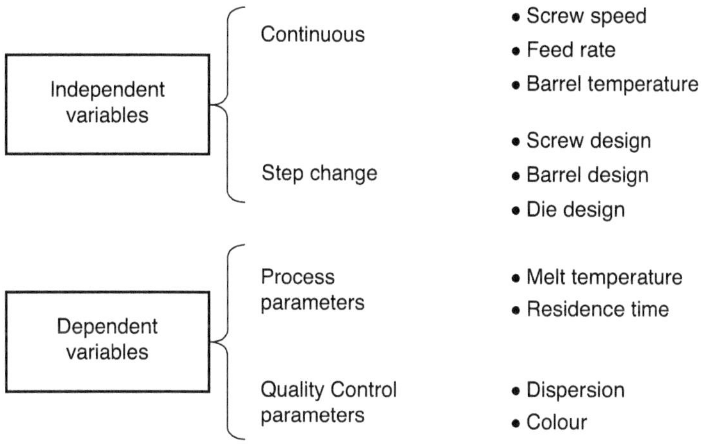

Figure 2.13 *Process values and parameters in TSE operation.*

whole system, weather the extruder screws turn fast or slow. The screw speed will influence the transport and mixing of the powder stream, as well as determine the amount of energy input to the system. This is an important parameter of the process for a given API/excipient combination in order to achieve a final product with desired quality attributes. During early research and development, an optimal formulation can be investigated and controlled precisely with the feeding systems. Screw speed variation can be used to achieve different mixing effects.

All relevant process values and resulting process parameters are depicted in Figure 2.13. For a successful set-up of a stable continuous extrusion process, it is necessary to understand the interdependencies of these parameters and values. Above all, the relevance of these parameters and values is specific to each distinct formulation and extruder set-up. As soon as the formulation composition varies or the extruder set-up changes, deviations from previously measured process values will be encountered.

The independent process parameters that can be set on the extruder by the operator, and some of the dependent process values as a result of their change, can be observed in Figure 2.13.

Process parameters can be distinguished between continuous and step-change parameters. Continuous parameters can be altered while the extruder is running, such as the screw speed or the feed rate of the dosing system. However, step-change parameters such as the alteration in barrel or screw design require a pause in the process.

The interdependencies of the process parameters and the process values is complex and, in many cases, not predictable to an exact value. Feedback loops, e.g. viscosity of the melt and the melt temperature are influencing parameters, force the user to gain an understanding of their dependencies to set up a robust process.

This can be achieved by conducting a set of relevant experiments to learn about the distinct process. To ensure that all relevant parameters are tested and their influences are characterized, Design of Experiments (DoE) approaches should be used to test in a systematic manner. The following sections introduce and describe some of the critical parameter/value interactions.

2.7.1 Filling Level

If material properties (e.g. inner friction) and screw design allow, the fed-in powder in the feeding section is conveyed onto the cold screws and pre-blended. Together with the powder, a considerable amount of air (either in particle cavities or adherent to the surface) is incorporated into the process. The screw can therefore only become filled with powder up to a certain degree as the entrained air prevents a higher intake of material.

If there is a restriction to the powder flow and material is compressed and subsequently molten, the filling degree in the barrel increases. This is normally the case when the conveyed material reaches the first mixing zone. If the material is molten, the degree of filling represents a full barrel.

As the residence time in a truly filled section is dependent on the degree of filling, the more restrictions we have the higher the residence time becomes (as shown in the simplified Figure 2.14). With an average higher filling level inside the complete barrel, the mean residence time increases.

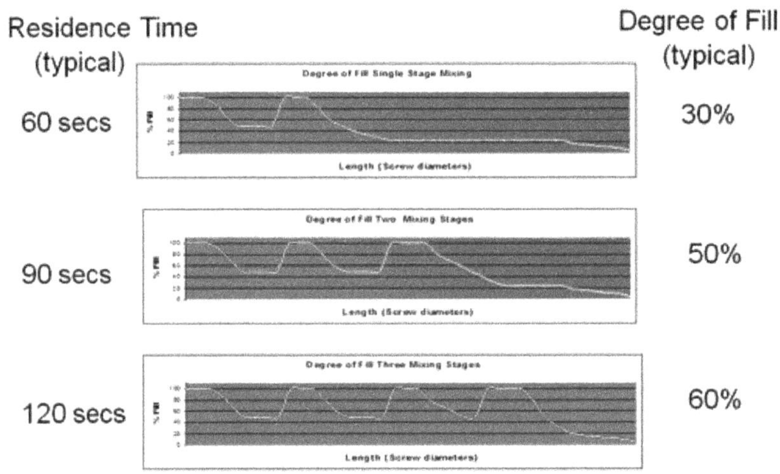

Figure 2.14 Average degree of filling along the screw.

2.7.2 Screw Speed

To investigate the influence of screw speed upon the melt temperature, a polypropylene was fed at a constant feed rate into a 24 mm TSE while screw speed was subsequently varied. The recorded melt temperatures are depicted in Figure 2.15.

It can be observed that the screw speed has a great influence on the melt temperature; the melt temperature increases with screw speed. As mentioned in Section 2.3.2, the specific energy input e is an important process value and should therefore be considered as characterizing the given process. Measured in units of kJ/kg, it is defined as the mechanical

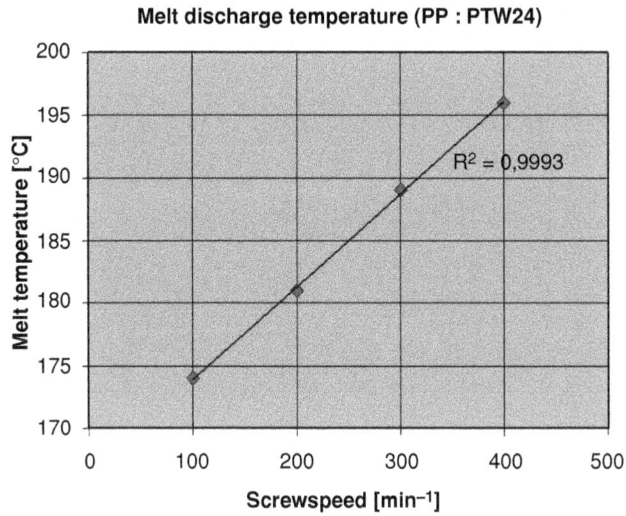

Figure 2.15 *Influence of screw speed on melt temperature.*

power transported via the screw shafts into a given mass of melt:

$$e = \frac{P_{\rm m}}{1000} \times \frac{3600}{\dot{m}}$$

where $P_{\rm m}$ is the mechanical power (see Section 2.3.1) and $\dot{m}$ (kg/h) is the throughput of the system. The specific energy brought into the system increases with increased screw speed, which explains the observation depicted in Figure 2.15.

The specific energy input is an important parameter for scale-up of an HME process. Related to the increase throughput of a production extruder, the specific energy per mass unit should be similar to the value obtained during laboratory-scale operations.

2.7.3 Feed Rate

When the feed rate is increased, the residence time decreases in the extruder. The effect on feed rate of residence time is greater than the effect of screw speed. This can be seen in Figure 2.16, where polypropylene was used on a 24 mm TSE to investigate this effect.

As can be seen, doubling the screw speed has less effect on residence time distribution than changing the feed rate. Extremely long residence times are therefore achieved by feeding at low rates.

2.7.4 Residence Time Distribution

Residence time is an important parameter and has an influence on the quality of the obtained extrudate. On the one hand, heat and/or shear-sensitive material can decompose over time; on the other hand, a minimum residence time is required for sufficient melting and mixing in order to obtain a homogeneous product.

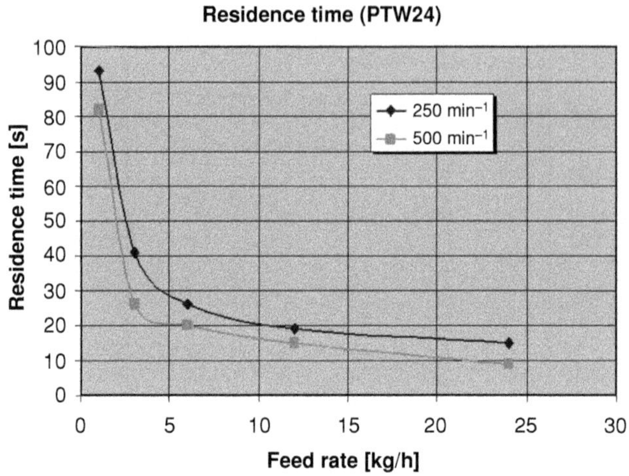

Figure 2.16 Influence of feed rate on residence time.

The residence time distribution (RTD) can be determined by introducing a tracer into the melt stream at a given time t_0. Over time, the concentration of the tracer is measured constantly at the output of the die. The tracer must therefore be detectable and the amount of tracer should not influence the system by e.g. increasing the torque. For each process, the RTD has a specific shape where the mean residence time can be calculated. Figure 2.17 depicts the RTD of a color tracer and its dependency upon screw speed and feed rate [3]. This illustrates the fact that a decrease in feed rate leads to a broader distribution, whereas change in screw speed for a constant feed rate does not influence the RTD very much.

The RTD can be influenced by the individual transport capabilities of the screw elements used. This effect is dominant in partially filled screws, whereas in a fully filled condition the feed rate plays a more significant role.

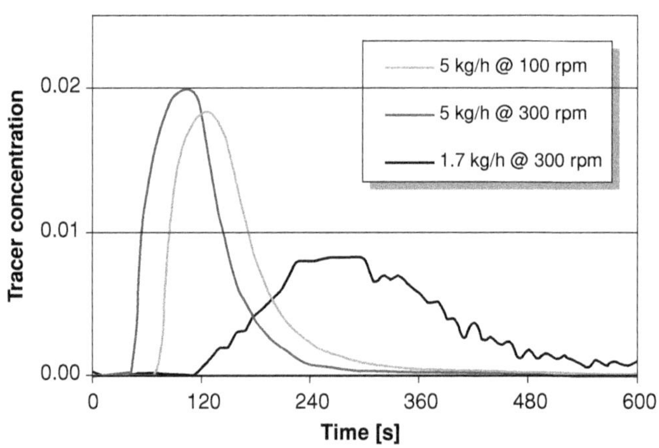

Figure 2.17 RTD depending on throughput and rpm. For a better understanding of the figure, please refer to the color section.

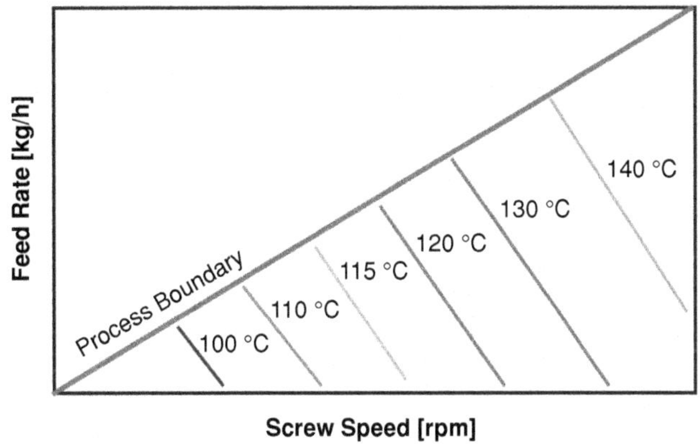

Figure 2.18 *Effect of feed rate and screw speed on melt temperature.*

2.7.5 Effect of Screw Speed and Feed Rate on Melt Temperature

With information on the effect of screw speed and feed rate on the melt temperature and RTD (Figures 2.16 and 2.17), we can represent the effects of these process parameters on the process values as shown in Figures 2.18 and 2.19.

The melt temperature will increase with increased screw speed, but will decrease with increased feed rate (Figure 2.18). Alternatively, residence time reduces with increased screw speed and increased feed rate (Figure 2.19).

The process boundary is also depicted in Figures 2.18 and 2.19. This boundary is the physical limit which a process cannot overcome, even when process parameters are changed accordingly. One reason for a boundary can be the torque constraint. At a certain viscosity and filling level, the power of the drive unit is limited to process the melt and the extruder has to stop in order to avoid damage to the screws or gearbox. Another reason for this boundary can be the processing of powders with a very low bulk density. The entrained air in these

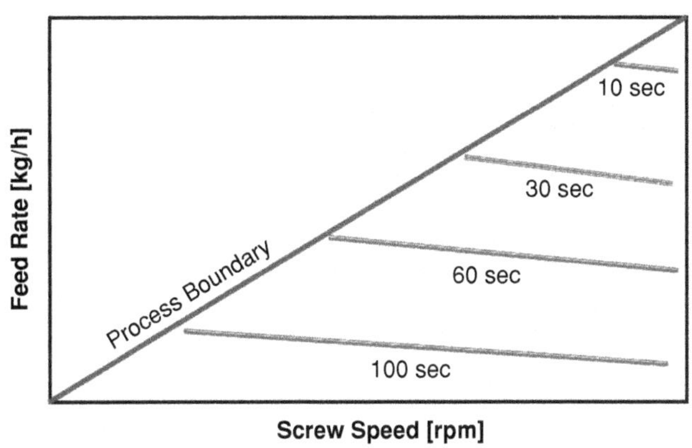

Figure 2.19 *Effect of feed rate and screw speed on residence time.*

particles prevents a complete filling of the screw; increased feed rates and screw speeds lead to an overflow of the feeding section and thus limit the maximum throughput of the process. It is important to note that these boundaries are specific to a distinct formulation and screw/barrel design. The characteristic of these boundaries are determined for each new process.

2.8 Setting up an HME Process using QbD Principles

When setting up a stable extrusion process, use can be made of the process parameters and value dependencies derived from Sections 2.7.1–2.7.5 and hence the Quality by Design (QbD) approach promoted by the US Food and Drug Administration (FDA).

An understanding of the manufacturing process allows manufacturing equipment to be designed to deliver the desired quality of the final product. All critical sources of variability of the final product quality are identified and can be explained. The impact of process parameters is defined, and thus the variability can be managed by the process [4].

When applied to process equipment, the *knowledge space* is defined from the understanding of the limitations of the equipment and characteristics of the materials being processed. The *design space* is defined from an understanding of the critical and non-critical product parameters, and experiments are conducted to determine the relationships between different process parameters.

Even when using Design of Experiment (DOE) techniques, a large number of experiments are required to define the design space based on the effects of different process parameters on product quality attributes. When using a continuous process such as HME, the individual feed streams allow the formulation changes to be rapidly made and a minimum sample size produced per experiment. Material usage and experimental time can be significantly reduced.

Finally, the *control space* defines the operating window within which all critical process parameters can be controlled to deliver the required product quality attributes. To complete a full QbD cycle, one has to work his way from the outside (Knowledge space) to the required operation window (control space).

2.8.1 Understanding Knowledge Space

Thorough understanding of a manufacturing process, when critical attributes are defined and controlled and the influence of process variables is known, enables a product to be delivered with required and constant quality. When applying the term knowledge space to an extruder, it mirrors the understanding of the limitations of the equipment (e.g. the maximum screw speed that can be applied and the maximum feed rate of the dosing system). These parameters have to be linked to the characteristics of the material being processed and the critical quality attributes (e.g. the degradation temperature of used polymer and/or API). These maximum values and the material processed determine the process space in which experiments can be performed.

2.8.2 Defining Design Space

An empirical approach of evaluating the design space requires a sophisticated experimental set-up. Even when using DoE techniques, a large number of experiments have to be performed. It is critical to include all relevant parameters sets of a process in such an evaluation. A simple linear extrapolation of obtained results for only a few operating parameters may result in an inacceptable product quality.

2.8.3 Determining Control Space

Consider starting the process at point 'A' with a low screw speed and a low feed rate to gently fill the screw with material (Figure 2.20). This will result in a low melt temperature and long residence time; a material bed will build up inside the extruder barrel, the mixing zones will become slowly filled and plastification of the polymer will begin.

When the feed rate is increased to point 'B', the residence time will reduce. As point 'B' is close to the process boundary limit at this applied screw speed, only an increase in screw speed to point 'C' will enable us to apply higher feed rates and therefore increase the throughput of the whole system. The process is however well away from its boundary; we can therefore increase feed rate to point 'D'.

We can follow this 'staircase' approach until arriving at the limitations of the system at maximum feed rate and screw speed, as indicated by point 'F'. For this particular formulation under these process conditions, barrel/screw design and operating temperature, point 'F' represents the maximum achievable throughput.

If the material properties are measured at each of the staircase points, it is possible to detect differences in the critical quality attributes of the final product due to the combination of melt temperature (screw speed) and RTD (feed rate) applied. A specific limit for melt temperate and RTD can therefore be defined where the quality attributes of the final product are within the required range.

If we now combine all our data, the control space becomes evident, i.e. the operation window within which all critical process parameters can be controlled to deliver the required product quality attributes.

The process boundary line defines one boundary of the process. When the product in question requires a certain temperature, then the maximum temperature line defines the second boundary. If residence time is also found to be a critical quality parameter,

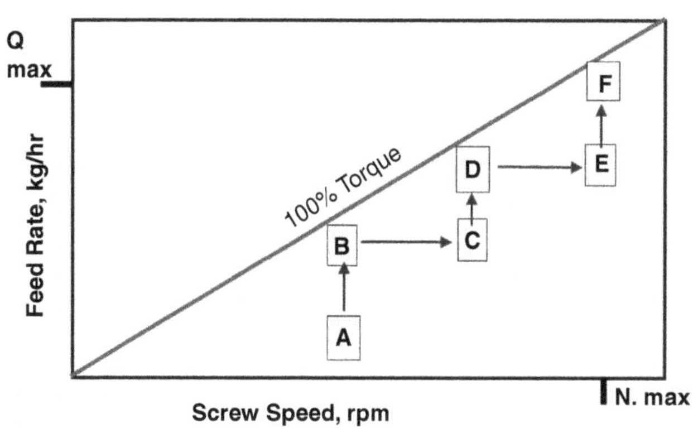

Figure 2.20 *Evaluating the design space.*

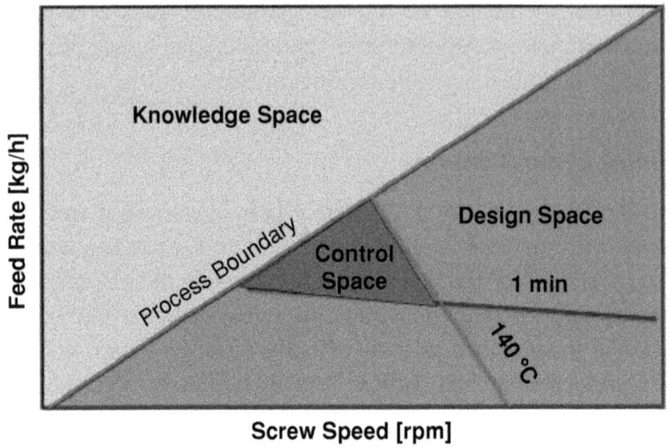

Figure 2.21 *Determining the control space. For a better understanding of the figure, please refer to the color section.*

the maximum residence time defines the third boundary. In the example described by Figure 2.21, point 'D' falls within the control space and hence delivers the acceptable product quality at the maximum throughput of the system.

2.9 Summary

Today's pharmacists have a powerful tool at their disposal to obtain a solid dispersion, in the form of the twin-screw extruder. Certain design attributes have to be considered, as they are different from the original TSEs that are available for plastic compounding. Once the interdependencies of process parameters and the resulting values of final product quality attributes and process conditions are developed, it is possible to produce final drug dosage forms in a continuous fashion. This will assist in the development of drug production, where product quality and conformance are ensured through the design of effective and efficient manufacturing procedures.

References

(1) Kolter, K., Karl, M., Nalawade, S. and Rottmann N. (2010) *Hot-Melt Extrusion with BASF Pharma Polymers. Extrusion Compendium.* BASF SE, Ludwigshafen.
(2) Kohlgrüber, K. (2008) *Co-rotating Twin-Screw Extruders: Fundamentals, Technology, and Applications.* Carl Hanser Verlag, Munich.
(3) Geilen, T. (2011) Doppelschencken extrusion. Presentation at APV Seminar No. 6382, HME Masterclass, Karlsruhe.
(4) Swanborough, A. (2008) Benefits of continuous granulation for pharmaceutical research, development and manufacture. Application Note LR-63, Thermo Fisher Scientific, Karlsruhe.

3

Hot-melt Extrusion Developments in the Pharmaceutical Industry

Ana Almeida, Bart Claeys, Jean Paul Remon and Chris Vervaet
Laboratory of Pharmaceutical Technology, Ghent University, Belgium

3.1 Introduction

Starting from the plastic and rubber industry in the second half of the 19th century, passing through the food industry, hot-melt extrusion (HME) has over the last years also been introduced as a manufacturing technique in the pharmaceutical industry. As a process of converting raw materials into a product of uniform shape and density by forcing it through a die under controlled conditions (temperature, feed rate and pressure), HME is a well-established manufacturing technology with a plethora of technical solutions already available in other fields. This has resulted in the availability of equipment (extruders, downstream auxiliary equipment and monitoring tools for evaluation of performance and product quality) to support the introduction of this technique for pharmaceutical applications, using single- and twin-screw extruders. Historically, single-screw extruders were an economical option for melt processing (commonly used to produce films, pipes and sheets). Although a relatively simple process, single-screw extrusion does not offer the mixing capability of a twin-screw extruder and is therefore not the preferred approach for the production of pharmaceutical formulations. Moreover, as twin-screw extrusion allows separate batch operations in a single continuous process and has a shorter residence time, lower process temperature and higher output, this set-up offers more versatility and is able to accommodate various pharmaceutical formulations.

Hot-melt Extrusion: Pharmaceutical Applications, First Edition. Edited by Dennis Douroumis.

Based on the advantages offered by HME, the interest in this technique within the pharmaceutical industry has grown over the last 10–15 years; this is reflected in the increasing number of scientific reports and patents describing the pharmaceutical applications of HME. HME also received a boost as a pharmaceutical manufacturing technique when the Food and Drug Administration (FDA) encouraged drug-markers to use continuous manufacturing processes.

Despite the availability of HME equipment and the benefits of this technique, HME is not yet a common manufacturing tool for drug delivery systems. Although it is extensively used to manufacture medical devices (often in combination with injection molding for shaping of the HME material into e.g. stents, syringes, valves), the number of drug products on the market which are manufactured via HME is limited. The best-known examples include implants (e.g. Implanon®, Nexplanon®) and vaginal rings (e.g. Nuvaring®), providing controlled delivery of hormones for contraception. For oral drug delivery HME is used to manufacture Kaletra® tablets, a protease-inhibitor combination product for the treatment of human immunodeficiency virus (HIV). This formulation, developed by Soliqs based on the Meltrex technology, uses a polyvinylpyrrolidone/vinylacetate matrix to enhance the bioavailability of lopinavir and ritonavir to reduce the dosing frequency and to improve storage stability. Using the Meltrex technology, Soliqs also developed a sustained-release formulation of verapamil (the first directly shaped HME product on the market), a fast-onset ibuprofen system and an antiretroviral tablet containing ritonavir (Norvir®) which (in contrast to the conventional formulation) does not require refrigeration during storage.

To streamline the development of drug products, the regulatory bodies also stimulated the investment in new drug delivery platforms and the innovation of pharmaceutical plants to enhance pharmaceutical productivity, highlighting continuous production, Quality by Design (QbD) and Process Analytical Technologies (PAT) as important tools to simplify, control and understand the manufacturing process. As a continuous process, HME fits perfectly within this framework. PAT tools (e.g. spectroscopic techniques, rheology measurements) can play a crucial role for real-time quality evaluation and understanding of the extrusion process of pharmaceutical dosage forms.

3.2 Advantages of HME as Drug Delivery Technology

Due to its versatility in embracing a wide spectrum of applications, HME offers many advantages over conventional pharmaceutical production methods.

- HME can be operated as a continuous process, ensuring optimal reproducibility (consistent product flow at relatively high throughput rates). As a consequence, it requires less offline testing compared to batch processes, online or inline PAT tools can be easily implemented and real-time release becomes a possibility.
- The limited number of processing steps (blending, melting, extrusion and shaping in a single-step process), the short residence time (maximum of a few minutes) and the reduction in labor forces (due to the extensive automation of the process) lead to a higher economic efficiency.

- HME is a solvent-free process and so organic solvents and water are not required, waiving the need for additional production stages (e.g. time-consuming drying steps), which makes the process environmentally friendly.
- There exists a wide range of dosage forms which can be manufactured via HME (granules, pellets, tablets, films, sheets, rings, etc.), depending on the shape of the die and/or the post-processing technique (pelletizing, milling, calendering, injection molding, etc.). This offers excellent opportunities for product life cycle management using this technique.
- In order to fulfill the requirements of FDA and other regulatory authorities, companies responsible for the production of extrusion lines have adapted the knowledge acquired from the polymer industry to produce adequate GMP-compliant extrusion technology for pharmaceutical applications. An additional value when compared to other techniques is the self-cleaning capacity of extruders, minimizing cleaning procedures during changeover from one formulation or product to another.
- The intense mixing and agitation imposed by the rotating screw during extrusion (in function of screw design, throughput rate and barrel temperature) causes de-aggregation of suspended particles in the molten polymer, resulting in a uniform dispersion in the molten polymers.
- The drug release profile offered by HME products via proper selection of the polymers used during HME is highly versatile. Solid solutions (i.e. the dispersion of active pharmaceutical ingredients or APIs in a matrix at the molecular level) are an efficient approach in the delivery of poorly water-soluble, Biopharmaceutics Classification System (BCS) class II compounds because of the improved absorption and therapeutic efficacy. On the other hand, HME is also an excellent tool to create sustained-release formulations when, for example, a hydrophobic polymer is extruded with a highly water-soluble drug. In this case, the contact of the drug with the gastro-intestinal (GI) tract fluids is delayed and API is leached from the matrix much slower than compared to an equivalent compressed tablet.

The main disadvantage of HME is related to the thermal processing, limiting its application for thermolabile components. However, changes in the configuration of the equipment (screw configuration, twin-screw extruders) or the addition of plasticizers can reduce process temperature and residence time to avoid thermal degradation during processing. Another disadvantage is cost related, as the availability of HME equipment in pharmaceutical manufacturing plants is currently limited. Hence, high start-up costs due to investment in equipment, knowledge and facilities are still a significant barrier for the full-scale implementation of HME in the pharmaceutical industry. In addition, the perception within the pharmaceutical industry that authorities are still skeptical of a new technology and that queries about the development, validation and quality control will delay approval of the drug product are currently also a barrier to the widespread implementation of HME as a drug delivery platform.

3.3 Formulations used for HME Applications

Independent of the final dosage form (granules, pellets, mini-matrices, beads, tablets, films, implants, vaginal rings), the main applications of HME as drug delivery technology are:

Table 3.1 *Thermoplastic polymers commonly used to prepare immediate and sustained-release dosage forms via hot-melt extrusion.*

Immediate release
Polyethylene oxide (PEO)
Polyethylene glycol (PEG)
Polyvinylpyrrolidone (PVP)
Hydroxypropyl methyl cellulose (HPMC)
Hydroxypropyl cellulose (HPC)
Vinylpyrrolidone/vinylacetate copolymer (Kollidon® VA)
Dimethylaminoethyl methacrylate copolymer (Eudragit® E)
PEG 6000 / vinylcaprolactam / vinylacetate copolymer (Soluplus®)
Sustained release
Ethylcellulose (EC)
Ethylene vinyl acetate (EVA)
Polyvinyl acetate (PVA)
Poly(L-lactic acid) (PLA)
Poly(lactic-co-glycolic acid) (PLGA)
Polycaprolactone
Silicone
Ammonium methacrylate copolymer (Eudragit® RS/RL)
Lipid matrices (microcrystalline wax, stearic acid, carnauba wax, etc.)

(a) to enhance the dissolution rate and bioavailability of poorly water-soluble drugs and (b) to sustain drug release over an extended period of time.

To manufacture a specific HME dosage form, the drug is embedded in a polymer whose properties mainly determine the functionality of the end product (Table 3.1). Independent of the application, the basic requirements for pharmaceutical-grade polymers used as HME excipients are a thermoplastic behavior (as deformation of the formulation is essential), a suitable glass transition temperature (typically between 50 and 180°C, to allow processing at a relatively low temperature), a high thermal stability (ensuring a significant difference between T_g and degradation temperature), no toxicity (carrier is often the main ingredient and large doses will be administered to the patient) and a high or no solubilizing capacity (to ensure thermodynamic stability).

In addition to the API and polymers as main ingredients in the formulation, plasticizers are often required to enhance thermal processing. The use of other additives (e.g. bioadhesive polymers or drug-release modifiers via modulation of the matrix porosity or pH micro-environment) is optional depending on the application. Figure 3.1 summarizes the most relevant characteristics of polymers, API and additives for HME.

3.3.1 Active Pharmaceutical Ingredient

The ultimate goal of drug product development is to design a stable system that maximizes the therapeutic potential of the drug substance and facilitates its availability to patients. Consequently, it is essential to have an understanding of the physicochemical principles underpinning the behavior of such systems.

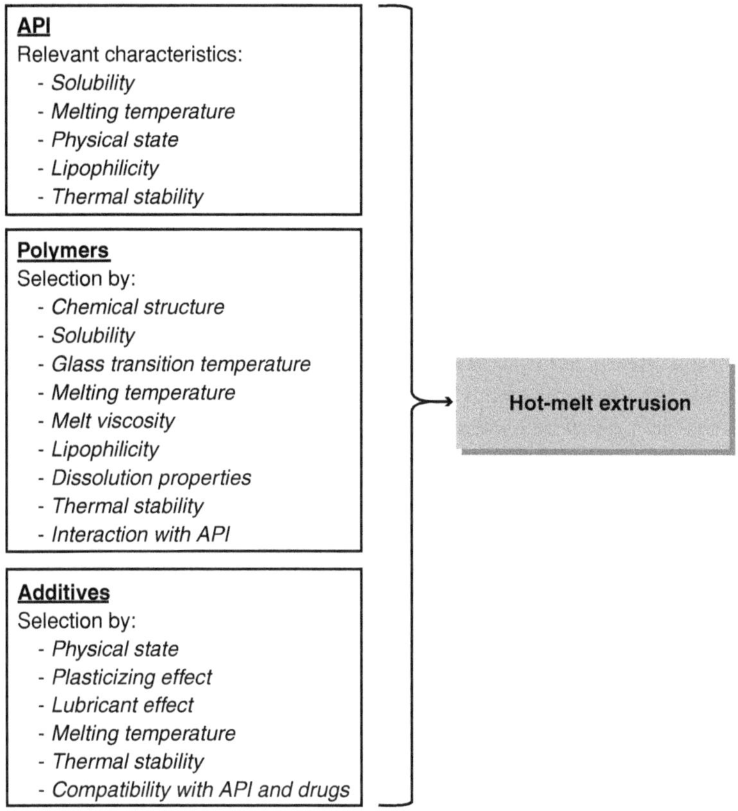

Figure 3.1 *Relevant characteristics of API, polymers and additives used for hot-melt extrusion.*

The majority of drugs are prepared in the crystalline state, characterized by a regular ordered lattice structure, which has unique advantages over the amorphous form in terms of physical (e.g. hygroscopicity) and chemical stability, processability and the availability of a diversity of forms (e.g. polymorphs, anhydrates, hydrates and solvates). Formulating crystalline drugs via HME is, in the majority of cases, related to the manufacture of sustained-release formulations. Polymers-based sustained-release matrices (using Eudragit® RL and RS as carriers) were processed by Quinten [1] via HME in combination with injection molding, using different metoprolol salts (tartrate, succinate and fumarate) as API. Drug release varied according to the salt form, due to changes in matrix hydration and permeability caused by different crystal lattices. Although the thermodynamic stability of crystalline products is superior to amorphous systems, a thorough understanding of possible polymorphic changes of the API during HME processing and/or subsequent storage of the dosage form is essential to ensure a stable dosage form.

Pharmaceutical delivery systems may also be prepared in an amorphous form using e.g. spray drying or hot-melt extrusion to prepare glassy drugs for enhanced dissolution behavior, or using HME polymers which are intrinsically at least partially amorphous at room or body temperature (e.g. polylactic acid or PLA, polyvinylpyrrolidone or PVP, polyethylene glycol PEG). Although the preparation of amorphous systems may be desirable for drug delivery purposes, there are a number of difficulties associated with their use since they are thermodynamically unstable and will tend to revert to the crystalline form on storage (devitrification). Although the onset of the devitrification process may be so slow as to be effectively irrelevant within the storage time of a product, an understanding of the nature and characterization of the glass transitional behavior is nevertheless essential in order to predict product stability of amorphous HME systems.

An in-depth characterization of amorphous systems using different techniques is essential to predict storage stability of these systems, as the physical structure of glassy materials is more difficult to characterize than crystalline systems. The mechanical properties and vapor sorption profiles of amorphous systems may be markedly different from the crystalline material, while the chemical reactivity of amorphous drugs may be greater. In addition, the behavior of the system below and above the glass transition temperature (T_g, at which the material changes on cooling from a liquid or rubbery state to a brittle state) will differ as the rate of crystallization is much higher above T_g. Hancock and Zografi [2] therefore suggested that T_g should be at least 50°C above the storage temperature to ensure that the product remains stable over its shelf life. A further consideration which is particularly pertinent to the commercial use of amorphous materials is the lack of a 'comfort factor' associated with such systems.

3.3.2 Solid Dispersions

The use of HME to produce solid dispersions in order to improve the dissolution properties of drugs has long been recognized. About five decades ago, Sekiguchi and Obi [3] made the first solid dispersion by melting sulfathiazole and urea, followed by cooling in an ice bath. The resulting eutectic mixture exhibited faster dissolution and better bioavailability than conventional formulations. Since then, solid dispersions have become one of the most-studied drug delivery technologies to solubilize and enhance the dissolution rate of BCS class II compounds. The term 'solid dispersion' refers to the dispersion of one or more active ingredients in a solid state carrier or matrix prepared by the melting (fusion), solvent or melting-solvent method. The concept of solid dispersion is therefore not only limited to poorly water-soluble drugs, but also includes water-soluble drugs dispersed in, for example, amorphous carriers which delay drug release.

A simple classification that has been commonly used to identify solid dispersions in pharmaceutical research is described in Table 3.2.

Table 3.2 *Classification of solid dispersions.*

	Solid dispersion		
	Glassy suspension	Crystalline suspension	Glassy solution
Drug	amorphous	crystalline	amorphous
Carrier	amorphous	amorphous	amorphous
DSC signals	2 T_gs (carrier and drug)	T_g (carrier) + T_m (drug)	1 T_g

The ideal type of solid dispersion for increasing dissolution is a solid glassy solution (often called solid solution) in which the amorphous drug has a lower thermodynamic barrier to dissolution together with a maximally reduced particle size (i.e. molecularly dispersed). In addition, the intimate presence of hydrophilic excipients can increase wetting and lead to supersaturation in the diffusion layer. Solid glassy solutions are thermodynamically stable provided that at storage temperature: the drug load is below the saturation concentration of the drug in the polymer; and that drug migration in the matrix is stopped via interactions with the polymer (e.g. via hydrogen bonds) and/or via immobilization in a carrier with a high glass transition temperature.

When a drug in the crystal form is dispersed in an amorphous polymer, a solid crystalline suspension is formed which is typically used for sustained-release purposes. While solid crystalline suspensions are thermodynamically stable, solid glass suspensions have a higher tendency for recrystallization due to clusters of amorphous drug present in the formulation. This classification indicates that the management of the drug release profile using solid dispersions is achieved by manipulation of the properties of the carrier and dispersed particles, whereby several parameters play an important role: molecular weight and composition of the carrier (e.g. varying the PEG and PEO content can yield immediate as well as sustained-release HME formulations [4, 5]), drug crystallinity, porosity (e.g. total porosity and pore size are key to modulate drug release from EVA mini-tablets prepared by HME [6]) and wettability (e.g. via the addition of hydrophilic polymers or surfactants).

The selection of a suitable polymer (or mixture of different polymers) as a carrier used in a solid dispersion is key to achieve the ultimate stable formulation. Since the polymer affects the dissolution characteristics of the dispersed drug, a water-soluble carrier results in a fast release of the drug from the extruded matrix while a poorly soluble or insoluble carrier reduces the release rate of the drug from the matrix. Moreover, to achieve a fast drug release from the matrix, it is generally necessary that the active drug be a minor component in the dispersion.

3.3.3 Bioavailability Improvement

The Biopharmaceutics Classification System (BCS) is widely used in drug development in order to promote the optimum candidate for development. The pillars (i.e. major factors governing bioavailability) of BCS are solubility and permeability, which are used to divide compounds into four classes as follows.

- Class I: High solubility and high permeability drugs, these product have ideal properties for oral absorption.
- Class II: Low solubility and high permeability drugs, a number of formulation strategies (including HME) have been developed to improve the delivery of BCS class II drugs. The different possibilities to improve the dissolution rate include increasing the surface area available for dissolution by decreasing the particle size, optimizing the wetting characteristics of the compound surface, decreasing the boundary thickness, ensuring sink conditions for dissolution and improving the apparent solubility of the drug under physiologically relevant conditions.
- Class III: High solubility but low permeability, pro-drug strategies are typically used for these compounds.

- Class IV: Low solubility and low permeability drugs, the development of this class of compounds can be risky and no in vitro/in vivo correlations are expected.

Improving the bioavailability of BCS class II drugs can be achieved via chemical (e.g. salt formation, incorporation of polar or ionizable groups in the main drug structure) or formulation approaches. The latter includes solubilization, particle size reduction and solid dispersions, among others.

Another important issue of solid glassy solutions is the fact that the drug is in the amorphous phase and tends to have a higher solubility. To obtain a stable formulation and avoid recrystallization, it is essential to select a suitable carrier which will exhibit specific interactions with functional groups of the drug. During extrusion, the drug should become molecularly dispersed in the polymer and the bounds established between drug and polymer must be strong enough to avoid the tendency of the drug to change to a more thermodynamically stable state via recrystallization. Several studies have identified ion-dipole interactions and intermolecular hydrogen bonding between drugs and polymers, inducing a higher miscibility and a better physical stability of the solid dispersions [7–9]. As supersaturation of the poorly soluble drug upon release from the polymer matrix increases the risk of drug recrystallization (negatively affecting bioavailability), hydrophilic polymers (e.g. PVP, PEG, HPMC) have been added to the formulation to inhibit crystal formation via drug/polymer interaction in solution or polymer adsorption on the initial crystal nucleus [10].

Below are some examples of studies illustrating the benefits of HME to improve the bioavailability of poorly water-soluble drugs.

Six et al. [11] compared the performance of itraconazole solid dispersions prepared by HME to a marketed form (Sporanox®). Although only a limited number of volunteers were used in this study and the existed a high variability in itraconazole pharmacokinetics, the solid dispersion of itraconazole (in combination with HPMC/Eudragit® E100 or Eudragit® E100/PVP-VA64) showed that HME was a valuable method to improve the bioavailability of itraconazole.

The use of PEO as matrix for immediate-release formulation manufactured by HME has been explored by different researchers, as this polymer is an ideal candidate for HME because of its broad processing window (stable up to 200°C) [12]. Li et al. [13] used PEO to improve the solubility of a poorly water-soluble drug, nifedipine. Although processed below the melting point of the drug, the API dissolved in the polymer (as visualized via hot-stage microscope) which resulted in a significant improvement of the drug dissolution rate compared to either pure nifedipine or a physical mixture of PEO and nifedipine. In another study, micronized particles of amorphous itraconazole (stabilized with PVP or HPMC) were melt extruded with PEO and a hydrophilic non-ionic surfactant (Poloxamer 407, a blockcopolymer of polyoxyethyleneoxide and polyoxypropyleneoxide) in order to disperse itraconazole into a hydrophilic polymer matrix [14]. Different analytical techniques demonstrated that HME processing did not alter the properties of the micronized itraconazole particles, and dissolution testing conducted at sink conditions revealed that the dissolution rate of the itraconazole improved by HME.

PVP was the focus of several early investigations into the feasibility of HME [15, 16] and more recently it is enjoying a revival, particularly for the preparation of solid solutions. Patterson et al. [17] reported the use of PVP in the formation of glass solutions using carbamazepine, dipyridamole and indomethacine comparing three different techniques: HME, spray drying and ball milling. HME formulations showed improved solubility in comparison with those prepared by spray drying, and the stability of solid solutions enhanced by the formation of hydrogen bonding between polymer and API.

3.3.4 Controlled Delivery Systems

In temporal control, drug delivery systems aim to deliver the drug over an extended period or at a specific time during treatment. Controlled release over an extended period is highly beneficial for drugs that are rapidly metabolized and eliminated from the body after administration. By controlling the delivery rate of the drug the duration of the therapeutic action can be sustained, improving patient compliance and reducing the incidence of adverse drug reactions. Controlled drug release is possible via different mechanisms using polymers with a variety of physicochemical properties: delayed polymer dissolution, diffusion controlled and reservoir systems. These systems can be manufactured via HME using poorly soluble or insoluble carriers to reduce drug release from the matrix (Figure 3.2).

In the first system, the carrier delays drug dissolution by reducing the rate of exposure to the dissolution medium as the polymer matrix degrades, erodes or dissolves at a slower

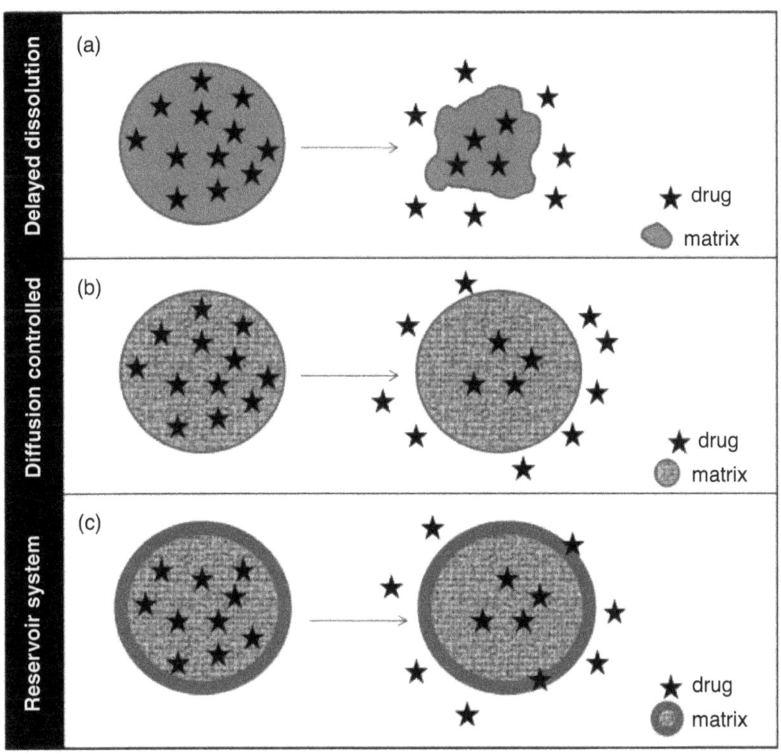

Figure 3.2 Schematic representation of controlled drug release systems: (a) delayed polymer dissolution; (b) diffusion controlled; and (c) reservoir systems.

rate than the drug (Figure 3.2a). Polymers that form a hydrogel when in contact with the dissolution medium (e.g. PEO, PEG, xanthan gum) are also responsible for retarding drug release. Despite its hydrophilicity, a matrix containing a high PEO content or a high molecular weight PEO will form a highly viscous gel at its surface when in contact with the dissolution medium, reducing diffusion of dissolved drug molecules across the gel barrier. (In contrast, formulations with a lower PEO content or low molecular weight PEO tend to increase drug dissolution based on the enhanced wettability.)

Zhang and McGinity [4] studied the effect of polyethylene oxide (PEO) as drug carrier on the release mechanism of chlorpheniramine maleate (CPM) from matrix tablets prepared by HME. The addition of PEG (which is completely miscible with PEO) weakened the cohesive interactions between PEO chains and reduced polymer friction and entanglement by increasing the interchain space between PEO molecules. The melt viscosity therefore decreased significantly and the processability improved. The molecular weight of PEO, the drug load and the inclusion of PEG all affected the processing conditions and drug release properties of the extruded tablet. Drug release from the matrix tablet was controlled by erosion of the PEO matrix and the diffusion of the drug through the swollen gel layer at the surface of the tablets. CPM was dispersed at a molecular level in the PEO matrix at low drug loading level, while recrystallization of CPM after HME was observed at high drug concentrations.

In diffusion-controlled systems, drug diffusion is delayed by an insoluble polymer (Figure 3.2b). The drug will only leach from the matrix after penetration of the dissolution medium into the matrix, dissolution of the drug in the solvent and diffusion of the dissolved drug through the matrix via the pore network. Several authors have described this type of system as one of the most common ways of producing sustained-release HME formulations. Examples of polymers used for HME where the main drug release mechanism is governed by diffusion include: ethylcellulose (EC) [18]; hydroxypropyl methyl cellulose (HPMC) [19]; polymethacrylate polymers [20]; and ethylene vinyl acetate [6].

Ethylcellulose, a polymer with known sustained-release properties when formulated in tablets manufactured by direct compression, has thermoplastic properties making this compound suitable for hot-melt extrusion. However, due to the high extrusion temperature required for HME, ethylcellulose has been commonly combined with plasticizers or other polymers (e.g. HPMC) to improve thermal processing. After HME, ethylcellulose forms an insoluble sustained-release matrix, where drug diffusion is the predominant release mechanism as drug is leached from the dosage form via the porous network formed when drug crystals are dissolved from the inert matrix.

De Brabander et al. [21] assessed ibuprofen release from ethylcellulose matrices combined with hydrophilic polymers (HPMC and xantham gum). The obtained mini-matrices provided a flexible system to tailor the drug release by changing the viscosity, substitution type and concentration of HPMC. Substituting HPMC for xanthan gum yielded formulations having a nearly zero-order drug release without burst effect and complete drug release within 24 h. In addition, the incorporation of xanthan gum resulted in a longer sustained-release effect, allowing a lower concentration of hydrophilic polymer to be used. Rheological and drug diffusion studies of hydrated HPMC and xanthan gum compacts elucidated the difference in the release-controlling ability of both polymers. The higher ability of xanthan gum to control drug release in comparison to HPMC originated from their different hydrophilicity, hydration properties and swelling behavior. A gradual increase in liquid uptake and swelling prevailed for the HPMC systems while the maximum liquid uptake and swelling was reached within 2 h for the xanthan gum formulation, quickly forming a viscous gel around the matrix core.

Another example of a diffusion-controlled system manufactured via HME is ethylene vinyl acetate (EVA) matrices [4]. The low T_g of EVA polymers (around $-25°C$, independent of the vinyl acetate content) allowed easy processing via HME, and its hydrophobic chains ensured sustained-release capacity. A porous network similar to an ethylcellulose matrix is formed when the dispersed crystalline drug is dissolved and released by means of percolation, leaving behind an empty porous EVA carcass. However, a specific feature of EVA matrices is the elastic rearrangement once part of the drug is leached from the tablet. While the matrix is initially (i.e. after HME) structurally supported by drug crystals, the structure partially collapses during drug release, reducing the number of pathways available for release of the remaining drug fraction. In addition to the elastic rearrangement, scanning electron microscopy (SEM) and x-ray tomography experiments showed a further reduction of the porosity of the EVA40 matrix due to molecular rearrangement of EVA40 (40% vinyl acetate content) during dissolution.

EVA polymers have also been used for the production of controlled-release reservoir systems via hot-melt extrusion. Based on this technology, contraceptive controlled release systems have been developed: e.g. Implanon®, an implant designed to release progestagen over a period of three years, or NuvaRing® [22], a contraceptive vaginal ring releasing both progestagen and estrogen over a period of 21 days. The principle of a reservoir system is that the drug is incorporated in a bulk polymer that is surrounded by a permeable membrane (Figure 3.2c). As a consequence of the concentration difference over the membrane, the drug dissolved in the core will diffuse through the membrane. The release rate of a reservoir system is controlled by the drug concentration in the core, permeability properties of the polymeric membrane and surface and thickness of the membrane. While conventional reservoir systems are prepared via film coating of solid dosage forms (e.g. tablets), co-extrusion can be used to manufacture reservoir systems via HME. Co-extrusion consists of extruding two (or more) materials through a single die with two (or more) orifices arranged so that the extrudates merge and weld together. The main advantage of this manufacturing technique is the capacity of producing bi-phasic or multi-phasic drug formulations.

Quintavalle et al. [23] prepared a sustained-release dosage form of theophylline via co-extrusion of a bilayered cylinder. The release rate was determined by the dimensions of the co-extrudate and the composition of the inner (PEG-based) and outer (microcrystalline-based) cylinder.

3.3.5 Plasticizers

In addition to the drug and polymer, the incorporation of plasticizers in the formulation is often essential to allow processing via HME. Plasticizers occupy sites along the polymer chain and prevent chain–chain interactions, significantly reducing the frictional forces between chains and hence providing polymer chain mobility. They therefore lower the T_g of the polymer, reducing the processing temperature during HME processing. Table 3.3 lists the plasticizing agents typically used for HME.

Using a plasticizer in a formulation is essential in order to increase the processability by lowering the extrusion temperature (effectively limiting the risk of thermal degradation),

Table 3.3 *Commonly used plasticizers during hot-melt extrusion of pharmaceutical formulations.*

Phthalate esters (dimethyl, diethyl, dibutyl, dioctyl phthalate)
Citrate esters (triethyl, tributyl, acetyl triethyl, acetyl tributyl citrate)
Fatty acid esters (butyl stearate, glycerol monostearate)
Sebacate esters (dibutyl sebacate)
Vitamin E TPGS
Polyethylene glycol, propylene glycol, polyethylene oxide
Triacetin
Surfactants (polysorbates, docusate sodium, polyethylene glycol monostearate)
Carbon dioxide

improve flow during extrusion and reduce the brittleness of the end product. The selection of the plasticizer is based on thermal analysis of polymer/plasticizer blends, chemical stability (including during storage) and possible molecular interactions between plasticizer and drug. Entwistle and Rowe [24] studied polymer/plasticizer interactions by measuring the intrinsic viscosity of polymer solutions in pure (liquid) plasticizers. EC and hydroxypropylmethyl cellulose were investigated in a series of dialkyl phthalates (dimethyl, diethyl, dibutyl and dioctyl phthalate) and in a series of liquid glycols. The intrinsic viscosity of the polymer/plasticizer solutions was correlated with the mechanical properties of cast films; tensile strength, elongation at rupture and work done in stressing-to-failure were at a minimum when the intrinsic viscosity was at a maximum. This correlation held only within a homologous series of plasticizers and none was found for plasticizers of different chemical structures.

Verreck et al. [25] explored the possibilities of pressurized carbon dioxide as a temporary plasticizer during hot-melt extrusion of polyvinylpyrrolidone-co-vinyl acetate, Eudragit® E100 and ethylcellulose. Carbon dioxide was effective as plasticizer for all three polymers, reducing the processing temperature during the hot-melt extrusion process. Furthermore, due to the foaming action of the volatile plasticizer at the extrusion die, the specific surface area and porosity of the HME matrix increased resulting in enhanced dissolution.

Although a plasticizer is intentionally added to the API/polymer blend in most formulations, APIs themselves have also been effective plasticizers during HME [26, 27]. De Brabander et al. [27] identified ibuprofen as plasticizer for hot-melt extruded ethylcellulose. Thermal analysis of EC extrudates containing 0–20% (w/w) ibuprofen showed the compatibility between drug and polymer (single glass transition temperature), and a drop of T_g in function of ibuprofen concentration in the solid solutions indicated the plasticizing effect of ibuprofen. The plasticizing efficiency was of the same magnitude as for the traditionally used plasticizers. Infrared spectroscopy of the molecular dispersions confirmed chemical interactions via hydrogen bonds between ibuprofen and ethylcellulose. Methylparaben, ibuprofen and chlorpheniramine maleate also had a plasticizing effect on Eudragit® RS 30D [28], reducing T_g and decreasing the tensile strength of the films containing higher levels of ibuprofen and methylparaben.

In conclusion, the selection of drug, polymer(s) and plasticizers in a formulation intended for HME is mainly related to the following issues: thermal stability of drug, polymer and other added substances (directly related to their thermal, chemical and physical properties); processability of the polymer (T_g, thermal degradation, viscosity, molecular weight); drug/polymer miscibility; compatibility between components; extrudate stability; and functionality of the final dosage form.

3.4 Characterization of Extrudates

The objective of processing a formulation via HME is to transform thermoplastic materials into homogeneous extrudates with a specific shape. In order to manufacture a homogeneous system via HME, the materials require a substantial energy uptake which is provided by elevated temperatures, high shear forces and pressure to enable the intense mixing of drug and carrier(s) during processing. To characterize the physical nature of extrudates at molecular or microscopic level, several methods can be used. An overview of the most common techniques for physicochemical characterization of HME-processed formulations (and some examples to illustrate their applications) is provided in the following sections.

3.4.1 Thermal Analysis

Knowledge of the thermal behavior of the drug and polymers incorporated in the formulation is an essential aspect of dosage form development for HME, as product performance in terms of dissolution, bioavailability and stability highly depends on its thermal properties.

The most common application of differential scanning calorimetry (DSC) for HME formulations is the assessment of drug crystallinity following HME processing and during storage, in order to determine its impact on drug release and bioavailability.

During development of an HME formulation thermal analysis is essential to determine the process conditions during HME, based on the thermal stability of the individual components as determined via DSC and/or thermal gravimetric analysis (TGA) (Table 3.4). The extrusion temperature should be around 20–30°C higher than the glass transition temperature (T_g) of the polymer to ensure good flow properties during HME, but below the thermal

Table 3.4 *Thermal properties of some thermoplastic polymers used for hot-melt extrusion (adapted from Kolter [30]).*

Polymer	T_g (°C)	Process temperature (°C): Without plasticizer	Process temperature (°C): + 10% PEG 1500	Degradation temperature (°C)
Kollidon® VA 64	101	155–200	120–155	230
Soluplus®	70	120–200	90–120	250
Kollidon® 12 PF	90	115–165	75–115	225
Kollidon® 17 PF	138	170–180	140–170	175
Kollidon® SR	152	140–180	90–140	210
Kollicoat® IR	208[a]	160–185	140–160	200
Kollicoat® Protect	205[a]	160–185	140–160	200

[a]Melting temperature

degradation temperature of any of the ingredients. The extrusion temperature of Eudragit® E for instance is limited to 180°C, as TGA indicated a 30% weight loss which started at 180°C due to the loss of the dimethylaminogroup of the polymer [29].

DSC also assists in the design of HME formulations by linking the thermal data with the miscibility of the individual components in a formulation (drug/polymer(s), polymer/polymer, plasticizer/polymer). De Brabander et al. [27] illustrated that the miscibility between ethylcellulose and ibuprofen influenced their thermal properties. Melting of the pure polymer led to partial recrystallization afterwards; when combined with ibuprofen, partial recrystallization was no longer observed. The incorporation of the drug resulted in a homogeneously dispersed system where recrystallization of ethylcellulose was inhibited. Experimentally, miscibility can be demonstrated by changes in melting endotherms (e.g. decrease of melt endotherm onset temperature and heat of fusion) and based on the glass transition temperature of the mixtures: a single concentration-dependent T_g lying between the glass transition temperature of the individual components illustrates miscibility.

Thermal analysis has also been used to identify an anti-plasticizing effect of specific components on HME polymers. For example, indomethacine formed a one-phase system with Eudragit® E [31]. However, T_g of the binary mixture was higher compared to pure polymer. Similar observations were made between chitosan and polyacryclic acid (due to complex formation and possibly hydrogen bonds) [32]. Although an anti-plasticizing effect can be beneficial towards stability, it results in a higher process temperature and possibly limits its application.

A new dimension to the thermal analysis was introduced with the modulated DSC (MDSC) technique. This method is based on the same principle as conventional DSC; however, the linear heating rate is superimposed with a sinusoidal wave modulation [33]. This enables the method to dissociate the total heat flow into a reversing component (heat capacity component showing melting and glass transition temperatures) and non-reversing component (kinetically controlled events, e.g. crystallization, evaporation, curing), improving resolution and enhancing sensitivity. MDSC therefore allows analysis of mixtures with overlapping thermal events, e.g. determination of T_g (increase of heat capacity, reversing component) independently of an accompanying enthalpy relaxation (non-reversing component) as described by Janssens et al. [34] or an exothermic event such as crystallization.

3.4.2 Atomic Force Microscopy

Atomic force microscopy (AFM), a method which can study the surface microstructure of hot-melt extrudates, can be used to visualize phase separation and/or non-homogeneity of HME samples. Six et al. [35] identified topographic differences in Eudragit® E dispersions containing different itraconazole concentrations. The polymer alone showed no significant surface discontinuities, whereas at higher drug loadings more surface roughness was observed, possibly indicating phase separation. Qi et al. [36] employed pulsed force mode atomic force microscopy (PFM-AFM), which reduces lateral shear forces between the tip and the sample, to map phase separation at a submicron scale. In formulations with high drug loadings of felodipine, amorphous felodipine co-existed with the crystalline form. The amorphous phase was however mainly concentrated at the centre of the extrudates, due to expansion of the polymer after extrusion.

3.4.3 Residence Time

The material residence time in the extrusion barrel is, besides the extrusion temperature, an important parameter in obtaining homogeneous extrudates with sufficient thermal stability. A high screw speed and powder flow rate significantly reduce the mean residence time. In order to provide intense mixing during HME, a high screw speed is preferred but a too-high screw speed can result in insufficient exposure to the heating zones and insufficient melting and/or dissolving of the drug substance as the residence time becomes too short. Monitoring of the residence time is possible via off-line or on-line detection of a tracer which is incorporated in the formulation (possibly in combination with a transparent barrel to improve visualization).

3.4.4 Spectroscopic Techniques

Non-destructive spectroscopic techniques (UV-VIS, IR, Raman, NIR) can be used for quantitative (e.g. concentration) as well as qualitative (e.g. drug crystallinity, identification of polymer/polymer or polymer/API interactions) analysis of HME formulations.

Wang et al. [37] used UV-VIS for in-line monitoring of the thermal degradation (molar mass reduction and color formation) during extrusion of poly(L-lactic acid) (PLA). This showed that degradation of extruded PLA depended on the processing parameters (temperature, torque and screw speed).

Qi et al. [38] used infrared spectroscopy to verify the homogeneity at the surface and in a cross-section of a paracetamol/Eudragit® E extrudate. Attenuated total reflectance-Fourier transform infrared spectroscopy (ATR-FTIR) spectra showed a higher intensity in the cross-section spectra compared to the extrudate surface. Besides homogeneity, IR allows interactions between API and carrier to be detected as shown by Jeung and Mishra [32]: IR-spectra identified complex formation and possible hydrogen bonding between chitosan and polyacrylic acid during the melt process.

Saerens et al. [39] evaluated Raman spectroscopy as an in-line PAT tool to monitor the API concentration and the solid state of the formulation during HME. A similar study was conducted by Tumuluri et al. [40] with clotrimazole and ketoprofen as APIs. Raman spectroscopy has also been used in combination with HME to identify interactions between drug and polymers based on peak shifts in the spectra of e.g. metoprol tartrate/Eudragit® RS PO [39] and to determine the residence time distribution [41]. Raman spectroscopy is also a valuable tool to study drug distribution in a matrix: at low processing temperatures, the high viscosity of PEO limited drug diffusion and resulted in poor homogeneity; the Raman spectra however showed that higher processing temperatures yielded a uniform drug distribution due to the lower PEO viscosity.

The application of in-line NIR spectroscopy to monitor the extrusion process was demonstrated by Fischer et al. [42] by successfully determining the ethylene vinyl acetate (EVA) content in a polypropylene matrix. Using NIR as a non-destructive on-line monitoring tool of critical parameters during HME improves process efficiency and product quality as adjustments to the process can be made in real time when deviations in the process are detected during HME. This was shown by Coates et al. [43] for polyethylene (PE)/polypropylene (PP) blends where rapid and accurate determination of the ethylene content is essential.

3.4.5 X-ray Diffraction (XRD)

As X-ray patterns provide a specific fingerprint of a molecule, this technique can be used to monitor a single component in a complex HME formulation. XRD is commonly used to determine the solid state of the drug in an HME formulation immediately after processing and during storage, and to assess the impact on dissolution and bioavailability. For example, the absence of sharp and narrow peaks of crystalline drug in the diffractogram of a solid glassy solution is imperative to ensure a fast release and high bioavailability of a poorly water-soluble BCS class II drug. XRD has also been used to establish maximum drug solubility in polymers [44] and to monitor crystallization (type, orientation) induced by shear flow after processing of the melt [45].

3.4.6 Microscopy

While hot-stage microscopy can be used to visualize the thermal events in a formulation, scanning electron miscroscopy (SEM) is used to identify the microscopic structure of a dosage form manufactured via HME and allows characterization of phenomena such as differences of crystal growth in the bulk and at the surface of a dosage form. Bruce et al. [46] investigated the effect of hydrophilic polymers on crystal growth in melt extrudates via scanning electron microscopy. Drug recrystallization was lower in extrudates containing polycarbophil and PVP compared to formulations without additive.

3.4.7 Drug Release

Dissolution testing of HME formulations using compendial dissolution methods is an essential part to evaluate the effect of HME processing on the functionality of the end product, e.g. fast release of an amorphous poorly water-soluble drug embedded in a hydrophilic polymer, or sustained release of a highly water-soluble crystalline drug in a hydrophobic matrix.

In addition, determination of the wettability (via the contact angle as a measure of surface energy) and porosity of an HME dosage form also provides valuable information on drug release. Porosity of solid dosage forms can be determined via conventional techniques (helium or mercury porosimetry), but X-ray tomography (computed tomography or CT-scan) provides an in-depth view of the porosity of a solid structure. Based on the total porosity (expressed as the percentage of pore volume to its total volume) and the pore distribution before and after dissolution, Almeida et al. [6] showed via X-ray tomography that elastic rearrangement of the EVA matrix reduced the pore size which sustained drug release over a longer period (Figure 3.3). Quinten et al. [47] used X-ray tomography to visualize the internal matrix structure of an injection-molded tablet containing L-HPC and EC, showing an anisotropic skin-score microstructure with pores mainly localized opposite to the injection point. The importance of porosity in relation to drug release was also defined by Verreck et al. [25] using a pressurized carbon dioxide to foam the extrudate at the die exit. The higher porosity (and increased specific surface area) resulted in faster dissolution rates.

3.5 Hot-melt Extruded Dosage Forms

The appeal of HME for the pharmaceutical industry is certainly based on the wide array of dosage forms that can be manufactured via this technique for a large number of applications.

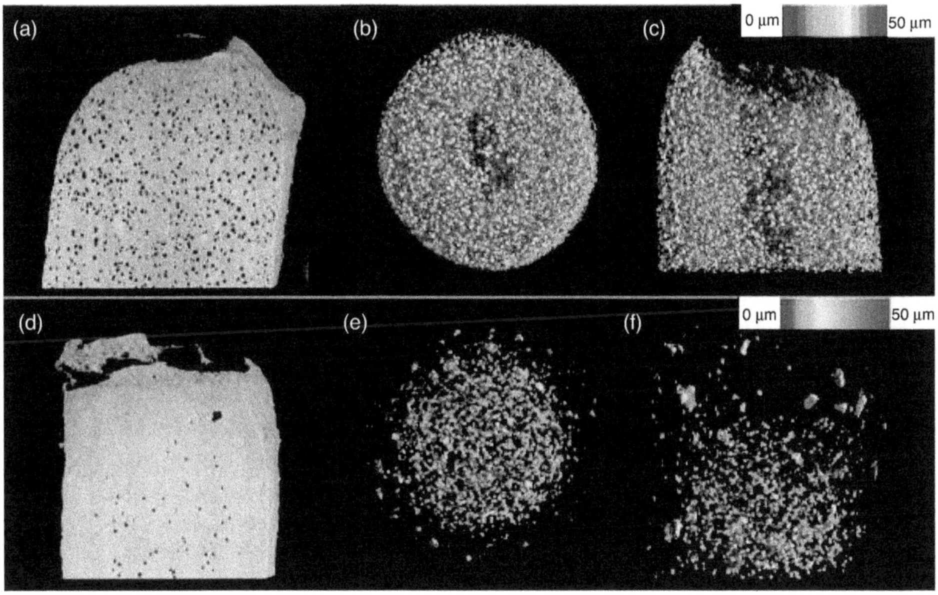

Figure 3.3 *X-ray tomography renderings of EVA40 matrices with 50% of metoprolol tartrate (MPT). (a, c) Axial and (b) radial cross-sections before dissolution. (d) Axial cross-section after 24 h dissolution and (e) radial and (f) axial cross-sections after 72 h dissolution. Black spots (a, d) indicate pores. The color scale used in (b), (c), (e) and (f) represents the pore size (maximum opening) where blue represents small pores and red represents larger pores. S. Almeida et al. 2011, reproduced with the permission of Elsevier. For a better understanding of the figure, please refer to the color section.*

Proper selection of the thermoplastic polymer and additives in the formulation defines the application (e.g. sustained release, immediate release, taste-masking, bioadhesion, etc.), while the type of dosage form manufactured via HME is determined by post-processing of the material (using a die with a specific design, calendering, injection molding, pelletizing, milling and tabletting, etc.). The following sections highlight the most common dosage forms and pharmaceutical applications of HME.

3.5.1 Oral Drug Delivery

As previously illustrated, processing of thermal polymers via HME is a valuable method to manufacture immediate-release (using solid solutions/dispersions formulated with hydrophilic polymers, e.g. PVP, PEO, HPC) and sustained-release (using hydrophobic matrices based on e.g. EC, EVA, Eudragit® RS/RL) dosage forms which have been presented as granules, tablets, pellets, rods or mini-tablets.

Miyagawa et al. [48] and Sato et al. [49] prepared controlled release granules via HME, containing carnauba wax as matrix former. The formulation was easily processed into strong granules (even below the melting point of the wax) and diclofenac release from the granules depended on the additives in the formulation (HPC, Eudragit® L, sodium chloride) due to the physicochemical properties (i.e. swelling and solubility) of these dissolution rate controlling agents.

Tablets of an HME formulation can be formed via calendering or injection molding of the hot thermoplastic material, or via milling of the cooled extrudate followed by mixing with external excipients (disintegrant, lubricant) and tableting.

Direct pelletization of the cylindrical strand at the die exit (die face cutting) is an efficient means to formulate an HME formulation as a multi-particulate dosage form (pellets, mini-tablets), especially important for sustained-release HME formulations. For a better control of the drug release, Young et al. [50] processed the cylindrical matrices obtained after pelletizing into spherical particles, which could subsequently be compressed into rapidly disintegrating tablets without affecting the drug release profile from the pellets [51].

The wide range of available thermoplastic polymers and/or specific manufacturing techniques allows oral HME formulations to be designed for a myriad of applications as follows.

- Taste masking: via embedding of the drug in a polymer or lipid matrix during HME, or via ionic interactions induced during thermal processing of drug and polymer [52].
- Enteric formulations: the enteric polymers traditionally used for film-coating purposes (e.g. Eudragit® L/S, HPMC-AS, CAP) have the potential to formulate enteric matrix formulations based on their thermoplastic properties. Andrews et al. [53] and Schilling et al. [54] identified that plasticized matrices of Eudragit® L and S provided sufficient gastric protection when processed via HME. A specific application of HME for the manufacturing of a gastro-protective dosage form was presented by Mehuys et al. [55], manufacturing hollow pipes of enteric polymers (PVAP and HPMC-AS). After filling these pipes with a drug-containing powder mixture and thermal sealing of both ends, this system had excellent resistance against the gastric acidic environment.
- Gastro-retentive systems based on the buoyancy of porous matrices. Fukuda et al. [56] described a controlled-release matrix tablet (using Eudragit® RS and Eudragit® E as polymers) with a porous structure due to the thermal decomposition (CO_2 formation) of sodium bicarbonate during HME.
- Targeted drug delivery via HME processing of pH-sensitive polyacrylate-based polymers, e.g. colon delivery of 5-ASA using Eudragit® S100 [57].
- Orally disintegrating dosage forms using highly soluble sugar alcohols (sorbitol, mannitol, xylitol) as drug carriers for rapid dissolution in the oral cavity [58].
- Injection molding is an important manufacturing process to create a variety of complex shapes with high-dimensional precision. This can be used as an extension of HME as the molten material obtained from an extruder is directly transferred by means of an injection step into a closed and shape-specific mould cavity. After solidification, the article with a shape duplicating the cavity is recovered by opening the mold to release the product, eliminating any need for a post-processing step. Due to mould design flexibility, this technique can be used not only to manufacture tablet-shaped items but also a variety of structures including medical devices, implants, tissue-engineered scaffolds and vaginal rings [59]. Quinten et al. [47] evaluated this technique to manufacture sustained-release matrix tablets using ethylcellulose or acrylate polymers as matrix former, in combination with hydrophilic polymer to modify drug release.

  Obviously injection molding is not limited to sustained-release applications, as all thermoplastic polymers used for HME have the potential to be processed via injection molding. Even complex drug delivery systems with a drug-containing core and a protective outer shell (ethylcellulose and cetostearylalcohol) can be manufactured (Egalet system) to obtain zero-order drug release via controlled erosion of the core (polyethylene glycol monostearate and PEO) [59].

3.5.2 Films

Adhesive films have been used for drug delivery via the transdermal, transmucosal or transungual route. While casting from organic or aqueous solvents was the main technique to manufacture these films, the disadvantages of film casting (long process time, high cost, environmental toxicity, waste) have promoted HME to manufacture film-based drug delivery systems.

The possibility of producing film via HME technology was initially demonstrated by Aitken-Nichol et al. [60] using an acrylic film (Eudragit® E) and lidocain (which plasticized the acrylic film). However, the main body of work in relation to HME films was done by the group of Repka, focusing on HPC and PEO as bioadhesive carriers. Processing HPC film via HME was facilitated using a variety of conventional (e.g. triethyl citrate or TEC, PEG, acetyltributylcitrate) and non-traditional (chlorpheniramine maleate, Vitamin E TPGS) plasticizers [26,61]. Crowley et al. [62] investigated the properties of transdermal PEO films containing either guaifenesin or ketoprofen. Both drugs plasticized the polymer during thermal processing; however, miscibility of ketoprofen and PEO was better as crystals of guaifenesin were detected on the film surface (ketoprofen could be dissolved in PEO up to a 15% concentration).

Transmucosal drug delivery via films (mainly via the buccal route) is advantageous for drugs with a high first-pass metabolism. This application requires thin, flexible and bioadhesive films sufficiently strong to withstand the mechanical stress in the oral cavity to sustain drug release over a longer period. Again PEO and HPC have been used for this purpose, providing zero-order release of clotrimazole over a period of up to 10 h (the release rate depends on the molecular weight of the polymers) [63]. PEO films are also a tool to formulate thermolabile drugs in a transmucosal drug delivery system: Δ^9-tetrahydrocannabinol and its prodrug were successfully incorporated in an HME-processed PEO film, using suitable plasticizers and additives to minimize drug degradation [64].

A specific application of HME films is the treatment of onychomycosis (nail infections) where targeted local drug delivery via a bioadhesive film improves patient compliance as oral systemic antifungal therapy requires long-term treatment, suffers from systemic side effects and has a low success rate. PEO and HPC films manufactured via HME are effective tools for this application based on their bioadhesion to nails and the incorporation of poorly water-soluble antifungal drugs as an amorphous phase (e.g. up to 20% ketoconazole) [65]. Sustained release of itraconazole from HPC film could be modified by tuning the hydration of the film based on the molecular weight and crystallinity of the HPC polymers [66], while bioadhesion of HPC films to nails improved with the addition of tartaric acid (due to modifications at the nail surface and the plasticizing effect of tartaric acid) [67].

3.5.3 Vaginal Rings and Implants

Vaginal rings are flexible, torus-shaped, elastomeric drug delivery devices that provide long-term release of substances to the vagina for local or systemic effect. They are designed to be self-inserted and removed, and are positioned in the upper third of the vagina adjacent to the cervix. The concept of drug delivery to the human vagina using vaginal ring devices initially focused on steroid-releasing silicone elastomer rings, resulting in the market introduction of a ring for estrogen replacement therapy (Estring®). Polyethylene vinyl acetate-based vaginal rings have also been commercialized (Nuvaring®, Femring®). The Nuvaring, extensively studied by Van Laarhoven et al. [68, 69], consists of a coaxial fiber (reservoir system prepared via co-extrusion) using EVA 28 and EVA 9 in the core and membrane, respectively, to provide sustained drug release over a period of 21 days. Although the drugs (etonogestrel and ethinyl estradiol, incorporated in the core) completely dissolved in the polymer melt during HME, the process conditions, drug concentration, storage temperature and storage time determined the tendency for drug recrystallization (hence drug-release kinetics) in the EVA matrix after cooling to room temperature.

More recently, Johnson et al. [70] reported about a segmented polyurethane intravaginal ring for sustained delivery of antiretroviral agents (dapivirine and tenofovir) to prevent the male-to-female sexual transmission of the human immunodeficiency virus. Due to the different hydrophilicity of both drugs, tenofovir and dapivirine were separately formulated into polymers with matching hydrophilicity via hot-melt extrusion using a hydrophilic water swellable and a hydrophobic non-water swellable polyurethane grade. The drug-loaded rods were joined together to form dual segment intravaginal polyurethane rings. Within their polymeric segments dapivirine and tenofovir were amorphous and crystalline, respectively. In vitro release of both drugs from the segmented device was sustained over 30 days.

A silicone-based intravaginal ring with inserts was proposed by Morrow et al. [71] to allow controlled release of hydrophilic and/or macromolecular drugs (e.g. peptides, proteins, antibodies). As their poor permeation through hydrophobic elastic polymers limited their application in a conventional silicone ring, a drug-free ring with openings to insert small drug-containing rods was prepared. These inserts were manufactured via tabletting or thermal processing and used polymers (e.g. HPMC or silicone combined with hydrophilic compounds) to control drug release over periods varying from several hours to several weeks.

EVA-based matrices have also been used for contraceptive implants (e.g. Implanon®, Nexplanon®). A single rod of a sustained-release hormonal implant is inserted just under the skin of a woman's upper arm, releasing its drug content over a 3-year period.

Implants processed via HME and formulated with poly(D,L lactide-co-glycolide) (PLGA) have been studied for a number of applications, e.g. controlled release of gentamycin for local treatment of osteomyelitis [72] or incorporation of a protein (lysozyme) in an HME matrix with full recovery of the biological activity [73]. The factors controlling drug release from the biodegradable PLGA matrix were polymer degradation and erosion, which were controlled by the physical properties of the polymer (e.g. molecular weight and viscosity). Chemical modification of the PLGA structure via the formation of a block-copolymer with PEO also modified the drug release pattern from the matrix as swelling was observed due to relaxation of the PEO blocks [74].

Polycaprolactone implants were manufactured via injection molding for sustained release of 5-fluorouracil. Drug release was diffusion-controlled and dependent upon the dimensions and drug load of the system [75].

3.6 A View to the Future

Today's challenges during drug development have exponentially increased due to increased competition by cheap generics, reduced healthcare budgets in many countries, increased requirements for safety and efficacy, counterfeiting activities with criminal intent, increased pressure to supply cheap drugs and poor chemical and physical stability of many new drug compounds. These factors have had a tremendous negative impact on the time, cost and success rate of pharmaceutical drug development and are forcing the pharmaceutical industry to consider new efficient and versatile manufacturing techniques. As hot-melt extrusion allows the continuous production of a variety of dosage forms for multiple applications, this drug delivery platform offers the opportunity to improve the efficacy within the pharmaceutical industry.

Despite more than a century of experience in the plastics and rubber industry and having been used since 1930 in the food industry, HME is a relatively new technology in the pharmaceutical industry; it has however been widely advocated as a valuable method to produce drug delivery systems. Replacement of traditional batch processes with a consistent continuous process, solvent-free manufacturing of solid molecular dispersion, applicable for high dose as well as potent compounds, good content uniformity based on the intense mixing capacity, potential of automation and reduction of labor costs are just some of the advantages that have made HME worthy of consideration for pharmaceutical applications.

Barriers to the implementation of HME in the pharmaceutical industry are related to the thermal processing (and associated risk of thermal degradation) and the significant capital investment initially required (as availability of HME equipment is currently limited at pharmaceutical manufacturing sites). However, proper selection of extruder design (screw assemblies, die, etc.) and formulation (polymers, plasticizers and/or processing aids) will reduce the material residence time in the extruder, minimizing the negative impact on product quality and degradation of drug and polymers. Introduction of HME as a single-step continuous process can significantly reduce manufacturing costs (as batch processors are prone to out-of-spec manufacturing based on significant batch-to-batch variability), warranting the initial capital investment in HME technology.

Another barrier to the use of HME in pharmaceutical manufacturing is the limited number of thermoplastic polymers available for development of HME formulations. This complicates the development process as freedom-to-operate might not be guaranteed based on the complex intellectual property landscape regarding HME and thermoplastic polymers.

As a large fraction of the new chemical entities have a low bioavailability due a low aqueous solubility, the application of HME technology in the pharmaceutical industry has tended to focus on formulations which increase the efficacy of these compounds via the formation of solid dispersions. HME is the preferred manufacturing method for this purpose as solvent-based systems suffer from many issues (residual solvent, handling of large solvent volumes, possible phase separation during evaporation, solvent must dissolve API and polymer). Although solid dispersions offer a means to considerably enhance product performance, the stability of these systems remains a challenge. Although this is not a specific issue of HME but of solid solutions/dispersions in general (independent of the manufacturing method), the availability of new thermoplastic polymers specifically designed for HME applications might provide additional benefits to dissolve and stabilize APIs in polymeric matrices. An interesting development in this respect was the launch of

Soluplus® in 2009, a thermoplastic polymer suitable for HME (T_g 70°C). This copolymer of PEG, vinylcaprolactam and vinylacetate is designed to provide excellent solubilization capacity for poorly aqueous drugs in solid dispersions.

Although solid dispersions will probably remain the major focus of HME within the pharmaceutical industry, the broad scope of HME applications (oral solid dosage forms, implants, bioadhesive films, stents, etc.) supports HME as a versatile processing technology for drug delivery systems, with an exciting future within the pharmaceutical industry.

References

(1) Quinten, T. (2010) *Evaluation of injection molding as a pharmaceutical production technology for sustained-release matrix tablets.* PhD thesis, Ghent University.

(2) Hancock, B.C. and Zografi, G. (1997) Characteristics and significance of the amorphous state in pharmaceutical systems. *Journal of Pharmaceutical Sciences*, **86**, 1–12.

(3) Sekiguchi, K. and Obi, N. (1961) Studies on absorption of eutectic mixture. I. A comparison of the behavior of eutectic mixture of sulfathiazole and that of ordinary sulfathiazole in man. *Chemical & Pharmaceutical Bulletin*, **9**, 866–872.

(4) Zhang, F. and McGinity, J.W. (1999) Properties of sustained-release tablets prepared by hot-melt extrusion. *Pharmaceutical Development & Technology*, **4**, 241–250.

(5) Crowley, M.M., Zhang, F., Koleng, J.J. and McGinity, J.W. (2002) Stability of polyethylene oxide in matrix tablets prepared by hot-melt extrusion. *Biomaterials*, **23**, 4241–4248.

(6) Almeida, A., Possemiers, S., Boone, M.N., De Beer, T., Quinten, T., Van Hoorebeke, L., Remon, J.P. and Vervaet, C. (2011) Ethylene vinyl acetate as matrix for oral sustained release dosage forms produced via hot-melt extrusion. *European Journal of Pharmaceutics & Biopharmaceutics*, **77**, 297–305.

(7) Zografi, G. and Taylor, L.S. (1997) Spectroscopic characterization of interactions between PVP and indomethacin in amorphous molecular dispersions. *Pharmaceutical Research*, **14**, 1691–1698.

(8) Schachter, D.M., Tirol, G.C. and Xiong, J.C. (2002) Solid state NMR perspective of drug-polymer solid solutions: a model system based on poly(ethylene oxide). *International Journal of Pharmaceutics*, **281**, 89–101.

(9) Vasanthavada, M., Tong, W.Q., Joshi, Y. and Kislalioglu, M.S. (2005) Phase behavior of amorphous molecular dispersions – II: Role of hydrogen bonding in solid solubility and phase separation kinetics. *Pharmaceutical Research*, **22**, 440–448.

(10) Janssens, S. and Van den Mooter, G. (2009) Review: physical chemistry of solid dispersions. *Journal of Pharmacy & Pharmacology*, **61**, 1571–1586.

(11) Six, K., Daems, T., de Hoon, J., Van Hecken, A., Depre, M., Bouche, M.P., Prinsen, P., Verreck, G., Peeters, J., Brewster, M.E. and Van den Mooter, G. (2005) Clinical study of solid dispersions of itraconazole prepared by hot-stage extrusion. *European Journal of Pharmaceutical Sciences*, **24**, 179–186.

(12) Verhoeven, E., De Beer, T.R.M., Schacht, E., Van den Mooter, G., Remon, J.P. and Vervaet, C. (2009) Influence of polyethylene glycol/polyethylene oxide on the release

characteristics of sustained-release ethylcellulose mini-matrices produced by hot-melt extrusion: in vitro and in vivo evaluations. *European Journal of Pharmaceutics & Biopharmaceutics*, **72**, 463–470.

(13) Li, L., AbuBaker O. and Shao, Z.Z.J. (2006) Characterization of poly(ethylene oxide) as a drug carrier in hot-melt extrusion. *Drug Development & Industrial Pharmacy*, **32**, 991–1002.

(14) Miller, D.A., McConville, J.T., Yang, W., Williams, R.O. and McGinity, J.W. (2007) Hot-melt extrusion for enhanced delivery of drug particles. *Journal of Pharmaceutical Science*, **96**, 361–376.

(15) Gruenhagen, H. (1996) Polymer/drug-melt extrusion: therapeutic and technological appeal. *Pharmaceutical Technology Europe*, **11**, 22–27..

(16) Forster, A., Hempenstall, J., Tucker, I. and Rades, T. (2001) Selection of excipients for melt extrusion with two poorly water-soluble drugs by solubility parameter calculation and thermal analysis. *International Journal of Pharmaceutics*, **226**, 147–161.

(17) Patterson, J.E., James, M.B., Forster, A.H., Lancaster, R.W., Butler, J.M. and Rades, T. (2007) Preparation of glass solutions of three poorly water soluble drugs by spay drying, melt extrusion and ball milling. *International Journal of Pharmaceutics*, **336**, 22–34.

(18) Verhoeven, E., De Beer, T.R.M., Van den Mooter, G., Remon, J.P. and Vervaet, C. (2008) Influence of formulation and process parameters on the release characteristics of ethylcellulose sustained-release mini-matrices produced by hot-melt extrusion. *European Journal of Pharmacology & Biopharmacology*, **69**, 312–319.

(19) Young, C.R., Dietzsch, C., Cerea, M., Farrell, T., Fegely, K.A., Rajabi-Siahboomi, A. and McGinity, J.W. (2005) Physicochemical characterization and mechanisms of release of theophylline from melt-extruded dosage forms based on a methacrylic acid copolymer. *International Journal of Pharmaceutics*, **301**, 112–120.

(20) Yang, R., Wang, Y.J., Zheng, X., Meng, J., Tang, X. and Zhang, X.F. (2008) Preparation and evaluation of ketoprofen hot-melt extruded enteric and sustained-release tablets. *Drug Development & Industrial Pharmacy*, **34**, 83–89.

(21) De Brabander, C., Vervaet C. and Remon, J.P. (2003) Development and evaluation of sustained release mini-matrices prepared via hot melt extrusion. *Journal of Controlled Release*, **89**, 235–247.

(22) van Laarhoven, J.A.H., Kruft, M.A.B. and Vromans, H. (2002) In vitro release properties of etonogestrel and ethinyl estradiol from a contraceptive vaginal ring. *International Journal of Pharmaceutics*, **232**, 163–173.

(23) Quintavalle, U., Voinovich, D., Perissutti, B., Serdoz, F., Grassi, G., Dal Col, A. and Grassi, M. (2008) Preparation of sustained release co-extrudates by hot-melt extrusion and mathematical modelling of in vitro/in vivo drug release profiles. *European Journal of Pharmaceutical Sciences*, **33**, 282–293.

(24) Entwistle, C.A. and Rowe, R.C. (1979) Plasticization of cellulose ethers used in the film coating of tablets. *Journal of Pharmacy & Pharmacology*, **31**, 269–272.

(25) Verreck, G., Decorte, A., Li, H., Tomasko, D., Arien, A., Peeters, J., Rombaut, P., Van den Mooter, G. and Brewster, M. (2006) The effect of pressurized carbon dioxide as a plasticizer and foaming agent on the hot melt extrusion process and extrudate properties of pharmaceutical polymers. *J. Supercritical Fluids*, **38**, 383–391.

(26) Repka, M.A. and McGinity, J.W. (2000) Influence of vitamin E TPGS on the properties of hydrophilic films produced by hot-melt extrusion. *International Journal of Pharmaceutics*, **202**, 63–70.
(27) De Brabander, C., Van den Mooter, G., Vervaet, C. and Remon, J.P. (2002) Characterization of ibuprofen as a nontraditional plasticizer of ethyl cellulose. *Journal of Pharmaceutical Sciences*, **91**, 1678–1685.
(28) McGinity, J.W. and Wu, C.B. (1999) Non-traditional plasticization of polymeric films. *International Journal of Pharmaceutics*, **177**, 15–27.
(29) Lin, S.Y., Yu, H.L. and Li, M.J. (1999) Formation of six-membered cyclic anhydrides by thermally induced intramolecular ester condensation in Eudragit E film. *Polymer*, **40**, 3589–3593.
(30) Kolter, K. (2011) Excipients for hot-melt extrusion improving bioavailability. Presented at the APV experts' workshop on hot melt extrusion and its use in the manufacturing of pharmaceutical dosage forms. April 12–13, Tarrytown, NY.
(31) Chokshi, R. and Zia, H. (2004) Hot-melt extrusion technique: a review. *Iranian Journal of Pharmaceutical Research*, **3**, 3–16.
(32) Jeung, S. and Mishra, M.K. (2010) Hot melt reactive extrusion of chitosan with poly(acrylic acid). *International Journal of Polymeric Materials*, **60**, 102–113.
(33) Coleman, N.J. and Craig, D.Q.M. (1996) Modulated temperature differential scanning calorimetry: A novel approach to pharmaceutical thermal analysis. *International Journal of Pharmaceutics*, **135**, 13–29.
(34) Janssens, S., de Armas, H.N., Roberts, C.J. and Van den Mooter, G. (2008) Characterization of ternary solid dispersions of itraconazole, PEG 6000, and HPMC 2910 E5. *Journal of Pharmaceutical Sciences*, **97**, 2110–2120.
(35) Six, K., Murphy, J., Weuts, I., Craig, D.Q.M., Verreck, G., Peeters, J., Brewster, M. and Van den Mooter, G. (2003) Identification of phase separation in solid dispersions of itraconazole and Eudragit E100 using microthermal analysis. *Pharmaceutical Research*, **20**, 135–138.
(36) Qi, S., Belton, P., Nollenberger, K., Gryczke, A. and Craig, D. (2011) Compositional analysis of low quantities of phase separation in hot-melt-extruded solid dispersions: a combined atomic force microscopy, photothermal Fourier-transform infrared microspectroscopy, and localised thermal analysis approach. *Pharmaceutical Research*, **28**, 2311–2326.
(37) Wang, Y., Steinhoff, B., Brinkmann, C. and Alig, I. (2008) In-line monitoring of the thermal degradation of poly(l-lactic acid) during melt extrusion by UV-vis spectroscopy. *Polymer*, **49**, 1257–1265.
(38) Qi, S., Gryczke, A., Belton, P. and Craig, D.Q.M. (2008) Characterisation of solid dispersions of paracetamol and Eudragit E prepared by hot-melt extrusion using thermal, microthermal and spectroscopic analysis. *International Journal of Pharmaceutics*, **354**, 158–167.
(39) Saerens, L., Dierickx, L. Lenain, B., Vervaet, C., Remon, J.P. and De Beer, T. (2011) Raman spectroscopy for the in-line polymer-drug quantification and solid state characterization during a pharmaceutical hot-melt extrusion process. *European Journal of Pharmaceutics & Biopharmaceutics*, **77**, 158–163.
(40) Tumuluri, V.S., Kemper, M.S., Lewis, I.R., rodduturi, S., Majumdar, S., Avery, B.A. and Repka, M.A. (2008) Off-line and on-line measurements of drug-loaded hot-melt

extruded films using Raman spectroscopy. *International Journal of Pharmaceutics*, **357**, 77–84.

(41) Ward, N.J., Edwards, H.G.M., Johnson, A.F., Fleming, D.J. and Coates, P.D. (1996) Application of Raman spectroscopy for determining residence time distributions in extruder reactors. *Applied Spectroscopy*, **50**, 812–815.

(42) Fischer, D., Bayer, T., Eichhorn, K.J. and Otto, M. (1997) In-line process monitoring on polymer melts by NIR-spectroscopy. *Journal of Analytical Chemistry*, **359**, 74–77.

(43) Coates, P.D., Barnes, S.E., Sibley, M.G., Brown, E.C., Edwards, H.G.M. and Scowen, I.J. (2003) In-process vibrational spectroscopy and ultrasound measurements in polymer melt extrusion. *Polymer*, **44**, 5937–5949.

(44) Martínez-Ohárriz, M.C., Rodríguez-Espinosa, C., Martín, C., Goñi, M.M., Tros-Ilarduya, M.C. and Sánchez, M. (2002) Solid dispersions of diflunisal–PVP: polymorphic and amorphous states of the drug. *Drug Development & Industrial Pharmacy*, **28**, 717–725.

(45) Somani, R.H., Hsiao, B.S., Nogales, A., Fruitwala, H., Srinivas, S. and Tsou, A.H. (2001) Structure development during shear flow induced crystallization of i-PP: in situ wide-angle x-ray diffraction study. *Macromolecules*, **34**, 5902–5909.

(46) Bruce, C., Fegely, K.A., Rajabi-Siahboomi, A.R. and McGinity, J.W. (2007) Crystal growth formation in melt extrudates. *International Journal of Pharmaceutics*, **341**, 162–172.

(47) Quinten, T., Gonnissen, Y., Adriaens, E., De Beer, T., Cnudde, V., Masschaele, B., Van Hoorebeke, L., Siepmann, J., Remon, J.P. and Vervaet, C. (2009) Development of injection moulded matrix tablets based on mixtures of ethylcellulose and low-substituted hydroxypropylcellulose. *European Journal of Pharmaceutical Sciences*, **37**, 207–216.

(48) Miyagawa, Y., Okabe, T., Yamaguchi, Y., Miyajima, M., Sato, H. and Sunada, H. (1996) Controlled-release of diclofenac sodium from wax matrix granule. *International Journal of Pharmaceutics*, **138**, 215–224.

(49) Sato, H., Miyagawa, Y., Okabe, T., Miyajima, M. and Sunada, H. (1997) Dissolution mechanism of diclofenac sodium from wax matrix granules. *Journal of Pharmaceutical Sciences*, **86**, 929–934.

(50) Young, C.R., Koleng, J.J. and McGinity, J.W. (2002) Production of spherical pellets by a hot-melt extrusion and spheronization process. *International Journal of Pharmaceutics*, **242**, 87–92.

(51) Young, C.R., Dietzsch, C. and McGinity, J. (2005) Compression of controlled release pellets produced by a hot-melt extrusion and spheronization process. *Pharmaceutical Development & Technology*, **10**, 133–139.

(52) Douroumis, D. (2011) Orally disintegrating dosage forms and taste-masking technologies 2010. *Expert Opinion on Drug Delivery*, **8**, 665–675.

(53) Andrews, G.P., Jones, D.S., Diak, O.A., McCoy, C.P., Watts, A.B. and McGinity, J.W. (2008) The manufacture and characterisation of hot-melt extruded enteric tablets. *European Journal of Pharmaceutics & Biopharmaceutics*, **69**, 264–273.

(54) Schilling, S., Lirola, H., Shah, N., Malick, A.W. and McGinity, J.W. (2010) Influence of plasticizer type and level on the properties of Eudragit S100 matrix pellets prepared by hot-melt extrusion. *Journal of Microencapsulation*, **27**, 521–532.

(55) Mehuys, E., Remon, J.P. and Vervaet, C. (2005) Production of enteric capsules by means of hot-melt extrusion. *European Journal of Pharmaceutical Sciences*, **24**, 207–212.
(56) Fukuda, M., Peppas, N.A. and McGinity, J.W. (2006) Floating hot-melt extruded tablets for gastro-retentive controlled drug release system. *Journal of Controlled Release*, **115**, 121–129.
(57) Bruce, L.D., Shah, N.H., Malick, A.W., Infeld, M.H. and McGinity, J.W. (2005) Properties of hot-melt extruded tablet formulations for the colonic delivery of 5-aminosalicylic acid. *European Journal of Pharmaceutics & Biopharmaceutics*, **59**, 85–97.
(58) Sherry, R. (2006) Granules comprising a NSAID and a sugar alcohol made by melt extrusion. US patent 20070254028.
(59) Quinten, T., De Beer, T., Remon, J.P. and Vervaet, C. (2011) Overview of injection molding as a manufacturing technique for pharmaceutical applications. *In Injection molding. Process, Design and Applications.* Kauffer, P.H. (ed.), Nova Publishers, New York.
(60) Aitken-Nichol, C., Zhang, F. and McGinity, J.W. (1996) Hot melt extrusion of acrylic films. *Pharmaceutical Research*, **13**, 804–808.
(61) Repka, M.A., Gerding, T.G., Repka, S.L. and McGinity, J.W. (1999) Influence of plasticizers and drugs on the physical-mechanical properties of hydroxypropylcellulose films prepared by hot melt extrusion. *Drug Development & Industrial Pharmacy*, **25**, 625–633.
(62) Crowley, M.M., Fredersdorf, A., Schroeder, B., Ucera, S., Prodduturi, S. and Repka, M.A. (2004) The influence of guaifenesin and ketoprofen on the properties of hot-melt extruded polyethylene oxide films. *European Journal of Pharmaceutical Sciences*, **22**, 409–418.
(63) Repka, M.A., Prodduturi, S. and Stodghill, S.P. (2003) Production and characterization of hot-melt extruded films containing clotrimazole. *Drug Development & Industrial Pharmacy*, **29**, 757–765.
(64) Repka, M.A., Munjal, M., ElSohly, M.A. and Ross, S. (2006) Temperature stability and bioadhesive properties of d9-tetrahydrocannabinol incorporated hydroxypropyl cellulose polymer matrix systems. *Drug Development & Industrial Pharmacy*, **32**, 21–32.
(65) Mididoddi, P.K. and Repka, M.A. (2007) Characterization of hot-melt extruded drug delivery systems for onychomycosis, *European Journal of Pharmaceutics & Biopharmaceutics*, **66**, 95–105.
(66) Trey, S.M., Wicks, D.A., Mididoddi, P.K. and Repka, M.A. (2007) Delivery of Itraconazole from Extruded HPC Films. *Drug Development & Industrial Pharmacy*, **33**, 727–735.
(67) Mididoddi, P.K., Prodduturi, S. and Repka, M.A. (2006) Influence of tartaric acid on the bioadhesion and mechanical properties of hot-melt extruded hydroxypropyl cellulose films for the human nail. *Drug Development & Industrial Pharmacy*, **32**, 1059–1066.
(68) van Laarhoven, J.A.H., Kruft, M.A.B. and Vromans, H. (2002) Effect of supersaturation and crystallization phenomena on the release properties of a controlled release device based on EVA copolymer. *Journal of Controlled Release*, **82**, 309–317.

(69) van Laarhoven, H., Veurink, J., Kruft, M.A. and Vromans, H. (2004) Influence of spinline stress on release properties of a coaxial controlled release device based on EVA polymers. *Pharmaceutical Research*, **21**, 1811–1817.
(70) Johnson, T.J., Gupta, K.M., Fabian, J., Albright, T.H. and Kiser, P.K. (2010) Segmented polyurethane intravaginal rings for the sustained combined delivery of antiretroviral agents dapirivine and tenofovir. *European Journal of Pharmaceutical Sciences*, **33**, 203–212.
(71) Morrow, R.J., Woolfson, A.D., Donnelly, L., Curran, R., Andrews, G., Katinger, D. and Malcolm, R.K. (2011) Sustained release of proteins from a modified vaginal ring device. *European Journal of Pharmaceutics & Biopharmaceutics*, **77**, 3–10.
(72) Gosau, M. and Muller, B.W. (2010) Release of gentamicin sulphate from biodegradable PLGA-implants produced by hot melt extrusion. *Pharmazie*, **65**, 487–492.
(73) Ghalanbor, Z., Korber, M. and Bodmeier, R. (2010) Improved lysozyme stability and release properties of poly(lactide-co-glycolide) implants prepared by hot-melt extrusion. *Pharmaceutical Research*, **27**, 371–379.
(74) Witt, C., Mader, K. and Kissel, T. (2000) The degradation, swelling and erosion properties of biodegradable implants prepared by extrusion or compression moulding of poly(lactide-co-glycolide) and ABA triblock copolymers. *Biomaterials*, **21**, 931–938.
(75) Hou, J., Li, C., Cheng, L. Guo, S., Zhang, Y. and Tang, T. (2011) Study on hydrophilic 5-fluorouracil release from hydrophobic poly(ϵ-caprolactone) cylindrical implants. *Drug Development & Industrial Pharmacy*, **37**, 1068–1075.

4

Solubility Parameters for Prediction of Drug/Polymer Miscibility in Hot-melt Extruded Formulations

Andreas Gryczke
Postfach 1207, 64549 Riedstadt

4.1 Introduction

One of the main challenges that drug development faces is the limited drug solubility and bioavailability of synthesized drug molecules in the pharmaceutical industry. Different strategies are employed to enhance both solubility and permeability of drug molecules, where the majority of these aim to increase the wettability of the drug by reducing drug particle size. Solid dispersions prepared by the melt-extrusion process is among the approaches that can be used to reduce the drug particle size to its absolute minimum by creating a molecular dispersed system. To obtain a stable solid system where the drug is molecularly dispersed in the polymer, selection of the appropriate polymer is a prerequisite. The successful development of a solid dispersion depends on the drug–polymer miscibility, the manufacturing process and the processing parameters. This book focuses on hot-melt extrusion, a technique broadly used in other industrial sectors; in particular, we examine the relationship between hot-melt extrusion and the compounding process in the plastic industry. Similarly to the extrusion process in the plastic industry, thermoplastic polymers are preferred in pharmaceutical industry as the main matrix where fillers and plasticizers are also added via a compounding-extrusion step, including the drug molecule.

Hot-melt Extrusion: Pharmaceutical Applications, First Edition. Edited by Dennis Douroumis.

In this chapter we aim to:

- define the different types of solid dispersions obtained from a melt-extrusion process;
- discuss types of interactions between drug and monomer/polymer;
- investigate mechanisms for stabilizing the drug in the polymeric carrier;
- address the question of what is a solvent and what is a solute in a solid dispersion; and
- demonstrate how to determine the drug–monomer, drug–drug and monomer–monomer interaction.

The models used to estimate the drug–monomer and eventually the drug–polymer miscibility are the Flory–Huggins and the solubility parameter models. In the same way as a plastic engineer applies these models to predict solubility of a polymer in different solvents or to predict plasticizer–polymer miscibility, these models can be used to predict polymer–drug miscibility. In general a drug molecule presents similar behavior to a plasticizer or a filler when incorporated into a polymer. The fundamentals and applications of the Hansen solubility parameter is well described in the literature [1], including the appropriate tools and equations required to predict the drug–polymer miscibility following a simplified approach. Nevertheless, there are several alternative methods to predict drug–polymer miscibility, e.g. molecular dynamic simulation using developed software.

A literature search for publications in the pharmaceutical field performed in early 2011 revealed 23,000 publications related to solid dispersions and more than 1500 publications implementing the Flory–Huggins theory, solubility parameters and cohesive energy density. However, in a similar search, approximately 140 publications were found to combine solid dispersions with the Flory–Huggins theory and the solubility parameter or the cohesive energy density. This is surprising, since it is commonly accepted that the majority of new active pharmaceutical ingredients (APIs) are often poorly soluble in water and possibly present low permeability, leading to insufficient absorption in the gastrointestinal (GI) tract. The reduced number of publications is also surprising, as it is expected that the solubility parameter concept and Flory–Huggins theory would be used more frequently since they can help to reduce the experimental workload and hence reduce cost.

In this chapter we describe approaches for selecting a set of polymers in order to prepare solid dispersions by estimating the drug–monomer and later the drug–polymer miscibility without conducting experimental trials in the first instance. Obviously, no theoretical model can replace the practical experience and real data generation. With the limited API amounts available in the beginning of early formulation development, this method can however help to exclude the majority of those excipients that would be likely to fail in later experiments.

4.2 Solid Dispersions

Before describing solid dispersion, let us take a look at aqueous dispersions and how solid dispersions differ from aqueous solutions (Figure 4.1).

Atkins [2] mentions that both the dispersion and the solution are homogeneous. The difference between a dispersion and a solution is defined as follows. In an aqueous dispersion, one component is dispersed in a separate phase in the other component and hence a very small sample taken from the dispersion could contain only component A or component B.

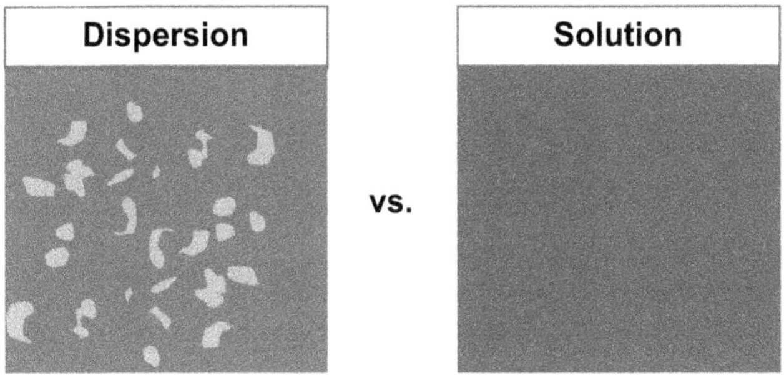

Figure 4.1 An aqueous dispersion and an aqueous solution.

In an aqueous solution, however, components A and B exist in one phase and hence a very small sample from a solution will always contain both components A and B. It is however questionable whether this definition can be strictly applied to the categories of solid dispersions defined below.

Particle size reduction of drug can be carried out mechanically, for example by milling. However, milling does not lead to the expected dissolution rate enhancement due to possible aggregation and agglomeration caused by the stronger van der Waals forces resulting from the increased surface energy [3]. Another possibility is to transform the drug to a non-crystalline state by dissolving it in a suitable solvent and then letting the drug recrystallize in controlled conditions. A melt technique (e.g. by hot-melt extrusion) can be employed to transform the drug into the amorphous state and let it recrystallize from there. Most likely, someone familiar with the hot-melt extrusion process has experienced the extrusion of a clear (amorphous) strand which became opaque after cooling at ambient conditions. Such a process might be optimized to obtain nano-sized particles, which are better able to dissolve in aqueous media due to their larger specific surface.

In this chapter, a solid dispersion is considered as a binary mixture of an amorphous polymer which acts as matrix for the drug molecule. Solid dispersions are also used to reduce the drug particle size to a minimum in the case of molecular dispersions. Solid dispersions can generally be obtained via a solvent method, a melting method or a combination of both, where the solute is firstly dissolved and then mixed as liquid solution with the polymeric matrix. Solid dispersions can be formed spontaneously through strong interactions (e.g. ionic) if the various components are miscible. Chiou and Riegelman [4] showed the relevance of the solid dispersion preparation method to the drug dissolution performance, using griseofulvin as a model poorly soluble drug. The same authors defined different classifications of solid dispersions based on the drug-release mechanism [5] while dispersions prepared by "traditional mechanical mixing" were excluded. The current edition focuses on hot-melt processing where amorphous polymers are present in solid dispersion matrices. The following classifications of solid dispersions will therefore be discussed in this chapter.

1. Solid crystalline suspension: the drug is suspended in a crystalline state within a polymeric (mostly amorphous) carrier. In a solid crystalline suspension, the drug and the carrier are located in two separate phases. The melting point of the drug can still be

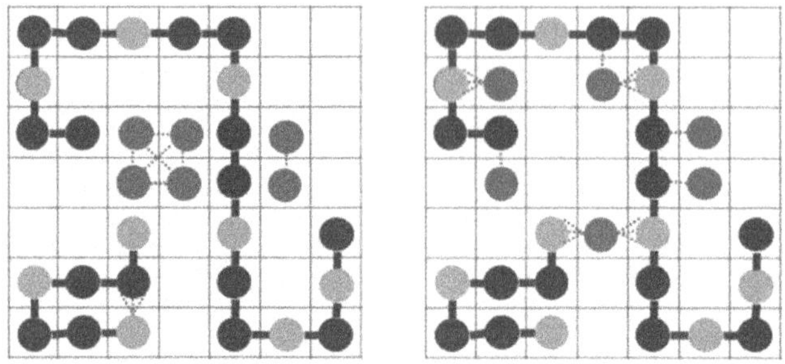

Figure 4.2 Lattice model of a solid glassy suspension (left) and a solid glassy solution (right). For a better understanding of the figure, please refer to the color section.

detected if the drug's crystal size is not too small to be dissolved in the carrier with increasing temperature [6]; the polymer glass transition temperature is also detectable. Crystalline suspensions are kinetically stable systems as the crystal is present at the lowest energy state for the drug. Solid crystalline suspensions usually appear opaque, although there are some exceptions reported in the literature.

2. Solid glassy suspension: the drug is molten and remains amorphous as a separate phase to the amorphous polymer. Prepared by melt processing, the cooled melt (for a binary system) will appear translucent in most cases. Solid glassy suspensions have a tendency to be kinetically unstable and tend to recrystallize under unfavorable circumstances if drug molecule migration cannot be avoided or sufficiently slowed down. By using analytical techniques, it is possible to determine a separate glass transition phase for both the drug and the polymer in such a system. In cases where the glass transition phase of the drug and the polymer are in close proximity, the discrimination between the solid glassy suspension and the solid glassy solution (Figure 4.2) needs to be further supported by suitable analytical tools. Solid glassy suspensions are usually seen as meta-stable or unstable, since the drug molecules can migrate inside the suspension which can cause mid- or long-term nuclei formation and crystal growth (ultimately resulting in recrystallization of the drug). Van Krevelen [7, 8] claims that the rate of crystallization is the result of nucleability (probability of nucleation) and transportability (migration or diffusion of the drug molecule through the polymer chains). The process is kinetically controlled and the timescale varies from a few seconds to several years. The relevant stability is therefore related to the transformation kinetics and needs to be investigated thoroughly.
3. Solid molecular dispersion: Chiou and Riegelman call this a solid glassy solution, where the drug is molecularly dispersed in the polymer matrix. For clarity, we avoid the use of 'solid solution' and use instead 'molecular dispersion'. Solid molecular dispersions prepared by a melt method usually appear to be translucent. A stable molecular dispersion can however be obtained at concentrations higher than nominal (e.g. 10–15% w/w). In addition, solid molecular dispersions prepared from binary systems without the influence of a third component (e.g. water) remain stable if the miscibility is not exceeded at ambient temperature.

The actual solubility of drug molecules in polymers is in many cases extremely low, due to the solid character of these substances, and has no practical use. However, the drug–polymer interactions are comparable to those of a plasticizer–polymer system. In a similar fashion to a plasticizer, the drug molecule is inserted between the polymer chains; this increases its flexibility and hence the polymer's mobility while the free volume of the system is increased. By considering a solid molecular dispersion as a 'solid solution', a polymer can become dissolved in a liquefied plasticizer; similarly, a polymer can be dissolved in a molten drug. A drug can act as a solvent for the polymer or vice versa only if the barrier of the cohesive energy is overcome through the polymer–drug interactions. The cohesive energy is defined as the energy required to break all interactions (van der Waals interactions, covalent bonds, hydrogen bonds and ionic bonds), allowing atoms or molecules to disconnect and result in solid to liquid/gas or liquid to gas transformation [9].

Figure 4.2 illustrates the difference between a solid glassy suspension and a solid molecular dispersion. The blue circles represent the monomers in a polymer chain while the green circles represent the drug molecules. The solid glassy suspension can be molecularly dispersed but the interaction forces between component A and component B are weaker than the forces for self-association between A–A and B–B, so that a migration of drug cannot be avoided or prevented sufficiently. If migration of drug cannot be avoided, a phase separation is likely to occur. This can lead to nuclei induction and nuclei growth which is kinetically controlled and generally leads to recrystallization or precipitation of the drug in the polymeric carrier. A solid molecular dispersion is created (although steric hindrance will be involved at some extent) if the interaction forces between drug and polymer (component A and component B) are stronger than the forces for self-association of A–A or B–B. Furthermore, the formation of hydrogen bonding prevents migration of the drug molecules within the polymer matrix. Hydrogen bonds are stronger than the weak dispersive and polar attractions. Single dispersive interaction is weak, but they sometimes appear in a large quantity even in aliphatic molecules. Hydrogen bonds dominate over polar attraction and dispersive interactions in terms of stabilization. The exact mechanism is not fully understood, but principally stabilization against crystallization takes place via steric hindrance and/or formation of hydrogen bonds between the drug and the polymer.

It has been demonstrated that the amorphous state of pharmaceuticals can provide faster dissolution rates with higher solution concentrations than their crystalline state; the formulation of stable solid molecular dispersions is therefore a viable alternative. The energy usually required to break up the crystalline structure of the drug before it dissolves is therefore not a limitation to the release of the drug from the molecular dispersion, leading to supersaturated solutions [10]. A challenging task for the galenical development is the formulation of dispersion that prevents precipitation of the drug molecules after entering the supersaturated stage. In conclusion, a solid dispersion can exist in any of the three categories described previously. Due to the relative miscibility between two molecules, the drug substance can be molecularly dispersed up to a certain range within the polymer matrix while the rest will be present in a crystalline stage. A combined dispersion where the drug exists simultaneously as both a molecular dispersion and a crystalline phase has been recently reported [11].

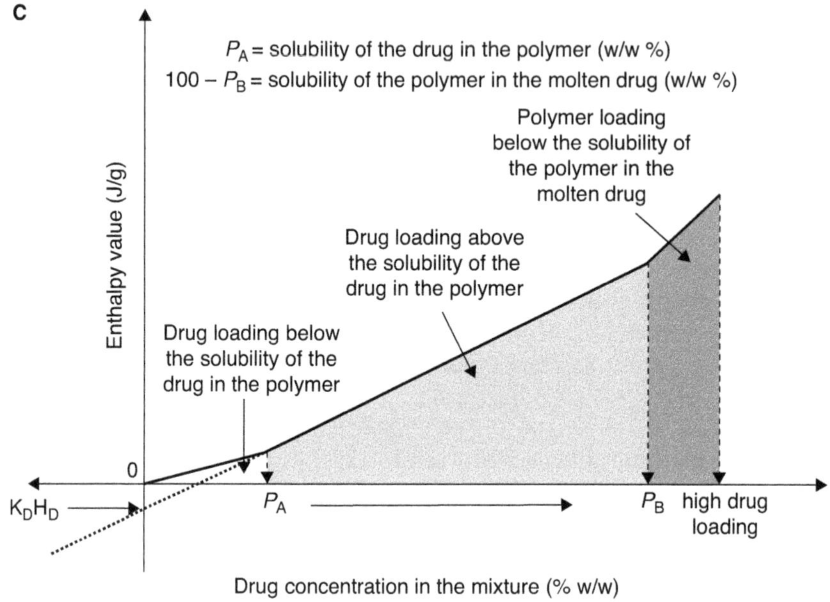

Figure 4.3 *Enthalpy versus drug concentration. S. Qi et al. 2010, reprinted with permission of Springer.*

The type of the produced solid dispersion depends mainly on the processing parameters (e.g. applied temperature, cooling rate) and the drug–polymer miscibility. This chapter focuses on the estimation of the drug–polymer miscibility prior to experimentation, which can serve as a valuable tool for the development of pharmaceutical dosage forms. The concept of a solubility parameter δ is well established in the areas of coating, paint and plastic industries, including pharmaceutics. The basic principles of the solubility parameter and the Flory–Huggins theory are discussed in the following, including the methodology to estimate the drug–polymer miscibility.

As mentioned previously, both components in a solid dispersion can be either the solute or the solvent. In a recent publication, Qi et al. [11] considered three regions of behavior of drug concentration within a polymer mixture plotted against the dissolution enthalpy, as shown in Figure 4.3. This included a low drug concentration region (where the drug loading is lower than the drug solubility in the polymer), an intermediate drug concentration region and a high drug concentration region (where the drug loading is higher than the polymer solubility in the molten drug).

In a simplistic approach, other researcher assumed that the drug acts as the solute and the polymer as the solvent; important information can however be overlooked using this limited approach. It is however possible to dissolve a certain amount of polymer in a drug-rich phase. Drug molecules are usually significantly smaller than polymer molecules and will bind non-covalently with the structural units of a polymer stronger than polymer functional groups interact with each other, thereby reducing the interactions between the polymer chains and softening the matrix [12, 13]. Good solvents for polymers are molecules having a composition similar to the structural units of the polymer [14]. It will be shown in Equation (4.27) how to tackle this when predicting the drug–polymer miscibility.

4.3 Basic Assumptions for the Drug–polymer Miscibility Prediction

The model(s) employed to predict drug–polymer miscibility are similar to those applied in polymer-solvent chemistry by calculating the interaction energies of a liquid solvent molecule with a solid polymer molecule for a solid–solid system. Further optimization of the predicted drug–polymer miscibility can be improved by using experimentally determined parameters as described in Marsac et al. [15], where experimental models were developed to allow for more quantitative estimates of the thermodynamics of mixing amorphous drugs with glassy polymers.

The model introduced in this chapter aims to facilitate the estimation of drug–polymer systems without conducting any experimentation. The intention of this chapter, however, is to allow the researchers to estimate miscibility *in silico*, that is, be one step ahead of any practical experiment.

It also focuses on the estimation of the drug–monomer miscibility due to the complexity of the possible intra-polymer interactions which cannot be covered in a simple model. This involves the estimation of the miscibility between the drug molecule and an interacting segment of a polymer chain, instead of considering the entire polymer entity. Practical approaches for testing the solubility of a drug molecule in the different monomers of a polymer showed acceptable correlation with the prediction of drug–polymer miscibility [16].

The model described here assumes certain interaction energy and, in some situations, the predicted solubility is underestimated. This can be observed when a stable amorphous system is obtained where the estimated interaction energy is low, but the system remains stable because of high viscosity and a low nucleation growth due to low drug molecule migration inside the polymeric matrix. The model can also yield misleading solubility values if stronger interactions between drug and polymer occur which are not reflected by the model. Similar results can be obtained if the appropriate process for the preparation of the solid dispersion is selected or if the method is not optimized. For example, the formation of hydrogen bonds is prevented due to the steric hindrance between the functional groups of the drug and the polymer. Selection of the appropriate manufacturing process can however overcome these issues and facilitate hydrogen-bonding interactions. If the model is applied to a binary mixture, the presence of other ingredients (such as water) can lead to an over- or underestimation of drug–polymer miscibility.

Furthermore, the current model does not take into account the efficiency and performance of the solid dispersion preparation method. In addition, it does not consider the influence of chemical effects such as the molecule or polymer chain conformation and neighboring effects due to the presence or absence of functional groups in the molecule. For example, the helical structure of a polyethylene glycol is not considered, even though it has significant influence on drug–polymer miscibility. The proposed model strictly assumes planar two-dimensional (2D) molecules resulting in a relatively lower precision compared to other approaches, such as molecular dynamic simulation. Nevertheless, in most cases the precision and accuracy of the estimated miscibility is adequate for the selection of a suitable polymer as inert carrier for the drug molecule to produce a solid glassy solution.

4.4 Solubility and the Flory–Huggins Theory

The miscibility estimation is used to predict the expected solubility of a component A in a component B, or vice versa. The thermodynamics of mixing in polymer systems has been discussed extensively [17, 18] and mixing is described by the Gibbs free energy:

$$\Delta G_{\mathrm{M}} = \Delta H_{\mathrm{M}} - T\Delta S_{\mathrm{M}} \tag{4.1}$$

where ΔG_{M} is the free energy of mixing, ΔH_{M} is the enthalpy of mixing and ΔS_{M} is the entropy of mixing. Solubility is assumed if ΔG_{M} becomes negative. In this equation, it becomes obvious for a polymer–drug system that the solubility depends most on the enthalpy term as, in most cases, the entropy term will not be large due to the large molecular weight of the polymer. This is why oligomers of two polymers might be soluble in each other, while the high molecular weight polymers are usually not miscible or poorly miscible. The combinatorial entropy of mixing is always positive; favorable but entropy contribution therefore depends on the size of the molecules (for high molecular-weight polymers, the entropy gain is insufficient to achieve miscibility).

The Flory–Huggins theory [19–21] was developed for a binary solution of a larger molecule (component 1, assumed to be the solvent) and a small molecule (component 2, assumed to be the solute). It describes a lattice model in which the structural units of a polymer and the solvent molecules are placed. Changes in entropy can be calculated based on the placement of the molecule units in the lattice. Based on the interactions between the structural units of a polymer and the solvent molecules, changes in the enthalpy can be calculated. The model does not consider compressibility of the polymer–solvent mixture and the entropy of mixing in Equation (4.1) as, with increasing polymerization of the polymer(s), the entropy becomes increasingly negligible. In Hansen's [1] approach for the estimation of the solubility parameter, the entropy term is kept constant at 0.34.

Hildebrand and Scott [22] introduced the following equation to calculate the enthalpy of mixing:

$$\Delta h_{\mathrm{M}} = \varphi_1 \varphi_2 (\delta_1 - \delta_2)^2 \tag{4.2}$$

where Δh_{M} is the enthalpy of mixing per unit volume, φ_1 and φ_2 are the volume fractions of components 1 and 2 and δ_1 and δ_2 are the solubility parameter for components 1 and 2.

Equation (4.2) can be also written in the form:

$$\Delta H_{\mathrm{M}} = V_0 \varphi_1 \varphi_2 (\delta_1 - \delta_2)^2 \tag{4.3}$$

where V_0 is the geometric mean of the volume of components 1 and 2 in the lattice.

Equation (4.3) predicts that $\Delta H_{\mathrm{M}} = 0$ if $\delta_1 = \delta_2$, which suggests that two substances are mutually soluble in each other if their structures are similar (i.e. 'like dissolves like'). The equation does not consider exothermic mixing, as the enthalpy term in Equation (4.3) cannot become negative. This is a major drawback of the proposed Hildebrand and Scott approach where polar systems and strongly interacting pharmaceutical species (e.g. hydrogen bonding) are not included.

The solubility parameter is defined as

$$\delta_{\text{Hildebrand}} = \sqrt{\frac{E_{\text{coh}}}{V_{\text{molar}}}} \tag{4.4}$$

where E_{coh} is the cohesive energy and V_{molar} is the molar volume of the component. E_{coh} is defined

$$E_{\text{coh}} = \Delta U_{\text{vap}} - p\Delta V \approx \Delta U_{\text{vap}} - RT \tag{4.5}$$

For low-molecular-weight molecules, the cohesive energy can be calculated from the heat of vaporization. For large-weight molecules such as polymers, indirect methods are used such as swelling or dissolving the polymer in a suitable solvent. Following the approach 'like dissolves like', the polymer's cohesive energy is assumed to be similar to the energy of the most suitable solvent. Several methods have been proposed by Fedors [23], Van Krevelen and Hoftyzer [7, 8], Hansen [1] and Stefanis and Panayiotou [24] to extend the solubility parameter concept to more polar strongly interacting species by including group contributions. Hansen and Panayiotou investigated the effect of polar attractions and hydrogen bonding in great detail and developed the 1D solubility parameter into a 3D parameter considering dispersive forces (which are always present even in aliphatic hydrocarbons), polar attractions and hydrogen bonding separately, namely:

$$E_{\text{coh}} = E_{\text{d}} + E_{\text{p}} + E_{\text{h}} \tag{4.6}$$

where E_{d} is the contribution of dispersion forces, E_{p} is the contribution of polar forces and E_{h} is the contribution of hydrogen bonding.

The 3D solubility parameter is defined:

$$\delta^2 = \delta_{\text{d}}^2 + \delta_{\text{p}}^2 + \delta_{\text{h}}^2 \tag{4.7}$$

Using Hansen's 3D solubility parameter system, the interaction between two molecules (originally polymer and solvent) could be interpreted more efficiently, since some effects are correlated strongly to hydrogen bonding or to polar forces. Hansen developed a group contribution method which allows the calculation of the 3D solubility parameter and only requires knowledge of compound's chemical structure. The partial solubility parameters, δ_{d}, δ_{p} and δ_{h} can be calculated using the combined group contribution methods of Van Krevelen–Hoftyzer and Fedors [23]. The method is especially useful in pharmaceutical development for drugs and polymers as it allows characterization of a material when there are no sufficient experimental data.

The 3D solubility parameters can be represented in a graphical form where the data are converted to a 2D plot.

Assuming that the value of δ_{d} of a given solvent is equal to that of a non-polar substance with similar chemical structure, the combined solubility parameter δ_{a} is introduced:

$$\delta_{\text{a}}^2 = \delta_{\text{p}}^2 + \delta_{\text{h}}^2 \tag{4.8}$$

Bagley et al. [25] introduced the Bagley solubility parameter δ_{v} and noted that the effects of δ_{d} and δ_{p} are thermodynamically similar whereas the effect of hydrogen bonding (δ_{h}) as a directed force is different in nature from both:

$$\delta_{\text{v}}^2 = \delta_{\text{d}}^2 + \delta_{\text{p}}^2 \tag{4.9}$$

Graphs plotting δ_h versus δ_v are commonly used to project a miscibility map of two substances.

Equation (4.3) can now be rewritten:

$$\Delta H_M = V_0 \varphi_1 \varphi_2 \left[(\delta_{d1} - \delta_{d2})^2 + (\delta_{p1} - \delta_{p2})^2 + (\delta_{h1} - \delta_{h2})^2 \right] \qquad (4.10)$$

The hydrogen bonds are especially important for the miscibility of two substances. The previously mentioned group contribution methods allow the calculation of hydrogen bonding energy. The limitation of such methods is that they do not consider different scenarios if certain hydrogen bond interactions are established between the two molecules. In addition, they do not consider the self-association between the A–A or B–B molecules. Hence, if the molecules A and B possess h-bond donors and h-bond acceptors, respectively, the predicted miscibility is misinterpreted through the solubility parameter model.

It is important to note that the solubility parameter is not valid for crystalline substances but only for amorphous structures, and the group contribution methods are estimated at an ambient temperature (25°C). The solubility parameter is temperature dependent and the hydrogen-bonding capability usually decreases with increasing temperature. Hansen [1] stated that a non-solvent for a substance can become a good solvent with increasing temperature and a boundary solvent can become a non-solvent. A good example is the poor solubility of caffeine in cold water. With increased temperature, water turns into a good solvent for caffeine. Under cooling conditions however, caffeine recrystallizes when the solution reaches a certain temperature. It worth mentioning that, for the purposes of hot-melt extrusion, the solubility parameters during the melting stage are different from those at ambient temperature; however, the predicted miscibility is successfully applied.

The solubility parameter has been used to predict the miscibility of a drug with excipients or carriers in solid dispersions. Greenhalgh et al. [26] classified compounds according to their difference in solubility parameters ($\Delta\delta_t$) between the drug and the polymer carriers as a tool to predict miscibility. The authors demonstrated that materials with $\Delta\delta_t < 7\,\text{MPa}^{0.5}$ are miscible while those with $\Delta\delta^t > 7\,\text{MPa}^{0.5}$ are immiscible. Forster et al. [27] concluded that compounds with $\Delta\delta_t < 2\,\text{MPa}^{0.5}$ are likely to be miscible and can form glassy solid solutions while those with $\Delta\delta_t > 10\,\text{MPa}^{0.5}$ are likely to be immiscible and unlikely to form glassy solid solutions.

The calculated solubility parameter provided by Greenhalgh and Foster is an efficient tool to predict the drug–polymer miscibility. The application of the Flory–Huggins theory allows a more precise prediction of drug–polymer miscibility by introducing important factors such as molecular size, which has a strong impact on the miscibility of two molecules. The application of the Flory–Huggins theory can be employed to different $\Delta\delta_t$ between a drug molecule and polymer grades of different molecular weight (e.g. polyethylene oxide). In contrast, the group contribution methods (e.g. that of Fedor) do not consider different molecular weights.

Flory introduced the dimensionless thermodynamic interaction parameter χ (generally referred to as 'Flory's Chi'). Huggins [20] calculated it based on the Hildebrand–Scatchard assumptions:

$$\chi = \chi_{\Delta_S} + \frac{V_{\text{Solvent}}(\delta_{\text{Polymer}} - \delta_{\text{Solvent}})^2}{RT} \qquad (4.11)$$

where $\chi_{\Delta s}$ is the entropy term which is kept constant at ~0.34 (according to Hansen [1]) for systems where a large-molecular-weight molecule such as a polymer is involved. Solubility is assumed if $\chi \leq \chi_{\text{critical}}$, where

$$\chi_{\text{critical}} = 0.5 \times \left(1 + \frac{1}{\sqrt{m}}\right)^2 \tag{4.12}$$

where m is the polymerization grade of the polymer. For large polymers, $\chi_{\text{critical}} = 0.5$.

By including the molecular size in the equation, the Flory's Chi approach is more accurate for the prediction of miscibility. The calculated Flory's Chi values can be used to determine whether two materials are miscible as follows:

- $\chi \leq 0$ implies good miscibility;
- $\chi = 0.5$ implies solvent or boundary solvent;
- $\chi \approx 1$ implies poor miscibility; and
- $\chi \geq 2$ implies immiscibility.

There are two limitations related to the Flory's Chi method as described by Equation (4.11):

- χ cannot become negative in Equation (4.12), i.e. exothermic mixing would be not captured by Equation (4.12); and
- hydrogen bonds with their stabilizing function in solid dispersions are not considered individually.

Below we describe the methodology followed by the Flory's Chi approach for the estimation of the solubility parameter and hence the prediction of the drug–excipient miscibility. The qualitative approach described herein differs from the commonly used approaches, as it provides a range of miscibility values rather than a single value.

Since the preparation process of the solid dispersion and its process parameters also influences the miscibility finally obtained (but cannot be included in the prediction calculations), it is better to predict an expected miscibility range rather than a fixed value. This approach is not suitable for predicting the intrinsic solubility of a component A in another component B however, but can be correlated with experimental values.

Due to the difficultly in predicting the various effects of the polymer chains, the model is initially used to estimate the miscibility between drug and monomers. Finally, the model introduces the method of determining the drug–polymer miscibility from the drug–monomer miscibility.

The cohesive energies of the drug and monomer are first calculated and then the partial contributions of dispersive forces, polar forces and hydrogen bonds are calculated separately.

$$E_{\text{coh}} = \delta^2_{\text{Fedors}} \tag{4.13}$$

The total solubility parameter can be estimated by Fedors' [23] method or similar group contribution approaches. However, it is important to ensure that the same method is employed for each molecule.

Equation (4.6) can be written as

$$E_{\text{d}} = E_{\text{coh}} - E_{\text{p}} - E_{\text{h}}, \tag{4.14}$$

enabling the calculation of the contribution of the dispersive forces. Prior to the calculation of the dispersive forces, it is necessary to calculate the contribution of polar forces and hydrogen bonds.

The contribution of polar forces can be calculated using the Hansen-Beerbower equation [1]:

$$E_p = \left(\sqrt{\frac{37.4\mu}{V_{molar}}}\right)^2 \tag{4.15}$$

where μ is the dipole moment and V_{molar} is the molar volume. If the dipole moment is not available from tables, it can be calculated using molecular dynamic algorithms such as those included in MOPAC (software is available online, Version 7).

The contribution of the hydrogen bonding energy is calculated via

$$E_h = N_k E_k \tag{4.16}$$

where E_k is the energy of the kth hydrogen bond and can be used to estimate the hydrogen bonding energy. In Table 4.1, the energy values have been extracted by the SOLPE software [28].

One of the main advantages of this approach is that the energy is calculated for dedicated hydrogen donor–acceptor pairs leading to increased precision of miscibility prediction. For example, it is possible to predict excellent miscibility for drug–polymer systems where the conventional approaches would suggest total immiscibility for these molecules. The Flory's Chi approach can even be employed for molecules that only have h-bond donors or h-bond acceptors to predict miscibility more precisely. The only limitation for the proposed method is that the actual hydrogen bond strength depends on neighboring effects to a considerable extent, which is not reflected here. For this reason, Gancia et al. [29] developed a much more comprehensive approach by calculating quantum-mechanical properties. This approach can also be used to calculate the hydrogen-bonding parameter with much greater confidence.

By calculating the specific hydrogen bonding considering electron donors and acceptors, it is possible to estimate the exchange energy density using the equation [1,30]:

$$A_{12} = \varepsilon_{11} + \varepsilon_{22} - 2\varepsilon_{12} \tag{4.17}$$

where A_{12} is the exchange energy density, ε_{11} and ε_{22} are the cohesive energy densities for self-association of molecule 1 and 2, respectively, and ε_{12} is the cohesive energy density

Table 4.1 *Energy value in kJ/mol for a variety of hydrogen bonds.*

	H-bonding acceptors							
H-bonding donors	O<	O =	N	S	F	C = C(π)	CN	Y
O–H	20.9	31.3	25.1	10	14.6	5	7.5	7
N–H	12.5	35.9	11.3	8	7.9	7	4.5	6
S–H	8	9	8	6	7	7	6	5
X>C–H	10.5	10.5	20.1	8	8	8	7	5
C–H	6	7	5	4	5	4	5	4

for the mixture of molecule 1 and 2. Equation (4.17) considers the fact that the breaking of each drug–drug and polymer–polymer bonds involves the simultaneous formation of two drug–polymer bonds [30].

According to Hildebrand and Scatchard [1,31–33] the heat of mixing and therefore the exchange energy density can be positive or zero only. They write Equation (4.17) in the form:

$$A_{12} = \left(\sqrt{\varepsilon_{11}} - \sqrt{\varepsilon_{22}}\right)^2 = (\delta_1 - \delta_2)^2 \tag{4.18}$$

where we again meet the general principle 'like dissolves like'.

The partial contribution of the hydrogen bonds can be calculated by considering Equations (4.10) and (4.18) [28]:

$$H_{12} = \varepsilon_{h(11)} + \varepsilon_{h(22)} - \frac{2U_{12}}{V_{12}} \tag{4.19}$$

where $\varepsilon_{h(11)}$ and $\varepsilon_{h(22)}$ are the terms for self-association of molecule 1 and 2 respectively and $2U_{12}/V_{12}$ is the term for interaction of molecule 1 and 2. $U_{12} = E_{\mathrm{h}}$ and is calculated using Equation (4.16).

By combining Equations (4.17–4.19), we obtain:

$$A_{12} = (\delta_{\mathrm{d1}} - \delta_{\mathrm{d2}})^2 + \left(\delta_{\mathrm{p1}} - \delta_{\mathrm{p2}}\right)^2 + H_{12} \tag{4.20}$$

The Huggins equation for the calculation of Flory's Chi interaction parameter χ can now be written:

$$\chi = \chi_{\Delta_S} + \frac{V_{\mathrm{Solvent}} A_{12}}{RT} \tag{4.21}$$

which can also be used to predict exothermic mixing and allow a higher degree of freedom for estimated miscibility scenarios. The interaction parameter χ can be obtained from Equation (4.21), which provides the basis for the miscibility estimation between drugs and monomers.

4.5 Miscibility Estimation of Drug and Monomers

The following example demonstrates the methodology to determine the miscibility of itraconazole with different monomers. Itraconazole is a trialzole antifungal agent with poor water solubility and is used against histoplasmosis or blastomycosis. Six et al. [34] investigated the itraconazole miscibility with a variety of polymers or polymer mixtures while Janssens et al. [35] examined the influence of the manufacturing methods on the miscibility of itraconazole and EUDRAGIT® E.

Initially, the molecules examined need to be fragmented by using the group contribution method in order to obtain the total cohesive energy density and to count possible hydrogen donors and acceptors. The total cohesive energy density and the molecular volume of the molecules can be obtained by implementing Fedors' method. The fragmentation of molecules can be carried out manually or automatically using the appropriate computer software. Furthermore, the molecular weight and the dipole moment of the molecules need

Table 4.2 *Number of hydrogen bonds suitable for self-association within Itraconazole.*

		H-bonding acceptors							
		O<	O =	N	S	F	C = C(π)	CN	Y
H-bonding donors	No.	–	–	–	–	–	–	–	–
O—H	–								
N—H	–								
S—H	–								
X > C—H	–								
C—H	–	6	7	5					4

to be obtained. If the dipole moment is not experimentally available, it can be estimated through molecular dynamic simulation software such as MOPAC.

It is particularly recommended to list all possible hydrogen bonds to consider self-association of the molecules, as demonstrated by Table 4.2.

By using Equations (4.13), (4.15) and (4.16), the total cohesive energy, the partial solubility parameter for polar forces and the partial solubility parameter for the hydrogen bonds are estimated as listed in Table 4.3.

As the model considers different types of possible hydrogen bonds, the solubility parameter for a molecule can capture different scenarios. We assume four different scenarios:

- Scenario 0: no hydrogen bonding is assumed;
- Scenario 1: the strongest single hydrogen bonding type is assumed;
- Scenario 2: the strongest plus the second strongest single hydrogen bonding type are assumed;
- Scenario 3: all possible hydrogen bonding types that add up to an acceptable maximum cohesive energy density (CED) are assumed.

Similarly, the same calculations can be carried out for the other molecules such as the monomers; the miscibility will be estimated in a later step. As the total cohesive energy of the molecule is constant in all three scenarios, the higher contribution of hydrogen bonds leads to a lower contribution of the dispersive forces. For the described model, this is acceptable as hydrogen bonds are considered to make the most significant contribution to the miscibility of a solid–solid system, such the drug–polymer solid dispersion.

Table 4.3 *Solubility parameters for Itraconazole.*

H-bonds assumed	Total δ_t	δ_d	δ_p	δ_h	δ_v
All H-bondings together	23.57	23.03	2.83	4.15	23.20
The strongest and second strongest	23.57	23.15	2.83	3.39	23.33
The strongest only	23.57	23.26	2.83	2.54	23.43
No H-bonding self-association	23.57	23.40	2.83	0.00	23.57

Table 4.4 *Hydrogen bond energies for interaction of itraconazole with vinyl acetate.*

		H-bonding acceptors							
		O<	O =	N	S	F	C = C(π)	CN	Y
H-bonding donors	No.	–	–	–	–	–	–	–	–
O–H	–								
N–H	–								
S–H	–								
X > C–H	–	**10.5**	**10.5**	**20.1**					**5**
C–H	–	6	7	5					4

In the next step, the possible interactions through hydrogen bonding between the drug and the corresponding monomer are explored. Table 4.4 illustrates a typical example of itraconazole (drug molecule) and the vinyl acetate as monomer, and highlights the favorable hydrogen bonds (in bold) which can be established between the drug and the monomer.

To obtain a first indication of whether hydrogen bonding could lead to strong interactions between the drug and monomer, the sum of energies of all possible hydrogen bonds of the drug–drug and the monomer–monomer self-association is compared to the sum of energies of all possible hydrogen bonds between drug and monomer. If the latter energy sum is greater than the sum of the two single molecules, this indicates a good probability for a given miscibility. As mentioned previously, the formation of hydrogen bonding depends on several factors such as the steric accessibility or the manufacturing process (parameters and conditions); the extent of the favorable hydrogen bonds that will eventually materialize is therefore difficult to predict. Every possibility of hydrogen bonding formation must therefore be investigated to obtain the dimensionless Flory's Chi interaction parameter. The number of possible hydrogen N_{h} bond scenarios can easily be calculated from:

$$N_{\mathrm{h}} = (N_{\mathrm{h(Drug-Drug)}} + 1) \times (N_{\mathrm{h(Monomer-Monomer)}} + 1) \times (N_{\mathrm{h(Drug-Monomer)}} + 1) \qquad (4.22)$$

In the case of itraconazole and the vinyl acetate monomer, Equation (4.22) gives 225 possible hydrogen bond constellations and Flory's Chi can be calculated for each constellation. The obtained Flory's Chi values are plotted in a normal distribution plot to obtain a mean Flory's Chi (Figure 4.4). The skewness and the kurtosis of the distribution are important indicators of the experimental Flory's Chi χ interaction parameter and the dependence of the drug–polymer miscibility on processing parameters. It is worth mentioning that the current approach does not consider the steric effects or the neighboring group or side chain effects.

The model is used to extract an interaction parameter that limits miscibility to a definite value similar to that obtained experimentally. However, due to the influence of the processing parameters on the experimental miscibility of a given drug–polymer system, it is unlikely that it will match the predicted value. This model therefore introduces an approach of predicting a range of possible drug–monomer miscibility values, which offers further interpretation options. The predicted drug–monomer miscibility can then be extrapolated

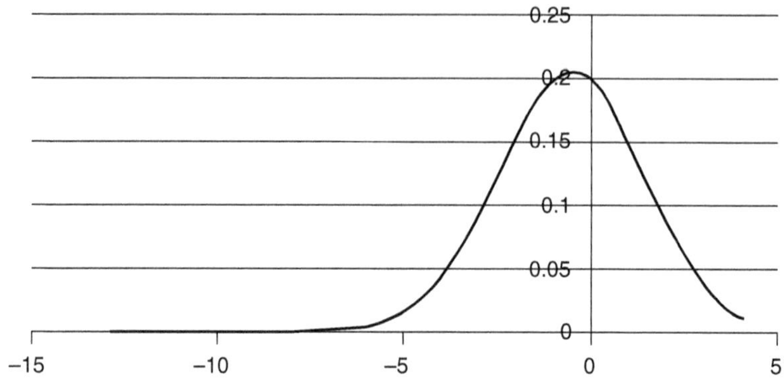

Figure 4.4 Normal probability distribution plot from Flory's Chi calculation for itraconazole and vinyl acetate, finding a skewness of −1.224 and a kurtosis of 0.986.

to the entire polymer molecule, allowing the selection of the suitable polymer from a given polymer range.

Further optimization of the model's flexibility can be achieved by representing the normal probability plot obtained from the Flory's Chi values for a drug–monomer combination in a box-whisker plot (Figure 4.5).

By taking in account the 25th and 75th percentiles instead of the mean Flory's Chi normal probability, a range of miscibility values can be predicted. The estimated group of values encompasses the imponderability of the processing conditions and chemistry aspects that influence the determined drug–monomer miscibility, and hence the drug–polymer miscibility.

The prediction of a miscibility range instead of the use of a single value is one of the key aspects introduced in this chapter. The spread of the predicted range depends strongly on the included hydrogen bonds. The dispersive and polar forces do not influence the spread or the range as they are non-directed and their formation is much less dependent on the molecule conformation compared to hydrogen bonds. The strong temperature dependence of the solubility parameter is mainly caused by the hydrogen-bonding capability of molecules. An increase in temperature decreases hydrogen-bonding capability.

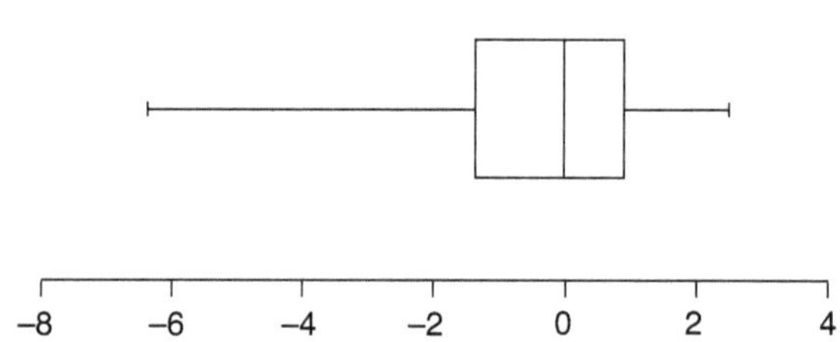

Figure 4.5 Box-whisker plot for calculated Flory's Chi values with min = −6.3, 75th percentile = −1.4, mean = −0.1, 25th percentile = −0.8 and max = 2.4 for itraconazole–vinyl acetate system.

From the obtained Flory's Chi values, the 25th percentile, the mean and the 75th percentile can be calculated. If all three values are smaller than χ_{critical}, then good miscibility of drug and monomer is anticipated. Furthermore, the influence of process parameters on the resulting miscibility might be low if the conformation of polymer chain does not hinder favorable interactions of drug molecule with the polymer segment.

As mentioned previously the Flory's Chi methodology provides a qualitative miscibility prediction suggesting poor, modest or excellent miscibility of a given drug–monomer combination, which can be expended to the entire polymer.

By calculating the Gibbs free energy it is feasible to obtain a quantitative determination of drug–polymer miscibility. The Gibbs free energy is defined [1]:

$$
\begin{aligned}
&f(\varphi) = G = \frac{1}{N}\varphi \ln \varphi + (1 - \varphi) \ln (1 - \varphi) + \chi\, \varphi (1 - \varphi) \\
&f(\varphi_0) \leq f(\varphi) \quad \forall \varphi \in [0, 1]
\end{aligned}
\tag{4.23}
$$

where G is Gibbs free energy, N is the polymerization grade of the polymer, ϕ is the volume fraction of the monomer and χ is the Flory's Chi interaction parameter.

By using Equation (4.23), the volume fraction is calculated when the Gibbs free energy reaches a minimum value. For negative Gibbs free energy, a drug–polymer system is considered miscible considering that the enthalpy is negative and the entropy is positive in Equation (4.23).

The molar volumes of the solute and solvent are estimated assuming that the volumes for both molecules do not differ by more than 40%. In fact, the monomers are not considered to be single free molecules but are covalently attached in the polymer chain. The molar volume of the polymer chain is therefore usually greater than the molar volume of the drug. The extrapolation of the drug–monomer miscibility to the drug–polymer leads to an approximate value which is acceptable for the current model. The molar ratio is given by:

$$
M_{\text{v(ratio)}} = \frac{M_{\text{v(solute)}}}{M_{\text{v(solvent)}}} \tag{4.24}
$$

where M_{v} is the molar volume.

For Equation (4.24), it is important to define which molecule is consider absorbed by the other component. The most common case is when the drug substance is viewed as the solute and the monomer as the solvent, but the drug can also be viewed as solvent for the monomer.

By incorporating the calculated molar ratio from Equation (4.24), the number of monomer molecules over a single drug molecule is estimated from:

$$
\frac{n_{\text{monomer}}}{n_{\text{drug}}} = \frac{-\varphi\, M_{\text{v(ratio)}}}{\varphi - 1} \tag{4.25}
$$

Equation (4.25) introduces a quantitative approach for the estimation of drug–monomer miscibility when the Gibbs free energy reaches a minimum.

Assuming that the interaction between drug and free monomer is comparable to the interaction between drug and monomer when the monomer is covalently bound into a polymer chain, the expected drug–polymer miscibility can be further estimated.

The relative polymer–molecule weight ratio strongly affects the drug–polymer interaction and hence the contribution of the monomer. Consequently the overall interaction is the result of the contribution of the weight fractions of the monomers [16].

The mass ratio of the expected drug–monomer miscibility is therefore calculated as:

$$w(\text{drug/monomer}) = \frac{x_{\text{monomer}} M_{w_{\text{drug}}}}{\frac{n_{\text{monomer}}}{n_{\text{drug}}} M_{w_{\text{monomer}}}} \tag{4.26}$$

where $n_{\text{monomer}}/n_{\text{drug}}$ is obtained from Equation (4.25), M_w is the molecular weight, and x_{monomer} is the monomer ratio in polymer with $\forall x_{\text{monomer}} \in [0, 1]$.

Finally, the expected drug–polymer miscibility is given by:

$$w(\text{drug/polymer}) = \sum_{i=1}^{n} w(\text{drug/monomer})_i \tag{4.27}$$

where n denotes the different monomers included in the polymer.

By applying the 25th, 75th and the mean percentiles of the Flory's Chi values from the box-whisker plot to Equations (4.23–4.27), the predicted drug–polymer miscibility range is finally estimated. The obtained miscibility reflects a certain flexibility which incorporates experimental variables such as the actual process for preparing solid dispersions and the processing parameters.

The results can be plotted in a graph as depicted in Figure 4.6, which illustrates the calculated miscibility for itraconazole and a range of polymers.

Figure 4.6 shows a bar graph as one possible way of plotting the calculated miscibility. The left end of each bar represents the lower end of the expected drug–polymer miscibility and is based on the 75th percentile Flory's Chi calculations. The right end of each bar represents the upper end of the expected drug–polymer miscibility and is based on the 25th percentile Flory's Chi calculations. The middle line in the bars represents the mean expected miscibility and is based on the mean Flory's Chi values.

The graph illustrates the spread of the predicted miscibility range and is easily read, whether the mean of expected miscibility level is orientated more to the upper or the lower end of the expected miscibility range.

The results depicted in Figure 4.6 are in good agreement with results reported by other research groups. For the miscibility of amorphous itraconazole with EUDRAGIT®, Janssens et al. [35] predict a miscibility of 7% w/w based on the approach developed by Marsac et al. [15]. The authors determined the drug–polymer miscibility level as a function of the preparation method for the amorphous solid dispersion. In contrast to the theoretical prediction, the obtained experimental miscibility for the stable amorphous solid dispersions was found to be 15% itraconazole prepared by a film-casting method and 27.5% prepared by spray drying. This study demonstrated the influence of the manufacturing technology of supersaturated solid dispersions. Six et al. [34] obtained stable amorphous solid dispersions with 15–20% itraconazole prepared by melt extrusion. The reported values for the miscibility of itraconazole and EUDRAGIT® are in excellent agreement with those predicted by the Flory's Chi approach as shown in Figure 4.6. The validity of the Flory Chi model is proved by another interesting study where Kolter et al. [36] reported a miscibility range of <25% for itraconazole with Kollidon® 12 PF, 40% with Kollidon® 17 PF and >50% with Kollidon 30 F or 90 F. Once more, these results are in agreement with the

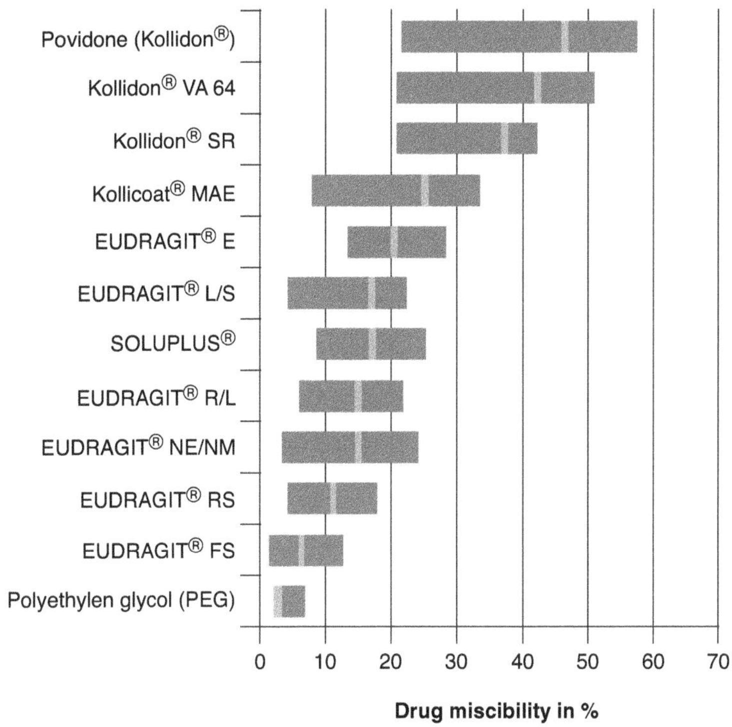

Figure 4.6 *Predicted miscibility of itraconazole in different polymers. For a better understanding of the figure, please refer to the color section.*

predicted values in Figure 4.6. Interestingly, a larger molecular weight of the homopolymer (polyvinylpyrrolidone) leads to a higher drug–polymer miscibility, suggesting that the conformation of the high-molecular-weight Kollidon® grades or their higher glass transition temperature ranges most likely prevent drug molecule migration at a greater extent. The reported miscibility values reported by Kolter et al. for Kollidon® VA 64 (40%) and Kollidon® SR (45%) are similar to those depicted in Figure 4.6 and fall within the predicted range. Nevertheless, the limitation of the proposed Flory Chi model is shown in the case for the predicted miscibility of Soluplus®. The Kolter et al. experimental value was found to be >50% for the itraconazole–Soluplus system, which is quite different from the values depicted in Figure 4.6. The effects that take place in a polymeric emulsifier such as Soluplus® have not yet been implemented in the model; the various drug interactions with the polymeric emulsifier are therefore not considered in the Flory Chi model.

4.6 Summary

This chapter describes the various types of solid dispersions manufactured by the melt-extrusion process of an amorphous polymer with a drug molecule. A solid crystalline suspension is obtained when the drug is incorporated in crystalline state in the amorphous polymer matrix, creating thermodynamic stable systems. An amorphous suspension is developed when the amorphous drug is incorporated in the polymer matrix, but cannot be immobilized through sufficient interactions with the polymer. The drug migration into the polymer matrix therefore leads to nuclei formation and subsequently to crystal formation of the drug (recrystallization). In a solid molecular dispersion, the drug is molecularly dispersed and is immobilized by sufficient non-covalent interactions with the polymer matrix. The behavior of the drug–polymer systems was considered similar to well-studied systems in the paint and plastic industry such as pigments, plasticizers and fillers.

Various methodologies for predicting the drug–polymer miscibility have been introduced, such as the Flory–Huggins and the solubility parameter as described by Hildebrand. In contrary to the models used in the literature where a single solubility value is determined, our approach introduces a model for the prediction of a range of miscibility values. The model reflects the influence of the processing parameters which are often underestimated; hence experimental results can often differ from the theoretical prediction. The model used to predict drug–polymer miscibility allows a greater flexibility on the predicted drug–polymer miscibility. System-integrated software can be further developed to process a wide set of drug–polymer systems in a time-efficient manner. Finally, the Flory Chi model was evaluated in comparison to reported examples by other research groups for solid dispersions of itraconazole in various polymers, and the predicted miscibility was discussed.

References

(1) Hansen, C.M. (2007) *Hansen Solubility Parameters: A User's Handbook*. CRC Press.

(2) Atkins, P.W. (1996). *Physikalische Chemie: Auflage v. 2 (p. 226)*. Wiley-VCH Verlag GmbH.

(3) Lin, S.L., Menig, J. and Lachman, L. (1968) Interdependence of physiological surfactant and drug particle size on the dissolution behavior of water insoluble drugs. *Journal of Pharmaceutical Sciences*, **57**, 2143–2146.

(4) Chiou, W.L. and Riegelman, S. (1970) Pharmaceutical applications of solid dispersion systems. *Journal of Pharmaceutical Sciences*, **59**, 937–942.

(5) Chiou, W.L., and Riegelman, S. (1971) Pharmaceutical applications of solid dispersion systems. *Journal of Pharmaceutical Sciences*, **60**(9), 1281–1302.

(6) Qi, S., Gryczke, A., Belton, P. and Craig, D.Q.M. (2008) Characterisation of solid dispersions of paracetamol and EUDRAGIT E prepared by hot-melt extrusion using thermal, microthermal and spectroscopic analysis. *International Journal of Pharmaceutics*, **354**(1–2), 158–167.

(7) Van Krevelen, D.W. and Hoftyzer, P. (1976) *Properties of Polymers. Their Estimation and Correlation with Chemical Structure*, 2nd ed. Elsevier, Amsterdam.

(8) Van Krevelen D.W. (1997) *Properties of Polymers: Their Correlation with Chemical Structure, Third Edition*. Elsevier, Amsterdam.

(9) Hancock, B.C., York, P. and Rowe, R.C. (1997) The use of solubility parameters in pharmaceutical dosage form design. *International Journal of Pharmaceutics*, **148**, 1–21.

(10) Leuner, C. and Dressman, J. (2000) Improving drug solubility for oral delivery using solid dispersions. *European Journal of Pharmaceutics & Biopharmaceutics*, **50**(1), 47–60.

(11) Qi, S., Belton, P., Nollenberger, K., Clayden, N., Reading, M. and Craig, D.Q.M. (2010) Characterisation and prediction of phase separation in hot-melt extruded solid dispersions: a thermal, microscopic and NMR relaxometry study. *Pharmaceutical Research*, **27**(9), 1869–1883.
(12) Radebaugh, G.W. (1992) Film coatings and film forming materials: Evaluation. *In* Swarbrick, J. and Boylan, J.C. (eds) *Encyclopedia of Pharmaceutical Technology*, Vol. 6. Marcel Dekker, New York and Basel, pp 1–28.
(13) Brabander, C.D., Mooter, G.V.D., Vervaet, C. and Remon, J.P. (2002) Characterization of ibuprofen as a nontraditional plasticizer of ethyl cellulose. *Journal of Pharmaceutical Sciences*, **91**, 1678–1685.
(14) Vollmert, B. (1988) *Grundriss der Makromolekularen Chemie Band IV Makromolekulare Aggregatzustaende*. E. Vollmert-Verlag, Karlsruhe.
(15) Marsac, P.J., Li, T. and Taylor, L.S. (2009) Estimation of drug-polymer miscibility and solubility in amorphous solid dispersions using experimentally determined interaction parameters. *Pharmaceutical Research*, **26**(1), 139–151.
(16) David, D., Rotstein, N. and Sincock, T. (1994) The application of miscibility parameter to the measurement of polymer-plasticizer compatibility. *Polymer Bulletin*, **33**(6), 725–732.
(17) Patterson, D. (1982) Polymer compatibility with and without a solvent. *Polymer Engineering & Science*, **22**, 64–72.
(18) Patterson, D. and Robard, A. (1978) Thermodynamics of polymer compatibility. *Macromolecules*, **11**, 690–695.
(19) Flory, P.J. (1941) Thermodynamic of high polymer solution. *Journal of Chemical Physics*, **9**, 660–661
(20) Huggins, M.L. (1941) Solution of long chain compound. *Journal of Chemical Physics*, **9**, 440.
(21) Flory, P.J. (1953) *Principles of Polymer Chemistry*. Cornell University Press.
(22) Hildebrand, J.H. and Scott, R.L. (1950) *The Solubility of Nonelectrolytes*. Reinhold.
(23) Fedors, R. (1974) A method for estimating both the solubility parameters and molar volumes of liquids. *Polymer Engineering & Science*, **14**, 147–154.
(24) Stefanis, E. and Panayiotou, C. (2008) Prediction of Hansen solubility parameters with a new group-contribution method. *International Journal of Thermophysics*, **29**(2), 568–585.
(25) Bagley, E.B., Nelson, T.P. and Scigliano, J.M. (1971) Three-dimensional solubility parameters and their relationship to internal pressure measurements in polar and hydrogen bonding solvents. *Journal of Paint Technology*, **43**, 35–42.
(26) Greenhalgh, D.J., Williams, A.C., Timmins, P. and York, P. (1999) Solubility parameters as predictors of miscibility in solid dispersions. *Journal of Pharmaceutical Sciences*, **88**, 1182–1190.
(27) Forster, A., Hempenstall, J., Tucker, I. and Rades, T. (2001) Selection of excipients for melt extrusion with two poorly water-soluble drugs by solubility parameter calculation and thermal analysis. *International Journal of Pharmaceutics*, **226**(1–2), 147–161.
(28) Schroeder, L.W. (1993–95) SOLPE: A program designed to predict Hansen Solubility Parameters and Flory's Chi Parameter From Molecular Composition. DYNACOMP, Inc.

(29) Gancia, E., Montana, J.G. and Manallack, D.T. (2001) Theoretical hydrogen bonding parameters for drug design. *Journal of Molecular Graphics Modeling*, **19**(3–4), 349–362.
(30) Gardon, J.L. (1966) The influence of polarity upon the solubility parameter concept. *Journal of Paint Technology*, **38**, 43–55.
(31) Hildebrand, J.H. and Scott, R. L. (1962) *Regular Solutions (Prentice Hall International Series in Chemistry)*. Prentice Hall.
(32) Hildebrand, J.H. and Scott, R.L. (1964) *The Solubility of Nonelectrolytes*. Dover Publications.
(33) Scatchard, G. (1931) Equilibria in non-electrolyte solutions in relation to the vapor pressures and densities of the components. *Chemical Reviews*, **8**, 321–333.
(34) Six, K., Daems, T., Hoon, J. de, Hecken, A.V., Depre, M., Bouche, M.-P., Prinsen, P., Verreck, G., Peeters, J., Brewster, M.E. and Mooter, G.V.d. (2005) Clinical study of solid dispersions of itraconazole prepared by hot-stage extrusion. *European Journal of Pharmaceutical Sciences*, **24**(2–3), 179–186.
(35) Janssens, S., De Zeure, A., Paudel, A., Humbeek, J.V., Rombaut, P. and Mooter, G.V.d. (2010) Influence of preparation methods on solid state supersaturation of amorphous solid dispersions: a case study with itraconazole and Eudragit E100. *Pharmaceutical Research*, **27**(5), 775–785.
(36) Kolter, K., Karl, M., Nalawade, S. and Rottmann, N. (2010) *Hot-melt Extrusion with BASF Pharma Polymers Extrusion Compendium*. Book brochure by BASF SE.

5

The Influence of Plasticizers in Hot-melt Extrusion

Geert Verreck
Janssen Research & Development, Belgium

5.1 Introduction

When applying hot-melt extrusion for pharmaceutical applications, a number of functional ingredients are introduced in the formulation in order to obtain: (1) acceptable and scalable processing conditions; (2) the required in vitro release characteristics of the drug substance; (3) adequate physical and chemical stability of the formulation during hot-melt extrusion and afterwards during long-term storage; and (4) the desired in vivo performance of the final dosage form. These functional ingredients comprise polymeric carriers, fillers, lubricants, stabilizing agents, plasticizers, etc. During hot-melt extrusion, the formulation is exposed to shear and frictional forces as well as to elevated temperatures and pressures. These forces and process conditions determine the processability of the formulation during the hot-melt extrusion process and downstream processing of the extrudate, and may cause thermal as well as mechanical degradation of the drug substance or ingredients of the formulation. In this respect, plasticizers can play a crucial role.

Plasticizers are typically ingredients with low molecular weight, either in the solid or the liquid/liquefied state. They add to the free volume of the main constituent of the formulation, i.e. the polymeric carrier, and thereby loosen the local liquid structure of the polymer. The theory of volume additivity was first derived for monomeric units in amorphous copolymers by Fox and Flory back in the early 1950s and, somewhat later, confirmed by Gordon and Taylor [1]. Further building on this ideal volume addivity for monomeric units, Kelley and Bueche described the molecular mobility of a polymer system plasticized with a

Hot-melt Extrusion: Pharmaceutical Applications, First Edition. Edited by Dennis Douroumis.

low-molecular-weight diluent. They applied the free volume equation of Williams, Landel and Ferry to derive the relationship between viscosity and glass transition temperature (T_g) as a function of the polymer-diluent concentration [2, 3]. This relation required the knowledge of glass transition temperatures of polymer and diluent and the thermal expansion coefficient of the diluent to be able to calculate the viscosity of the system. They experimentally confirmed their theoretically derived expressions for a number of polymer-diluent systems, indicating the usefulness of the free volume approach to explain segmental mobility of the polymer chain in presence of the diluent.

These empirical equations based on data fitting and relations build on the free volume hypothesis and classical thermodynamics as mentioned above, have contributed to the current understanding of the mechanism of plasticization for polymeric carriers. Based on these approaches, it has been known for some time that plasticizers decrease the glass transition temperature of amorphous polymers as a function of their concentration in case of compatible blends. In principal, this reduction in glass transition temperature during hot-melt extrusion results in an improved processability, an improved downstream processing and a reduction of the thermal degradation of any of the constituents of the formulation.

The purpose of this chapter is to provide an overview of the different applications of plasticizers used for pharmaceutical hot-melt extrusion, to describe their influence on the process and downstream processing and to give an overview of their influence on the performance of the pharmaceutical dosage form. Plasticizers can be divided into different classes: (1) the traditional plasticizers, ingredients of low molecular weight intentionally added to the formulation to obtain a desired property; (2) the non-traditional plasticizers, other low-molecular-weight ingredients present in the formulation, unintentionally generating plasticizing properties; and (3) specialty plasticizers.

5.2 Traditional Plasticizers

Traditional plasticizers are mostly low-molecular-weight ingredients added to the formulation to obtain a desired property either during the hot-melt extrusion process itself, during the downstream processing or in the final product. They can be present during hot-melt extrusion either in the liquid/liquefied state or in the solid state. For chemical applications, a large number of different plasticizers exist whereby the primary role consists of improving the flexibility and processability of the polymers [4]. Plasticizers reduce polymer–polymer chain secondary bonding and provide for more mobility. As such, they reduce parameters such as tensile strength, hardness, density, melt viscosity and glass transition temperature and, at the same time, increase parameters such as elongation at break, toughness and dielectric constant. Plasticizers can be internal or external. External plasticizers are not bound to the polymer chain by primary bonds and can therefore be lost by evaporation, migration or extraction. In contrast, internal plasticizers are inherently part of the polymer and therefore remain part of the final product. The ideal plasticizer should be highly compatible with the polymer, stable under the hot-melt extrusion conditions and sufficiently lubricating and stable when present in the final product. In addition, and even more important for pharmaceutical applications, they should comply with environmental, health and safety regulations. Although a lot of plasticizers exist for chemical applications, only limited choices of approved plasticizers for pharmaceutical industry are therefore available.

For pharmaceutical applications of hot-melt extrusion, plasticizers are most often added to the formulation either to improve the processing conditions during the manufacturing of the extruded dosage form or to improve the mechanical and physical properties of the final product. This final product is most often a tablet, a pellet or a polymeric film [5, 6]. Table 5.1 lists the most frequently used pharmaceutical traditional plasticizers, in combination with the used carriers and active substances, the plasticizer ranges that were investigated and its pharmaceutical applications.

5.3 Non-traditional Plasticizers

Non-traditional plasticizers are mostly low-molecular-weight ingredients present in the formulation either for other critical or non-critical functions, but contributing to the overall plasticization of the product mass in the melt extruder. The non-critical functions are most often residual materials present in the constituents of the formulation itself, such as residual solvents or water. Some polymeric materials are hygroscopic and may adsorb water during storage. If not removed prior to extrusion, and when extrusion is performed below the boiling point of the solvent, they can act as a plasticizer during the process. This may be beneficial for the product, but if viscosity becomes too low to produce an extrudable mass, these residual solvents have to be removed prior to melt extrusion. Alternatively, proper storage conditions may also help in preventing adsorption of water. In extreme cases, polymers may be stored under a nitrogen blanket or in vacuum containers.

Non-traditional plasticizers present in the formulation but having other critical functions are most often low-molecular-weight materials such as the active substance itself. One example of an active substance acting as a non-traditional plasticizer is ibuprofen, as described by De Brabander et al. [32]. The authors investigated the plasticizing effect of ibuprofen on ethylcellulose during hot-melt extrusion using a co-rotating intermeshing twin-screw extruder and compared their results to co-evaporates of the same composition. Ibuprofen is a highly crystalline drug substance with a melting endotherm at approximately 76°C. After quench-cooling the molten ibuprofen, a T_g could be measured at –43.6°C. Unprocessed ethylcellulose shows a T_g of 133.3°C. Thermal analysis of the co-evaporates showed that the glass transition temperature decreased with increasing ibuprofen concentration (0, 5, 10 and 20% w/w) and that a single T_g was observed for the different mixtures, indicating complete miscibility of ibuprofen with ethylcellulose (Table 5.2). These experimental T_g values were compared with those predicted by the Gordon–Taylor/Kelley–Bueche equation:

$$T_{gx} = \frac{T_{g_1} w_1 + T_{g_2} K w_2}{w_1 + K w_2}$$

where w_1 and w_2 are the weight fractions of ibuprofen and ethylcellulose, respectively, and T_{g1} and T_{g2} are the glass transition temperatures of ibuprofen and ethylcellulose, respectively. K is a constant based on the ratio of the differences in expansion coefficient ($\Delta\alpha$) at T_g of the active substance and the polymer. However, according to Simha and Boyer, when replacing volume fractions by weight fractions and assuming that $\Delta\alpha T_g$ is constant, K becomes [3]:

$$K \cong \frac{\rho_1 T_{g_1}}{\rho_2 T_{g_2}}$$

Table 5.1 *Overview of the most frequently used traditional plasticizers for pharmaceutical applications of hot-melt extrusion together with their carriers, active substances, concentration of plasticizers used and different applications.*

Plasticizer	Carrier	Active substance	Concentration of plasticizer	Application	Reference
Citric acid monohydrate	Eudragit® RS PO	–[a]	10–30 wt% (based on total weight)	Citric acid monohydrate was investigated as a solid-state plasticizer. It was observed that citric acid monohydrate served as an effective plasticizer during hot-melt extrusion. The tensile strength and elastic modulus of polymeric films reduced as a function of citric acid concentration while the elongation increased.	[7]
Dibutyl sebacate (DBS)	Ethylcellulose (EC)	Metoprolol tartrate, Propranolol hydrochloride, Theophylline monohydrate, Hydrochloro-thiazide	20–70 wt% (based on the polymer weight)	DBS was investigated as a plasticizer for EC to prepare mini-matrices with a sustained-release profile. Besides DBS, triethylcitrate (TEC), diethyl phthalate and triacetin were also investigated. All four plasticizers resulted in similar plasticizing properties for EC, but the more hydrophilic plasticizers resulted in a burst release. DBS was therefore selected for further investigation. A concentration of 50 wt% resulted in mini-matrices without cracks. In another study, DBS was used as plasticizer for EC to prepare hollow hot-melt extruded cylinders which would serve as a sustained-release barrier surrounding the drug-containing core.	[8–11]

	EC/Hydroxypropyl methylcellulose (HPMC), EC/Hydroxypropyl cellulose (HPC)	Metoprolol tartrate	20 wt% (based on the polymer weight)	In this work, DBS was added as a plasticizer for a polymer blend of EC/HPMC. This blend, together with the active substance, was first hot-melt extruded and subsequently injection-molded to obtain a monolithic tablet. Similar experiments were reported in a second study using DBS as plasticizer for a polymer blend of EC/HPC, melt extruded followed by injection molding.	[12, 13]
Diethyl phthalate	EC, HPMC, cellulose acetate butyrate, poly(vinyl chloride), poly(vinyl chloride-co-vinyl acetate, poly(ethylene-co-vinyl acetate), Eudragit® RS	Diltiazem hydrochloride, desacetyldiltiazem hydrochloride, oxprenolol hydrochloride, disopyramide phosphate	5–10 wt% (based on the polymer weight)	In this study, different polymers were investigated to evaluate sustained-release pellets prepared by hot-melt extrusion. Besides diethyl phthalate, other plasticizers were also used including triacetin, Polyethylene glycol 400 (PEG 400) and dioctyl phthalate.	[14, 15]
Glycerol	Maltodextrin	Pyroxicam	15–20 wt% (based on total weight)	Fast-dissolving films made of maltodextrin were prepared by hot-melt extrusion. Different plasticizers were evaluated including PEG400, glycerol, propylene glycol and esters of citric acid. Glycerol was selected as a suitable plasticizer based on an initial screening whereby different parameters such as stiffness, tensile stress, ductility, disintegration time, taste, etc. were investigated.	[16]
Methylparaben	Eudragit® RS PO	–[a]	5–25 wt% (based on the polymer weight)	Methylparaben was investigated as a solid-state plasticizer. It was observed that methylparaben was as effective as triethyl citrate during hot-melt extrusion. The T_g of the polymer as well as the viscosity were both reduced as a function of methylparaben concentration.	[17]

(continued)

Table 5.1 *Overview of the most frequently used traditional plasticizers for pharmaceutical applications of hot-melt extrusion together with their carriers, active substances, concentration of plasticizers used and different applications* (continued).

Plasticizer	Carrier	Active substance	Concentration of plasticizer	Application	Reference
	Eudragit® L100-55, Eudragit® L100, Eudragit® S100	Theophylline anhydrous	10–40 wt% (based on the polymer weight)	Hot-melt extruded enteric matrix pellets were prepared with a number of different polymers and plasticizers. Process-ability and in vitro release characteristics were evaluated. Differences in release rate as well as gastric protection were observed when water-soluble plasticizers were used versus less water-soluble materials.	[18, 19]
	Hydroxypropyl methyl cellulose acetate succinate (HPMC AS)	Theophylline anhydrous	10–40 wt% (based on the polymer weight)	Idem as above	[18, 19]
PEG 8000, PEG 3350, PEG 400	Eudragit® L100-55, Eudragit® L100, Eudragit® S100	Theophylline anhydrous	10–40 wt% (based on the polymer weight)	Idem as above	[18, 19]
	HPMC AS	Theophylline anhydrous	10–40 wt% (based on the polymer weight)	Idem as above	[18, 19]
	Polyethylene oxide (PEO) 1,000,000–7,000,000	Chlorpheniramine maleate	6–40 wt% (based on the total weight)	In this work, it was observed that adding PEG3350 to the hot-melt extrusion of PEO resulted in a lowering of the torque and, as a consequence, lower temperature settings could be used. This resulted in no discoloration of PEO and no decrease of the molecular weight of the polymer. Adding PEG3350 also resulted in a slightly faster drug release.	[20]

	HPC	Chlorpheniramine maleate, Hydrocortisone	1–5 wt% (based on the total weight)	In a first study, HPC films were prepared by hot-melt extrusion and a number of different plasticizers were evaluated including PEG400, PEG8000, TEC and acetyltributylcitrate. All materials possessed adequate plasticizing properties except PEG400, which showed physical-mechanical instabilities after 6 months of aging. In another study, PEG3350 was added to plasticize HPC films prepared by hot-melt extrusion. Other additives such as polycarbophil, carbomer, eudragit® E100 and sodium starch glycolate were also investigated for their influence on physical-mechanical and bioadhesive properties.	[21, 22]
Sugar alcohols: e.g. Sorbitol, Xylitol, Lactitol, Erythritol	Starches (corn, potato, rice, wheat, etc.)	Theophylline monohydrate	2.5–30 wt% (based on the total weight)	The influence of the solubility, melting point, particle size and concentration of plasticizer on the in-vitro characteristics of extrudates prepared for a controlled-release formulation was investigated. It was observed that mainly mechanical strength and dissolution profiles were influenced by the percentage plasticizer used. The authors also investigated the in-vivo performance of a formulation with 15 wt% Sorbitol as plasticizer. In another study using soluble starch, it was observed that sorbitol acted as an anti-plasticizer at low concentrations (< 5 wt%). As a consequence, different release rates were obtained as function of the plasticizer concentration.	[23–25]

(continued)

Table 5.1 *Overview of the most frequently used traditional plasticizers for pharmaceutical applications of hot-melt extrusion together with their carriers, active substances, concentration of plasticizers used and different applications* (continued).

Plasticizer	Carrier	Active substance	Concentration of plasticizer	Application	Reference
Surfactants (Polysorbate 80 and Dioctyl sodium sulfosuccinate)	Hydroxypropyl methylcellulose E5 (HPMC E5), Polyvinylpyrrolidone K30 (PVP K30), Vinylpyrrolidone / Vinylacetate (Plasdone® S-630)	Not mentioned in the publication, but physicochemical properties are given.	10 wt% (based on total weight)	The purpose of this work was to investigate surfactants as plasticizers during hot-melt extrusion of different polymers and to evaluate their influence on the physical stability of the amorphous drug. It was observed that this effect of the surfactants is minimal on the physical stability, although they lower the T_g of the system. It was therefore concluded that surfactants appear to be promising plasticizers.	[26]
Triacetin	Polyvinyl acetate phthalate (PVAP), HPMC AS	Hydralazine	10–50 wt% (based on the polymer weight)	In this study, enteric capsules were made by hot-melt extrusion. In order to do so, hollow cylinders were produced, filled with a model drug and side ends were closed. Besides triacetin, TEC and dioctyl phthalate were also investigated. For both polymers, triacetin proved to be the most suitable plasticizer for this application.	[27]
Triethylcitrate (TEC)	Eudragit® RS PO, Eudragit® RL PO	Theophylline anhydrous	5–10–15 wt% (based on the polymer weight)	The plasticizer was added to pre-plasticize a polymer blend using a single-screw extruder. The extrudates were cryogenically milled to obtain a fine powder. This powder was consecutively used to film-coat theophylline tablets. It was observed that the drug release rate decreased with increasing levels of plasticizer.	[28]

Eudragit® RS PO, Eudragit® RL PO	Diltiazem hydrochloride Chlorpheniramine maleate Indomethacin	4–12 wt% (based on the polymer weight)	The influence of the concentration of plasticizer on the drug-release characteristics was investigated. Drug release was increased with higher levels of TEC except for indomethacin. In another study it was shown that up to 12 wt% of TEC resulted in a T_g decrease of the polymer by 32.5°C. The drug release rate also increased with increasing concentration of plasticizer.	[29, 30]
Eudragit® L100-55	Chlorpheniramine maleate Theophylline anhydrous	20–30–40 wt% (based on the polymer weight)	The general idea is similar as described by Zheng et al. [29, 30]. Higher plasticizer concentrations were needed because of the higher T_g of the polymer. Similar observations were made in that the release properties were dependent on plasticizer content.	[31]
Hydroxypropyl methyl cellulose acetate succinate (HPMC AS)	Theophylline anhydrous	10–40 wt% (based on the polymer weight)	See under Methylparaben and references [18] and [19].	[18, 19]

[a] In the work described, there is no active substance used. In other words, the experiments are only related to carrier and plasticizer.

Table 5.2 *Glass transition temperatures (°C) as a function of ibuprofen (IBP) concentration (0, 5, 10 and 20% w/w) as a non-traditional plasticizer for co-evaporates and extrudates and with diethylphtalate (DEP) and dibutylsebacate (DBS) as traditional plasticizers at 10% w/w for ethylcellulose (EC) (data adapted from De Brabander et al. [32]).*

Concentration of plasticizer (% w/w)	IBP/EC co-evaporates	IBP/EC extrudates	DEP/EC co-evaporates	DBS/EC co-evaporates
0	128.6–129.3	126.3–127.8	128.6–129.3	128.6–129.3
5	105.8–109.2	103.7–108.4	–[a]	–[a]
10	84.6–87.1	78.3–78.7	89.3	83.1
20	52.8–66.2	57.6–64.5	–[a]	–[a]

[a] Experiment not performed.

where the densities of the amorphous drug and of the polymer are represented by ρ_1 and ρ_2, respectively.

Although the Gordon–Taylor relationship was originally derived for compatible polymer blends, it has also been used successfully for small organic molecules [33, 34]. When volume additivity holds, and when the heteromolecular interactions are of the same magnitude as the homomolecular interactions, values should coincide. However, the authors showed that measured values did not agree with the calculated values for the co-evaporates, indicating that heteromolecular interactions are not of the same magnitude as homomolecular interactions and thus volume additivity was not fulfilled. They observed a negative deviation from the calculated values and showed, via infrared spectroscopy, the presence of hydrogen bonding between active substance and polymer causing the deviation from ideal volume additivity.

They further evaluated similar concentrations of ibuprofen to ethylcellulose with hot-melt extrusion (temperature settings of the barrel were above the melting point of ibuprofen). Thermal analysis of the extrudates indicated that, during hot-melt extrusion, ibuprofen also decreased the T_g of ethylcellulose as a function of ibuprofen concentration in comparable ranges as the co-evaporates (Table 5.2). The plasticizing effect of ibuprofen on ethylcellulose was then compared with two traditional plasticizers: diethylphtalate and dibutylsebacate. At a 10% w/w concentration of these traditional plasticizers, similar T_g reductions were obtained as shown in Table 5.2. Based on this study, it could be concluded that ibuprofen acts as a non-traditional plasticizer for ethylcellulose in a comparable manner during co-evaporation as well as hot-melt extrusion and to a similar extent as traditional plasticizers.

Studies performed by Kidokoro et al. confirm that ibuprofen also acts as a non-traditional plasticizer for eudragit® RS [35]. This was evaluated by measuring the glass transition temperature of mixtures of ibuprofen and eudragit® RS with differential scanning calorimetry. A decrease of the glass transition temperature of 15°C was observed when 25% w/w ibuprofen was added to the polymer. They further measured the morphological properties of ibuprofen and eudragit® RS physical mixtures stored at 50°C for 1 day and identified the optimal ibuprofen concentration of 30% w/w based on film flexibility. The effect of ibuprofen as a plasticizer was obtained by increasing the polymer chain flexibility which resulted in a structural change in the tablets leading to a decrease in drug release. In the same study, they observed that theophylline was no plasticizer for eudragit® RS.

Another example where the active substance works as a non-traditional plasticizer during hot-melt extrusion is described by Aitken-Nichol et al. [36]. They investigated the use of hot-melt extrusion for preparing films for topical drug delivery using a single-screw extruder. The topical films consisted of eudragit® E100 as an amorphous polymeric carrier having a T_g of 40°C, a traditional plasticizer (different plasticizers were evaluated including TEC, triacetin and PEG 6000) and lidocaine HCl as the crystalline active substance with a melting point of approximately 77–79°C. The extrusion process was performed just above the melting point of the lidocaine HCl. The extruded films showed absence of crystallinity when analyzed by wide-angle X-ray diffraction, indicating miscibility between drug and polymer, and the T_g was depressed with increasing concentration of lidocaine HCl (5 and 10% w/w). However, the T_g decrease was more pronounced when solution-cast films were compared to the melt-extruded films. The authors mentioned that this difference could have been obtained due to a better intermolecular mixing in a solution compared to a high viscosity melt. Besides lidocaine HCl, they also looked into diphenhydramine HCl and observed no plasticizing effect of this active substance with eudragit® E100. They attributed this to the high melting point of the active substance (166–170°C) and the low processing conditions during hot-melt extrusion (i.e. between 80 and 130°C), meaning below the melting point of the drug. Since hot-melt extrusion was performed above the melting point of the drug, they suggested for lidocaine HCl that the molten active substance was solubilized in the polymer and, as such, led to a plasticizing effect.

A similar observation was made for itraconazole. Itraconazole is a crystalline active substance with a melting endotherm of approximately 168°C. Six et al. investigated the melting behavior of itraconazole and observed that a chiral nematic mesophase is formed at approximately 90°C when the active substance is cooled down from the molten phase [37]. Further cooling freezes the mesophase into the glassy state of the molecule. The T_g of itraconazole is located at 59°C. Verreck et al. and Six et al. evaluated the hot-melt extrusion of itraconazole with HPMC 2910 5 mPa.s (HPMC E5) at different drug loadings using a co-rotating intermeshing twin-screw extruder [38, 39]. HPMC E5 is an amorphous polymer with a T_g of 141°C. In one set of experiments, the extruder temperature settings were kept at 185°C (well above the melting point of the active substance). Itraconazole acted as a plasticizer for HPMC E5, as indicated by a decrease of the glass transition as a function of increasing itraconazole concentration. Up to 60% w/w itraconazole, drug and polymer initially seemed to be completely miscible as measured using modulated differential scanning calorimetry (m-DSC). Also, when extruded together with HPMC E5, formation of the monotropic mesophase could not be observed. In other words, HPMC E5 prevented the formation of this liquid crystalline phase of itraconazole. However, when the Gordon–Taylor equation was applied, experimental values deviated from calculated values, indicating non-ideal volume additivity between drug and polymer. The authors performed a further investigation of the system and revealed an amorphous phase-separated system existing of an amorphous itraconazole-rich phase and an amorphous HPMC E5-rich phase.

In another part of the investigation [38, 39], the authors also evaluated the effect of the different processing conditions (temperature 150–220°C, screw speed 200–400 rpm, feed rate 1–2 k/hr) during hot-melt extrusion at a 40% w/w itraconazole loading in HPMC E5. Under all conditions tested, itraconazole acted as a non-traditional plasticizer as long as the drug substance was transformed into the molten state during the extrusion process. Even when the temperature of the barrel was set at 150°C, well below the melting point of

Table 5.3 Experimental and calculated (according to the Gordon–Taylor equation) glass transition temperatures (°C) for itraconazole/PVPVA (data adapted from Six et al. [40]).

Concentration itraconazole (wt%)	Theoretical values (Gordon–Taylor)	Experimental values
0	108	108
10	101	103
20	94	96
40	82	84
60	72	73
80	65	64
100	59	59

itraconazole, a complete melt was still obtained in the extruder due to the shear created by the rotating twin screws.

These investigations show that itraconazole, although it forms a liquid crystalline mesophase upon cooling from the melt and is not completely miscible with HPMC E5, still acts as a non-traditional plasticizer for this cellulose polymer. Six et al. later investigated other polymeric carriers with itraconazole such as PVPVA 64 and eudragit® E100 [40–43]. It was also confirmed for these carriers that itraconazole worked as a non-traditional plasticizer. For instance, with PVPVA 64, complete miscibility was observed following the Gordon–Taylor equation and thus showing ideal volume additivity (Table 5.3).

It was also observed for PVP K30 that the active substance acted as a non-traditional plasticizer (the active substance was a compound manufactured by Novartis Pharmaceuticals Corp., Basel, Switzerland) [44]. Here, Lakshman et al. first transformed the crystalline active into the amorphous state via solvent evaporation and then performed the hot-melt extrusion process of a physical blend at 20–40 wt% drug load with PVP K30. At a 40 wt% drug load, no additional plasticizer was needed to prevent drug degradation. Below 40 wt%, sorbitol was added on top as a traditional plasticizer.

Based on the examples described above, it can be concluded that the main advantage of non-traditional plasticizers are their ability to reduce the extrusion temperature. The active substance itself can therefore be seen as an excellent processing aid during hot-melt extrusion.

5.4 Specialty Plasticizers

This class of low-molecular-weight materials can, depending on their physical state, also act as a plasticizer for polymeric carriers. More specifically, with specialty plasticizers the use of pressurized gases such as CO_2 is meant. By increasing the temperature and pressure towards the supercritical point, its isothermal compressibility approaches infinity and its density therefore changes dramatically. A pressurized gas can provide the solvent capacity of classical solvents, while providing higher diffusional capacity through its proximity to the gas state. CO_2 becomes a supercritical fluid when the critical temperature of 31°C and a critical pressure of 74 bar are reached, which are relatively mild conditions.

The combination of pressurized gases with polymer processing techniques has received increasing attention in polymer industry during the last 10 years [45]. The sorption of supercritical CO_2 into polymers results in swelling and changes of their mechanical and physical properties [46, 47]. The most important effect of carbon dioxide is the plasticization by reducing the glass transition temperature for a number of amorphous and semi-crystalline polymers. This plasticization occurs through two mechanisms [48]. First, carbon dioxide is absorbed between the polymer chains causing an increase of free volume and a decrease of chain entanglement. Second, carbon dioxide acts as a molecular lubricant that reduces melt viscosity. This has an impact on many polymer processing steps, such as viscosity reduction during polymer extrusion and polymer blending, foaming and changes in morphology due to induced crystallization.

A number of examples exist in polymer processing whereby pressurized CO_2 is injected into a single- and a twin-screw extruder to reduce the viscosity of the polymer and to create foam upon exiting the melt extruder when CO_2 is expanded to atmospheric conditions. In the last few years, the use of pressurized CO_2 during hot-melt extrusion for pharmaceutical drug delivery is also being investigated. Two major applications are therefore under consideration: the use of pressurized carbon dioxide as a plasticizer as well as its ability to form a foam upon expansion of the pressurized gas. Since CO_2 expands after exiting the die, it would not be present in the final product. In other words, the benefit would be that it serves as a temporary plasticizer present during the process but not in the final dosage form. In addition, also as a consequence of this expansion, foam is generated creating a porous structure with increased surface area which may be beneficial for increasing the dissolution rate.

Verreck et al. evaluated the injection of CO_2 using an intermeshing co-rotating twin-screw extruder for different polymers, including EC, PVPVA 64 and eudragit® E100 [49]. This type of extruder was chosen because of its excellent mixing capabilities [50]. However, due to its design, with this type of extruder the barrel cannot be completely filled with material as with a single-screw extruder. Injecting carbon dioxide could potentially lead to leakage of the gas, resulting in insufficient pressure build-up inside the barrel. The twin-screw configuration was therefore optimized in order to be able to inject and mix the carbon dioxide with the polymer melt at the appropriate pressures.

When using a twin-screw extruder, the use of an optimal screw design must be taken into account when optimizing the process as described by Lee et al. [51].

1. At the injection port of the carbon dioxide the pressure fluctuations should be minimized to obtain a stable injection; instead of kneading elements, transport elements should therefore be used at the site of injection.
2. Injected carbon dioxide should not be allowed to leak from upstream orifices, which is achieved by the creation of a melt seal using reversed screw elements.
3. The pressure downstream should be maintained at a sufficiently high level to ensure that the supercritical carbon dioxide remains dissolved in the polymer; this can be obtained by providing high die resistance.
4. Complete dissolution of carbon dioxide can be assured by using kneading elements to improve mixing downstream of the supercritical fluid introduction.

An extruder set-up and screw configuration were selected according to these suggestions, as shown in Figure 5.1 [49]. Using this set-up it was observed that CO_2 worked as a

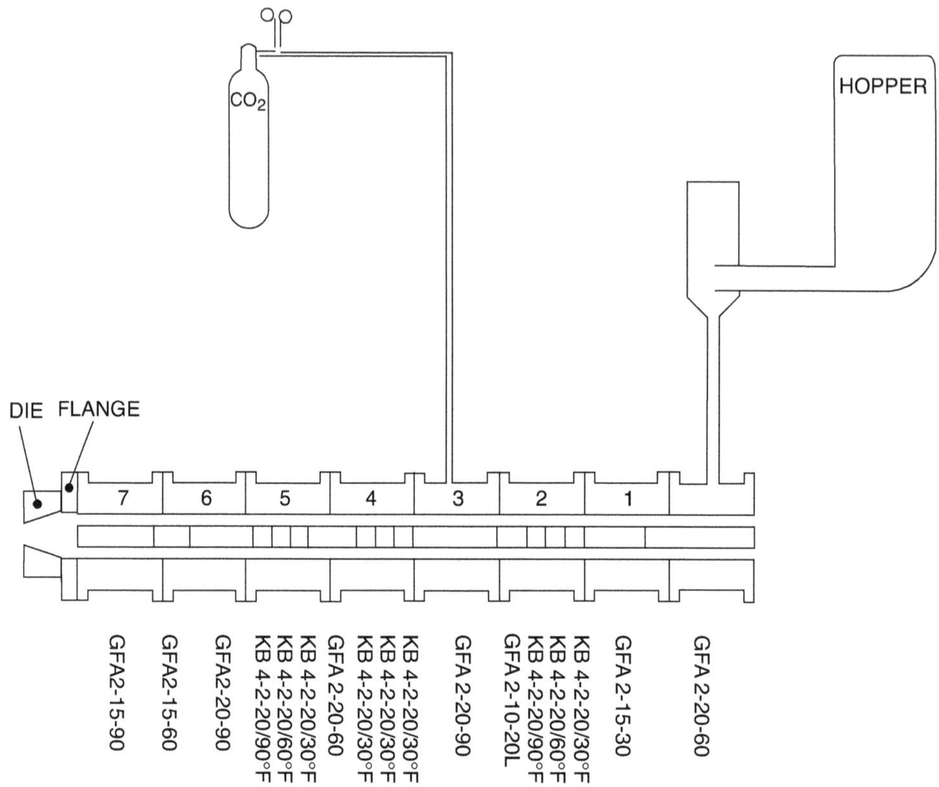

Figure 5.1 *Schematic set-up of an intermeshing co-rotating twin-screw extruder and screw configuration that allows for stable injection of pressurized CO_2. G. Verreck et al. 2003, reproduced with the permission of Elsevier.*

plasticizer reducing the temperature settings up to 15, 30 and 65°C for eudragit® E100, PVPVA 64 and EC, respectively. It was further observed that the morphology of the extrudates could be altered as a function of pressure and temperature. Due to the foam structures obtained, subsequent milling of the glassy extrudates was also improved.

In another study, hot-melt extrusion of solid dispersions consisting of the poorly soluble drug substance itraconazole with PVPVA 64 was investigated for the influence of injecting pressurized carbon dioxide [52]. It was confirmed that itraconazole works as a non-traditional plasticizer for PVPVA 64 (as described by Six et al. [40]) and that CO_2 was capable of even further reducing the temperature setting during extrusion, so that there was a combined plasticizing effect of the active substance and the pressurized gas. As with the plain PVPVA 64, the morphology was changed to a foam-like extrudate as a function of processing conditions. In vitro release of itraconazole could also be controlled as a function of the temperature and pressure.

Similar results were obtained when pressurized carbon dioxide was injected during hot-melt extrusion of itraconazole with EC as the polymeric carrier [53], i.e. drug release and foam morphology could be controlled as a function of pressure and temperature. With the latter solid dispersion, it was feasible to inject CO_2 even at supercritical conditions, allowing for a further reduction of the temperature settings of up to 65°C. The same coworkers then proved the concept with the thermally labile active substance p-amino salicylic acid (p-ASA); upon injection of supercritical CO_2 during hot-melt extrusion, significantly less degradation of p-ASA was observed [54].

In contrast with the work described above and the necessity of the use of a melt seal (by a reversing mixing screw element) prior to CO_2 injection to prevent leakage upstream, the work of Lyons et al. shows that this melt seal was not needed for their experiments [55]. They also describe the use of a co-rotating twin-screw extruder for the hot-melt extrusion process of the active substance Carvedilol with PEO 200.000 and Eudragit® E PO. Due to the flow behavior of these materials within the extruder barrel, leakage upstream was prevented without the need for a reversing mixing screw element. They also observed a drop in the melt viscosity when supercritical CO_2 was injected. The thermal properties of the extrudates were similar with or without carbon dioxide injection, and all samples exhibited a higher dissolution rate when treated with CO_2 during hot-melt extrusion. It was also observed that PEO crystallinity was increased for samples with carbon dioxide injection. Verreck et al. made the similar observation that the crystallinity of EC increased after CO_2 injection [52].

An alternative way to impregnate melt extrudate with CO_2 to obtain a porous structure is described by Andrews et al. [56]. Although the authors did not explicitly used carbon dioxide as a plasticizer during hot-melt extrusion, they do report on the interaction of CO_2 with a melt-extruded solid dispersion. They first performed hot-melt extrusion of Celecoxib and PVP K25 to obtain clear extrudate strands. In the next steps the strands are cut into tablets which are then exposed to CO_2 in a high-pressure vessel at 100 bar and 40°C for 24 hours. After those 24 hours, the chamber was depressurized and CO_2 evacuated. It was observed that extrudates after carbon dioxide treatment were still amorphous and T_g values were similar. FTIR and Raman spectra pre- and post-exposure to CO_2 were also identical, indicating the inert nature of this gas. The morphology was changed from a smooth surface pre-treatment to a porous material post-treatment. As a consequence, dissolution rates were significantly increased due to an increased surface area.

5.5 Conclusions

Based on the examples described in this chapter, it can be concluded that a number of different plasticizers exist to expand the application of hot-melt extrusion for pharmaceutical drug delivery. It is clear that the formulator can choose from either the traditional plasticizers or the specialty plasticizers and, in some cases, even a non-traditional plasticizer is present in the formulation. The major application of any of these categories is to: (1) optimize the processing conditions during hot-melt extrusion to prevent or minimize thermal degradation of the active substance and/or carrier; (2) tailor the extrudate properties during hot-melt extrusion or during post-die processing; and (3) modify or control the release properties of the final dosage form. Although it is obvious that a number of choices are already available, it is still an area of continuous focus to identify new materials with plasticizing properties to further foster the use of hot-melt extrusion for pharmaceutical applications.

References

(1) Gordon, M. and Taylor, J.S. (1952) Ideal copolymers and the second-order transitions of synthetic rubbers. I. Non-crystalline copolymers. *Journal of Applied Chemistry*, **2**, 493–500.

(2) Kelley, F.N. and Bueche, F. (1961) Viscosity and glass transition temperature relations for polymer-diluent systems. *Journal of Polymer Science*, **50**, 549–556.

(3) Simha, R. and Boyer, R.F. (1962) On a general relation involving the glass temperature and coefficients of expansion of polymers. *Journal of Chemical Physics*, **37**, 1003–1007.

(4) Rahman, M. and Brazel, C.S. (2004) The plasticizer market: an assessment of traditional plasticizers and research trends to meet new challenges. *Progress in Polymer Science*, **29**, 1223–1248.

(5) Crowley, M., Zhang, F., Repka, M.A., Thumma, S., Sampada, S.B., Battu, S.K., McGinity, J.W. and Martin, C. (2007) Pharmaceutical applications of hot-melt extrusion: part I. *Drug Development & Industrial Pharmacy*, **33**(9), 909–926.

(6) Breitenbach, J. (2002) Melt extrusion: from process to drug delivery technology. *European Journal of Pharmaceutics & Biopharmaceutics*, **54**, 107–117.

(7) Shilling, S.U., Shah, N.H., Malick, W., Infeld, M.H. and McGinity, J. (2007) Citric acid as a solid-state plasticizer for Eudragit® RS PO. *Journal of Pharmacy & Pharmacology*, **59**, 1493–1500.

(8) Verhoeven, E., De Beer, T.R.M., Van den Mooter, G., Remon, J.P. and Vervaet, C. (2008) Influence of formulation and process parameters on the release characteristics of ethylcellulose sustained-release mini-matrices produced by hot-melt extrusion. *European Journal of Pharmaceutics & Biopharmaceutics*, **69**, 312–319.

(9) Verhoeven, E., De Beer, T.R.M., Schacht, E., Van den Mooter, G., Remon, J.P., and Vervaet, C. (2009) Influence of polyethylene glycol/polyethylene oxide on the release characteristics of sustained-release ethylcellulose mini-matrices produced by hot-melt extrusion: in vitro and in vivo evaluations. *European Journal of Pharmaceutics & Biopharmaceutics*, **72**, 463–470.

(10) Mehuys, E., Vervaet, C. and Remon, J.P. (2004) Hot-melt extruded ethylcellulose cylinders containing a HPMC-Gelucire® core for sustained drug delivery. *Journal of Controlled Release*, **94**, 273–280.

(11) Mehuys, E., Vervaet, C., Gielen, I., Van Bree, H. and Remon, J.P. (2004) In vitro and in vivo evaluation of a matrix-in-cylinder system for sustained drug delivery. *Journal of Controlled Release*, **96**, 261–271.

(12) Quinten, T., De Beer, T., Vervaet, C. and Remon, J.P. (2009) Evaluation of injection moulding as a pharmaceutical technology to produce matrix tablets. *European Journal of Pharmaceutics & Biopharmaceutics*, **71**, 145–154.

(13) Quinten, T., Gonnissen, Y., Adriaens, E., De Beer, T., Cnudde, V., Masschaele, B., Van Hoorebeke, L., Siepmann, J., Remon, J.P. and Vervaet, C. (2009) Development of injection moulded matrix tablets based on mitures of ethylcellulose and low-substituted hydroxypropylcellulose. *European Journal of Pharmaceutics & Biopharmaceutics*, **37**, 207–216.

(14) Follonier, N., Doelker, E. and Cole, E.T. (1994) Evaluation of hot-melt extrusion as a new technique for the production of polymer-based pellets for sustained release

capsules containing high loadings of freely soluble drugs. *Drug Development & Industrial Pharmacy*, **20**(8), 1323–1339.
(15) Follonier, N., Doelker, E. and Cole, E.T. (1995) Various ways of modulating the release of diltiazem hydrochloride from hot-melt extruded sustained release pellets prepared using polymeric materials. *Journal of Controlled Release*, **36**, 243–250.
(16) Cilurzo, F., Cupone, I.E., Minghetti, P., Selmin, F. and Montanari, L. (2008) Fast dissolving films made of maltodextrins. *European Journal of Pharmaceutics & Biopharmaceutics*, **70**, 895–900.
(17) Wu, C., McGinity, J. (2003) Influence of methylparaben as a solid-state plasticizer on the physicochemical properties of Eudragit® RS PO hot-melt extrudates. *European Journal of Pharmaceutics & Biopharmaceutics*, **56**, 95–100.
(18) Shilling, S.U., Shah, N.H., Malick, W. and McGinity, J. (2010) Properties of melt extruded enteric matrix pellets. *European Journal of Pharmaceutics & Biopharmaceutics*, **74**, 352–361.
(19) Shilling, S.U., Lirola, H.L., Shah, N.H., Malick, A.W. and McGinity, J. (2010) Influence of plasticizer type and level on the properties of Eudragit® S100 matrix pellets prepared by hot-melt extrusion. *Journal of Microencapsulation*, **27**(6), 521–532.
(20) Zhang, F. and McGinity, J.W. (1999) Properties of sustained-release tablets prepared by hot-melt extrusion. *Pharmaceutical Development & Technology*, **4**(2), 241–250.
(21) Repka, M.A., Gerding, T.G., Repka, S.L. and McGinity, J.W. (1999) Influence of plasticizers and drugs on the physical-mechanical properties of hydroxypropylcellulose films prepared by hot melt extrusion. *Drug Development & Industrial Pharmacy*, **25**(5), 625–633.
(22) Repka, M.A. and McGinity, J.W. (2000) Physical-mechanical, moisture absorption and bioadhesive properties of hydroxypropylcellulose hot-melt extruded films. *Biomaterials*, **21**, 1509–1517.
(23) Henrist, D. and Remon, J.P. (1999) Influence of the formulation composition on the in vitro characteristics of hot stage extrudates. *International Journal of Pharmaceutical Sciences*, **188**, 111–119.
(24) Henrist, D., Lefebvre, R.A. and Remon, J.P. (1999) Bioavailability of starch based hot stage extrusion formulations. *International Journal of Pharmaceutics*, **187**, 185–191.
(25) Chamarthy, S.P. and Pinal, R. (2008) Plasticizer concentration and the performance of a diffusion-controlled polymeric drug delivery system. *Colloids and Surfaces A: Physicochemical & Engineering Aspects*, **331**, 25–30.
(26) Ghebremeskel, A.N., Vemavarapu, C. and Lodaya, M. (2006) Use of surfactants as plasticizers in preparing solid dispersions of poorly soluble API: stability testing of selected solid dispersions. *Pharmaceutical Research*, **23**(8), 1928–1936.
(27) Mehuys, E., Remon, J.P. and Vervaet, C. (2005) Production of enteric capsules by means of hot-melt extrusion. *European Journal of Pharmaceutical Sciences*, **24**, 207–212.
(28) Zheng, W., Cerea, M., Sauer, D. and McGinity, J.W. (2004) Properties of theophylline tablets powder-coated with methacrulate ester copolymers. *Journal of Drug Delivery Science Technology*, **14**(4), 319–325.
(29) Zhu, Y., Mehta, K.A. and McGinity, J.W. (2006) Influence of plasticizer level on the drug release from sustained release film coated and hot melt extruded dosage forms. *Pharmaceutical Development & Technology*, **11**, 285–229.

(30) Zhu, Y., Shah, N.H., Malick, W., Infeld, M.H. and McGinity, J.W. (2002) Solid-state plasticization of an acrylic polymer with chlorpheniramine maleate and triethyl citrate. *International Journal of Pharmaceutics*, **241**, 301–310.
(31) Sauer, D., Zheng, W., Coots, L.B. and McGinity, J.W. (2007) Influence of processing parameters and formulation factors on the drug release from tablets powder-coated with Eudragit® L100-55. *European Journal of Pharmaceutics & Biopharmaceutics*, **67**, 464–475.
(32) De Brabander, C., Van den Mooter, G., Vervaet, C. and Remon, J.P. (2002) Characterization of Ibuprofen as a non-traditional plasticizer of ethyl cellulose. *Journal of Pharmaceutical Sciences*, **91**(7), 1678–1685.
(33) Matsumoto, T. and Zografi, G. (1999) Physical properties of solid molecular dispersions of indomethacine with PVP and PVPVA in relation to indomethacine crystallization. *Pharmaceutical Research*, **16**, 1722–1728.
(34) Van den Mooter, G., Wuyts, M., Blaton, N., Busson, R., Grobet, P., Augustijns, P. and Kinget, R. (2001) Physical stabilization of amorphous ketoconazole in solid dispersions with polyvinylpyrrolidone K25. *European Journal of Pharmaceutical Sciences*, **12**, 261–269.
(35) Kidokoro, M., Shah, N.H., Malick, A.W., Infeld, M.H. and McGniity, J. (2001) Properties of tablets containing granulations of ibuprofen and an acrylic copolymer prepared by thermal processes. *Pharmaceutical Development & Technology*, **6**(2), 263–275.
(36) Aitken-Nichol, C., Zhang, F. and McGinity, J.W. (1996) Hot melt extrusion of acrylic films. *Pharmaceutical Research*, **13**, 804–808.
(37) Six, K., Verreck, G., Peeters, J., Binnemans, K., Berghmans, H., Augustijns, P., Kinget, R. and Van den Mooter, G. (2001) Investigation of thermal properties of glassy itraconazole: identification of a monotropic mesophase. *Thermochimica Acta*, **376**, 175–181.
(38) Verreck, G., Six, K., Van den Mooter, G., Baert, L., Peeters, J. and Brewster, M.E. (2003) Characterization of solid dispersions of itraconazole and hydroxypropylmethylcellulose prepared by melt extrusion, part I. *International Journal of Pharmaceutical Sciences*, **251**, 165–174.
(39) Six, K., Berghmans, H., Leuner, C., Dressman, J., Van Werde, K., Mullens, J., Benoist, L., Thimon, M., Meublat, L., Verreck, G., Peeters, J., Brewster, M. and Van den Mooter, G. (2003) Characterization of solid dispersions of itraconazole and hydroxypropylmethylcellulose prepared by melt extrusion, part II. *Pharmaceutical Research*, **20**(7), 1047–1054.
(40) Six, K., Verreck, G., Peeters, J., Brewster, M. and Van den Mooter, G. (2004) Increased physical stability and improved dissolution properties of itraconazole, a class II drug, by solid dispersions that combine fast- and slow-dissolving polymers. *Journal of Pharmaceutical Sciences*, **93**(1), 124–131.
(41) Six, K., Leuner, C., Dressman, J., Verreck, G., Peeters, J., Blaton, N., Augustijns, P., Kinget, R. and Van den Mooter, G. (2002) Thermal properties of hot-stage extrudates of itraconazole and eudragit E100. *Journal of Thermal Analysis & Calorimetry*, **68**, 591–601.
(42) Six, K., Murphy, J., Weuts, I., Craig, D.Q.M., Verreck, G., Peeters, J., Brewster, M. and Van den Mooter, G. (2003) Identification of phase separation in solid dispersions

of itraconazole and eudragit® E100 using microthermal analysis. *Pharmaceutical Research*, **20**(1), 135–138.
(43) Six, K., Verreck, G., Peeters, J., Augustijns, P., Kinget, R. and Van den Mooter, G. (2001) Characterization of glassy itraconazole: a comparative study of its molecular mobility below Tg with that of structural analogues using MTDSC. *International Journal of Pharmaceutics*, **213**, 163–173.
(44) Lakshman, J.P., Cao, Y., Kowalski, J. and Serajuddin, A.T.M. (2008) Application of melt extrusion in the development of a physically and chemically stable high-energy amorphous solid dispersion of a poorly water-soluble drug. *Molecular Pharmaceutics*, **5**(6), 994–1002.
(45) Tomasko, D.L., Li, H., Liu, D., Han, X., Wingert, M.J., Lee, J.L. and Koelling, K.W. (2003) A review of CO_2 applications in the processing of polymers. *Industrial & Engineering Chemistry Research*, **42**(25), 6431–6456.
(46) Kazarian, S.G. (2000) Polymer Processing with Supercritical Fluids. *Journal of Polymer Science, Series C*, **42**(1), 78–101.
(47) Kiran, E. (1994) Polymer formation, modifications and processing in or with supercritical fluids. *In* Kiran, E. and Sengers, J.M.H.L. (eds), *Supercritical Fluids*, Kluwer, Dordrecht.
(48) Chiou, J.S., Barlow, J.W. and Paul, D.R. (1985) Plasticisation of glassy polymers by CO_2. *Journal of Applied Polymer Science*, **30**, 2633–2642.
(49) Verreck, G., Decorte, A., Li, H., Tomasko, D., Arien, A., Peeters, J., Rombaut, P., Van den Mooter, G. and Brewster, M.E. (2006) The effect of pressurized carbon dioxide as a plasticizer and foaming agent on the hot melt extrusion process and extrudate properties of pharmaceutical polymers. *Journal of Supercritical Fluids*, **38**, 383–391.
(50) Mollan, M. (2003) Historical overview. *In* Ghebre-Sellassie, I. and Martin, C. (eds), *Pharmaceutical Extrusion Technology*, Marcel Dekker, New York.
(51) Lee, M., Tzoganakis, C. and Park, C.B. (1998) Extrusion of PE/PS blends with supercritical carbon dioxide. *Polymer Engineering & Science*, **38**, 1112–1120.
(52) Verreck, G., Decorte, A., Heymans, K., Adriaensen, J., Cleeren, D., Jacobs, A., Liu, D., Tomasko, D., Arien, A., Peeters, J., Rombaut, P., Van den Mooter, G. and Brewster, M.E. (2005) The effect of pressurized carbon dioxide as a temporary plasticizer and foaming agent on the hot stage extrusion process and extrudate properties of itraconazole with PVP-VA 64. *European Journal of Pharmaceutical Sciences*, **26**, 349–358.
(53) Verreck, G., Decorte, A., Heymans, K., Adriaensen, J., Liu, D., Tomasko, D., Arien, A., Peeters, J., Rombaut, P., Van den Mooter, G. and Brewster, M.E. (2007) The effect of supercritical CO_2 as a reversible plasticizer and foaming agent on the hot stage extrusion of itraconazole with EC 20 cps. *Journal of Supercritical Fluids*, **40**, 153–162.
(54) Verreck, G., Decorte, A., Heymans, K., Adriaensen, J., Liu, D., Tomasko, D., Arien, A., Peeters, J., Van den Mooter, G., and Brewster, M.E., (2006) Hot stage extrusion of p-amino salicylic acid with EC using CO_2 as a temporary plasticizer. *International Journal of Pharmaceutics*, **327**, 45–50.
(55) Lyons, J.G., Hallinan, M., Kennedy, J.E., Devine, D.M., Geever, L.M., Blackie, P. and Higginbotham, C.L. (2007) Preparation of monolithic matrices for oral drug delivery

using a supercritical fluid assisted hot melt extrusion process. *International Journal of Pharmaceutics*, **329**, 62–71.

(56) Andrews, G.P., Abu-Diak, O., Kusmanto, F., Hornsby, P., Hui, Z. and Jones, D.S. (2010) Physicochemical characterization and drug-release properties of celecoxib hot-melt extruded glass solutions. *Journal of Pharmacy & Pharmacology*, **62**, 1580–1590.

6

Applications of Poly(meth)acrylate Polymers in Melt Extrusion

Kathrin Nollenberger and Jessica Albers
Evonik Industries AG

6.1 Introduction

Polymethyl(meth)acrylate (PMMA) was developed and first brought to the market in 1933 by Röhm and Haas under the trademark Plexiglas®. It is a crystal-clear unbreakable organic glass of outstanding quality. A violin made of Plexiglas® was prized at the Paris World Exposition in 1937. PMMA was continuously developed in the following years for different applications including medical implants and prostheses. The high quality and excellent long-term stability under environmental stress, especially the impact of light, water and oxygen, is based on the rigid molecular structure of the polymer. It is composed of a continuous chain of carbon atoms as a backbone, which is additionally stabilized by methyl groups. The ester groups in poly(meth)acrylates are very resistant against hydrolysis [1].

In the 1950s the coating of oral solid dosage forms had become the main pharmaceutical application field of poly(meth)acrylates. The first patent application from 1952 by Trommsdorff and Grimm describes enteric coatings for solid dosage forms. The trademark EUDRAGIT® was registered in 1954. Beginning 1955 EUDRAGIT® L and EUDRAGIT® S became commercially available as organic solutions in isopropylic alcohol. Today poly(meth)acrylates are well known in the pharmaceutical industry and widely used for protective coatings such as taste masking, moisture and odor protection, furthermore for gastrointestinal targeting (delayed release) and extended drug release [1]. The polymers are available in different grades: as aqueous dispersions, organic solutions, granules and powders. Besides coatings they are also used as matrix formers either by direct compression or

Hot-melt Extrusion: Pharmaceutical Applications, First Edition. Edited by Dennis Douroumis.

Table 6.1 *Chemical name and compendial compliance of poly(meth)acrylates.*

Polymer	Physical properties	Chemical/IUPAC name	Ph. Eur.	USP/NF	JPE
EUDRAGIT® E 100	granules	Poly(butyl methacrylate-co-(2-dimethylaminoethyl) methacrylate-co-methyl methacrylate) 1:2:1	Basic Butylated Methacrylate Copolymer	polymer conforms to Amino Methacrylate Copolymer – NF	Aminoalkyl Methacrylate Copolymer E
EUDRAGIT® E PO	powder	Poly(butyl methacrylate-co-(2-dimethylaminoethyl) methacrylate-co-methyl methacrylate) 1:2:1	Basic Butylated Methacrylate Copolymer	polymer conforms to Amino Methacrylate Copolymer – NF	Aminoalkyl Methacrylate Copolymer E
EUDRAGIT® L 100-55	powder	Poly(methacrylic acid-co-ethyl acrylate) 1:1	Methacrylic Acid - Ethyl Acrylate Copolymer (1:1)	Methacrylic Acid and Ethyl Acrylate Copolymer (1:1)	Dried Methacrylic Acid Copolymer LD
EUDRAGIT® L 100	powder	Poly(methacrylic acid-co-methyl methacrylate) 1:1	Methacrylic Acid - Methyl Methacrylate Copolymer (1:1)	Methacrylic Acid and Methyl Methacrylate Copolymer (1:1)	Methacrylic Acid Copolymer L
EUDRAGIT® S 100	powder	Poly(methacrylic acid-co-methyl methacrylate) 1:2	Methacrylic Acid - Methyl Methacrylate Copolymer (1:2)	Methacrylic Acid and Methyl Methacrylate Copolymer (1:2)	Methacrylic Acid Copolymer S
EUDRAGIT® FS 30 D	30% aqueous dispersion	Poly(methyl acrylate-co-methyl methacrylate-co-methacrylic acid) 7:3:1	–	–	–
EUDRAGIT® RL 100	granules	Poly(ethyl acrylate-co-methyl methacrylate-co-trimethylammonioethyl methacrylate chloride) 1:2:0.2	Ammonio Methacrylate Copolymer, Type A	Ammonio Methacrylate Copolymer, Type A – NF	Aminoalkyl Methacrylate Copolymer RS

EUDRAGIT® RL PO	powder	Poly(ethyl acrylate-co-methyl methacrylate-co-trimethylammonioethyl methacrylate chloride) 1:2:0.2	Ammonio Methacrylate Copolymer, Type A	Ammonio Methacrylate Copolymer, Type A – NF	Aminoalkyl Methacrylate Copolymer RS
EUDRAGIT® RS 100	granules	Poly(ethyl acrylate-co-methyl methacrylate-co-trimethylammonioethyl methacrylate chloride) 1:2:0.1	Ammonio Methacrylate Copolymer, Type B	Ammonio Methacrylate Copolymer, Type B – NF	Aminoalkyl Methacrylate Copolymer RS
EUDRAGIT® RS PO	powder	Poly(ethyl acrylate-co-methyl methacrylate-co-trimethylammonioethyl methacrylate chloride) 1:2:0.1	Ammonio Methacrylate Copolymer, Type B	Ammonio Methacrylate Copolymer, Type B – NF	Aminoalkyl Methacrylate Copolymer RS
EUDRAGIT® NE 30 D	30% aqueous dispersion	Poly(ethyl acrylate-co-methyl methacrylate) 2:1	Polyacrylate Dispersion 30 Per Cent	Ethyl Acrylate and Methyl Methacrylate Copolymer Dispersion – NF	Ethyl Acrylate Methyl Methacrylate Copolymer Dispersion
EUDRAGIT® NM 30 D	30% aqueous dispersion	Poly(ethyl acrylate-co-methyl methacrylate) 2:1	Polyacrylate Dispersion 30 Per Cent	–	–

via wet granulation processes. Later the excellent biocompatibility of poly(meth)acrylates as well as their good dermal and mucosal tolerance were detected which opened up possibilities for their use in various medical applications such as ointments, gels, wound spray, patches or transdermal therapeutic systems [1].

In the late 1990s, poly(meth)acrylates have been introduced into melt extrusion for several reasons [2–7]. Being thermoplastic polymers, their physicochemical properties such as melt viscosity, glass transition temperature and temperature stability are ideal for the use in melt extrusion. Furthermore, their functionalities such as pH-dependent solubility or pH-independent swelling properties allow versatile applications.

Poly(meth)acrylates are synthetic polymers ensuring low batch-to-batch variations compared to natural-type polymers. They have very narrow specifications and are monographed in the European Pharmacopoeia [8], the USP/NF [9] and the Japanese Pharmaceutical Excipients [10]. For the entire range, drug master files exist at the US Food and Drug Administration (FDA).

This chapter focuses on those poly(meth)acrylates used in melt-extrusion applications suitable for the formulation of oral dosage forms, known under the brand EUDRAGIT® (Table 6.1). Other applications of poly(meth)acrylates in coatings, gels, ointments, transdermal systems and implants are described elsewhere in the literature. EUDRAGIT® L 30 D-55, EUDRAGIT® RL 30 D, EUDRAGIT® RS 30 D as well as organic solutions of the polymers are not mentioned in this chapter as their main application area are coatings.

6.2 Polymer Characteristics

6.2.1 Chemical Structure and Molecular Weight

The chemical structure of EUDRAGIT® polymers is based on poly(meth)-acrylates, whereas different monomers within the polymer chains provide the polymer its specific characteristics. In all EUDRAGIT® polymers the monomers are statistically ordered. EUDRAGIT® E (Figure 6.1) is a cationic copolymer based on dimethyl aminoethyl methacrylate, butyl methacrylate and methyl methacrylate. Since the dimethyl aminoethyl group is its functional unit, it rapidly dissolves by forming salts at acidic pH values below 5.

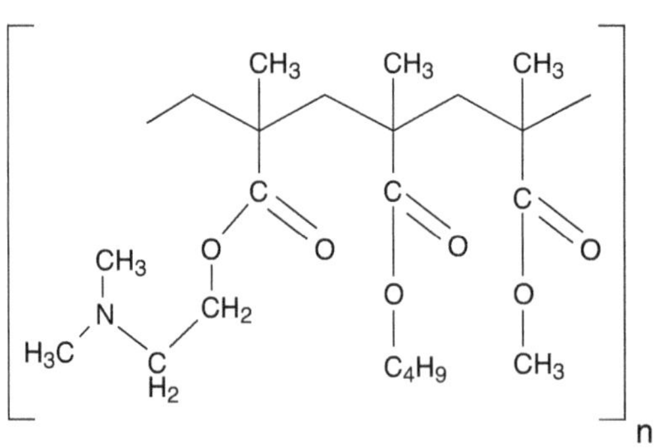

Figure 6.1 Chemical structure of EUDRAGIT® E.

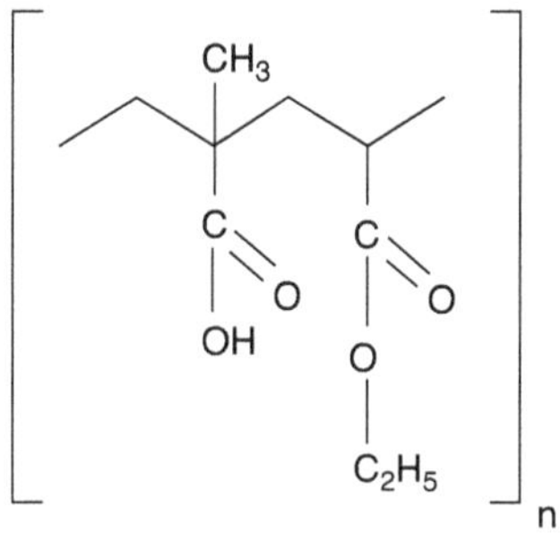

Figure 6.2 *Chemical structure of EUDRAGIT® L 100-55.*

Methacrylic acid copolymers with methyl methacrylate and ethylacrylate as ester components dissolve at basic pH values. The functionality is given by the carboxylic groups that are transformed to carboxylate groups starting in the pH range of 5–7. The dissolution pH of the polymers depends on the content of carboxylic groups. For EUDRAGIT® L 100-55 (Figure 6.2) and EUDRAGIT® L 100 (Figure 6.3), the ratio of the free carboxyl groups to the ester groups is approximately 1:1. For EUDRAGIT® S 100 (Figure 6.3), the ratio of the free carboxyl groups to the ester groups is approximately 1:2.

EUDRAGIT® FS 30 D (Figure 6.4) is the aqueous dispersion of an anionic copolymer based on methyl acrylate, methyl methacrylate and methacrylic acid. The ratio of the free carboxyl groups to the ester groups is approximately 1:10. Compared to the other anionic polymers, EUDRAGIT® FS has a much higher flexibility which allows extrusion at very low temperatures and without plasticizer.

Methacrylate ester copolymers are neutral polymers and are insoluble in water, diluted acids, buffer solutions or digestive fluids over the entire physiological pH range. The polymers EUDRAGIT® NE and EUDRAGIT® NM (Figure 6.5) are available as aqueous dispersions and can be used for melt extrusion processes by using a liquid dosing unit. The 30% aqueous dispersion of EUDRAGIT® NE contains 1.5% nonoxynol 100 as emulsifier and the 40% aqueous dispersion 2.0%. EUDRAGIT® NM contains 0.7% macrogol stearyl ether (20) as emulsifier.

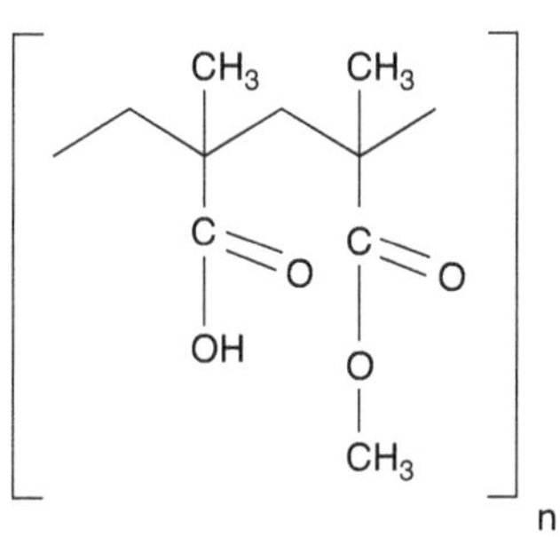

Figure 6.3 *Chemical structure of EUDRAGIT® L 100 and EUDRAGIT® S 100.*

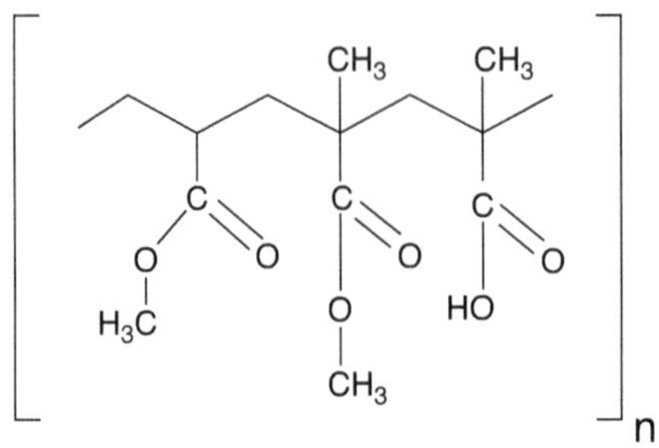

Figure 6.4 Chemical structure of EUDRAGIT® FS 30 D.

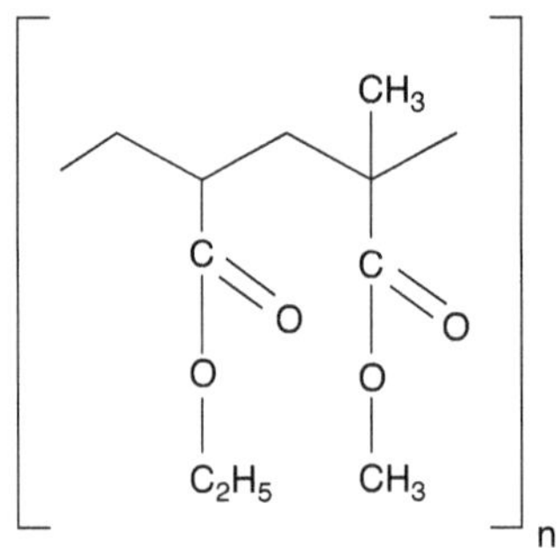

Figure 6.5 Chemical structure of EUDRAGIT® NE and EUDRAGIT® NM.

The permeability of methacrylic ester copolymers can be modified by a copolymerization with hydrophilic quarternary ammonium groups. These polymers are marketed as EUDRAGIT® RL and EUDRAGIT® RS. (Figure 6.6). Due to their pH-independent solubility, both polymers can be used for sustained-release applications.

For EUDRAGIT® RL the ratio of the ethacrylate and the methyl methacrylate groups to the quarternary trimethylammonioethyl chloride group is approximately 1:2:0.2. For EUDRAGIT® RS the ratio of the ethacrylate and the methyl methacrylate groups to the quarternary trimethylammonioethyl chloride group is approximately 1:2:0.1. Table 6.2

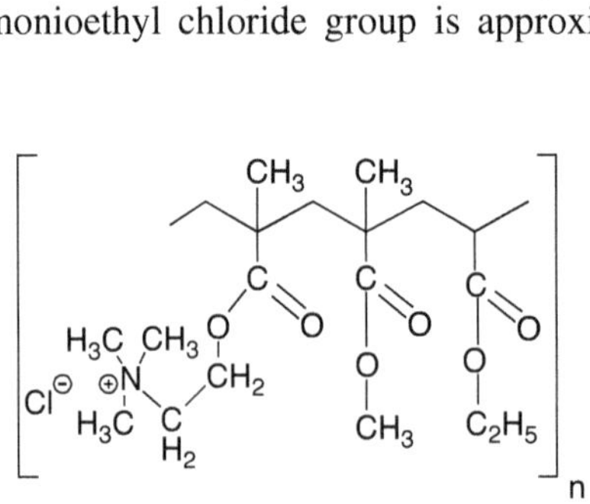

Figure 6.6 Chemical structure of EUDRAGIT® RL and EUDRAGIT® RS.

Table 6.2 Molecular weights determined by size exclusion chromatography.

Polymer	Molecular weight (g/mol)
EUDRAGIT® E	47,000
EUDRAGIT® L 100-55	320,000
EUDRAGIT® L	125,000
EUDRAGIT® S	125,000
EUDRAGIT® FS	280,000
EUDRAGIT® RL	32,000
EUDRAGIT® RS	32,000
EUDRAGIT® NE	750,000
EUDRAGIT® NM	600,000

shows the molar masses of the EUDRAGIT® polymers determined by size exclusion chromatography [11].

Due to their functional groups the poly(meth)acrylate copolymers EUDRAGIT® E and the anionic copolymers EUDRAGIT® L 100-55, L 100, S 100 and FS 30 D show a pH-dependent dissolution behavior (Figure 6.7).

The most important polymer properties for melt extrusion are the glass transition temperature, melt viscosity and thermostability. Energy is applied during the process in two ways: first by the heat of the barrels and secondly by the mechanical energy of the screws. Both can have a strong influence on the processability and stability of the polymers.

6.2.2 Glass Transition Temperature

The glass transition temperature T_g of an amorphous material is an important property. At this temperature, the rheological behavior changes from a glassy state to a more rubbery state. During this transition, the mobility of the polymer chains is increased [12].

In general, the processing temperature in the extruder needs to be above the glass transition temperature of the polymer to lower its melt viscosity, to soften the polymer and hence to improve the flow of the molten polymer in the extruder. Table 6.3 lists the mean glass transition temperatures of different EUDRAGIT® polymers [11].

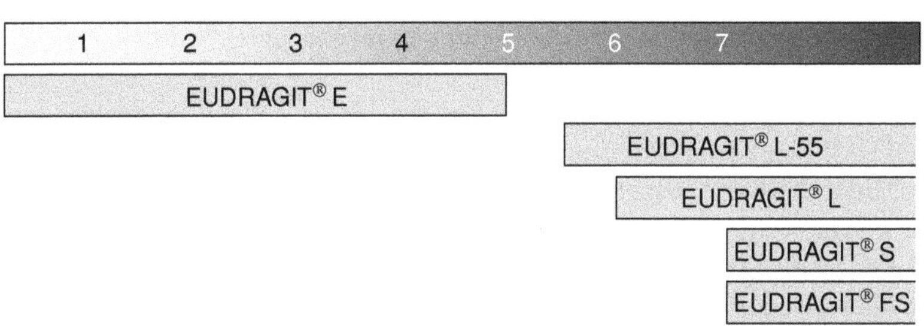

Figure 6.7 Dissolution pH of pH-dependent poly(meth)acrylate copolymers.

Table 6.3 Glass transition temperatures of EUDRAGIT® polymers given as interval measured at 20 K/min heating rate in the second heating cycle.

EUDRAGIT® grade	T_g ±5 (°C)
EUDRAGIT® E 100/E PO	~45
EUDRAGIT® L 100–55	~96
EUDRAGIT® FS 30 D	~43
EUDRAGIT® RL 100/RL PO	~63
EUDRAGIT® RS 100 / RS PO	~58
EUDRAGIT® NE 30 D/NE 40 D	~6
EUDRAGIT® NM 30 D	~9

The glass transition temperatures of EUDRAGIT® L 100 and EUDRAGIT® S 100 cannot be determined, as the thermal damaging of the functional group occurs around the glass transition temperature of the polymers which affects the value of T_g. Hence, these two polymers can only be extruded together with suitable plasticizers.

6.2.3 Plasticizers

The processability in melt extrusion can be improved by either adding plasticizers which lower the glass transition temperature and the melt viscosity of the polymer or by processing aids that act as lubricants and do not affect the glass transition temperature of the polymers.

Plasticizers increase the free volume in the polymer matrix, reducing the entanglement and friction of the polymer chains [13]. The solubility parameters of the polymer and the plasticizer should be similar to ensure good miscibility of the two components, leading to a higher effectiveness of the plasticizer.

In sustained-release applications, the influence of plasticizer type and level on the dissolution properties should not be neglected [14]. For solubility-enhancing formulations, hydrophobic plasticizers may have a negative effect on the increase in dissolution whereas surfactants can additionally improve the solubility enhancing effect. Five and ten percent sodium dodecyl sulfate (SDS) decreased the T_g of an API-EUDRAGIT® L 100 blend by 8.88°C and 11.25°C, respectively [15].

Due to the low glass transition temperature of EUDRAGIT® E, no plasticizer is required for melt extrusion processes. If a temperature-sensitive active pharmaceutical ingredient (API) is used, stearic acid is an efficient plasticizer for this polymer to enable extrusion at low temperatures <100°C.

In general, plasticizers are required for anionic poly(meth)acrylates except for EUDRAGIT® FS 30 D. Plasticizers commonly used in coating applications can also be used in melt extrusion, e.g. triethyl citrate (Figure 6.8), polyethylene glyocl 6000 or propylene glycole [16]. Bruce et al. prepared melt extruded tablets with EUDRAGIT® S 100 and 5-aminosalicylic acid for colon delivery. The process temperature was significantly decreased with the addition of triethyl citrate, but led to a faster drug release. Citric acid was proven to be an efficient solid-state plasticizer and decreased the T_g to 75.2°C and 103°C for the ratios citric acid:EUDRAGIT® S 100 1:1 and 1:4, respectively. As a higher amount

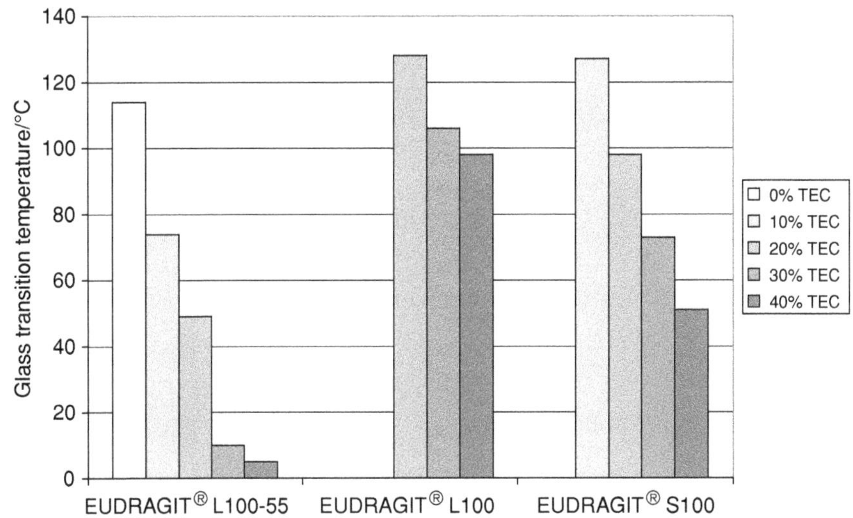

Figure 6.8 *Influence of triethyl citrate as plasticizer on the* T_g *of anionic poly(meth)acrylates EUDRAGIT® L 100-55, EUDRAGIT® L 100 and EUDRAGIT® S 100.*

of citric acid could decrease the release profile, a combination of triethyl citrate and citric acid was chosen and resulted in good processability and the desired release profile [17].

Based on the low T_g of EUDRAGIT® RL and EUDRAGIT® RS, those polymers can be processed without plasticizer. However, to enable even lower melt extrusion temperatures or less shear stress, plasticizers such as triethyl citrate, triacetin [3], dibutyl sebacate or polyethylene glycol 600 are suitable. As solid-state plasticizers, citric acid was identified for EUDRAGIT® RS as effective plasticizer. Citric acid monohydrate showed superior behavior than the anhydrous from, due to the higher solubility in the polymer [18].

Drug molecules can also act as plasticizers, reducing the processing temperature during melt extrusion. For example, chlorpheniramine maleate is reported to act as a solid-state plasticizer on EUDRAGIT® RS PO. The glass transition temperature is decreased by approximately 1.3°C for one percent of chlorpheniramine maleate [19].

Ibuprofen is an effective plasticizer and dramatically reduces the T_g of EUDRAGIT® RS 30 D [20] and EUDRAGIT® E in melt extrusion, as demonstrated in Figure 6.9.

The reduction of the glass transition temperature may lead to stickiness in the final product. Glidants such as talc and magnesium stearate can be added to improve processing.

6.2.4 Thermostability

Thermal stability of the polymers is as important for melt extrusion as the stability of the active ingredient. Degradation products can trigger the degradation of the drug and cause a high impurity level. Thermal stability as a function of temperature and time can be examined by thermogravimetric analysis and, if required, can be coupled with mass spectroscopy to identify the degradation products. Depolymerization of the poly(meth)acrylates only occurs at temperatures higher than 250°C. However, the

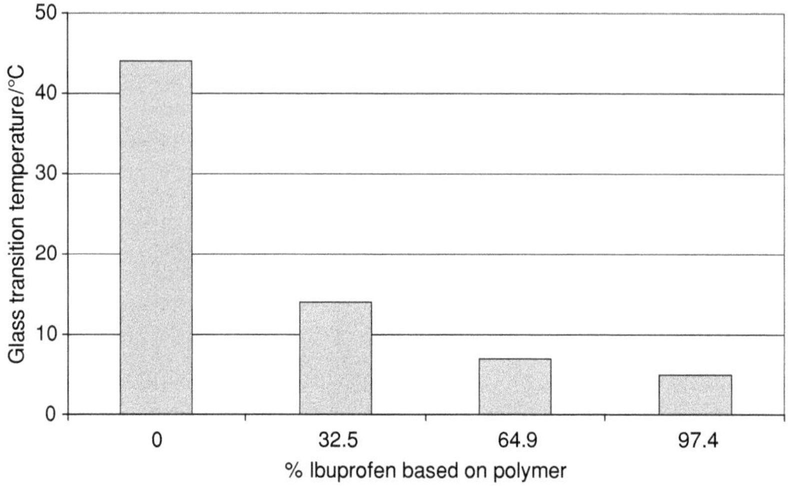

Figure 6.9 *Influence of ibuprofen on the glass transition temperature of EUDRAGIT® E.*

functional groups in the side chains are more sensitive to temperature and therefore, determine the thermostability of the polymer. It is not only the temperature which needs to be taken into consideration; the time during which the polymer is exposed to the temperature is also essential.

In Figure 6.10 the maximum temperatures for different poly(meth)acrylates are shown.

The maximum temperatures (T_{max}) were calculated for a time of 4.5 minutes, an average residence time during a melt extrusion process. The maximum damage of the functional group was allowed to be $<1\%$ ($<1.5\%$ for EUDRAGIT® RL/RS).

The neutral polymers EUDRAGIT® NE 30 D and EUDRAGIT® NM 30 D do not have a functional group and are therefore not prone to damage below 250°C.

The thermostability of EUDRAGIT® L 100 and EUDRAGIT® S 100 is not shown in the figure, as the two polymers cannot be extruded without the use of plasticizers due to their high melt viscosity and glass transition temperature.

In Figure 6.11 extrusion temperatures for EUDRAGIT® polymers are shown, taking into account the thermostability of the polymers and the use of plasticizers.

6.2.5 Viscosity

The viscosity of a polymer in its molten state affects the melt extrusion process in the melt temperature and the melting rate, the die flow and the output. Polymers with high viscosity require higher melt temperatures and higher melting rates in the screw. They also require discharge pressure and more power. The viscosities of the polymers are determined with a high-pressure capillary rheometer (Göttfert, Buchen, Germany) fitted with a 30 mm long by 1 mm diameter capillary die.

All EUDRAGIT® polymers show pseudoplastic behavior at high shears and are Newtonian at low shears (Figures 6.12–6.16). Carreau rheological models can be fitted to

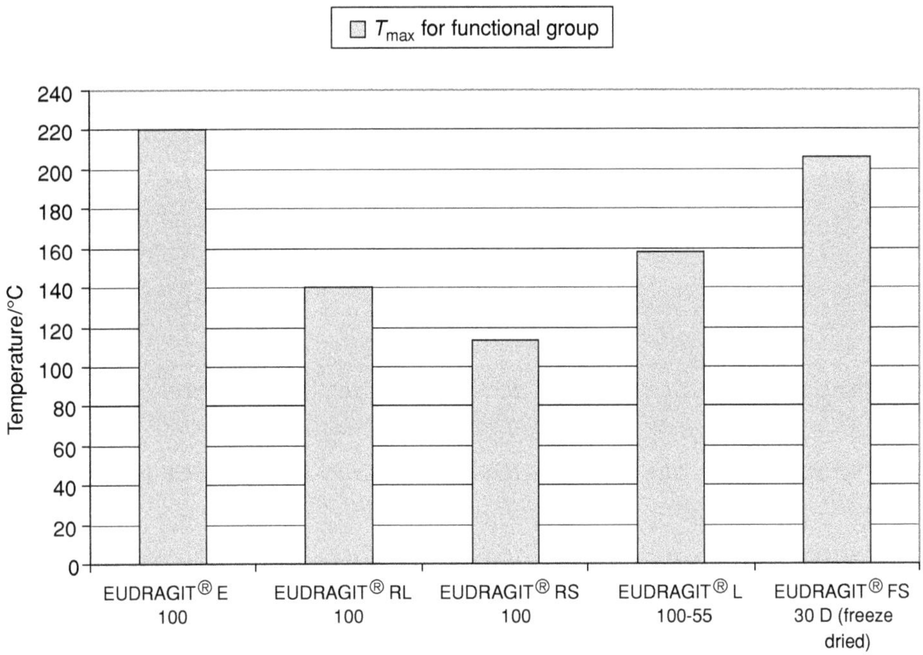

Figure 6.10 T_{max} *for functional groups of different EUDRAGIT® polymers at a residence time of 4.5 min.*

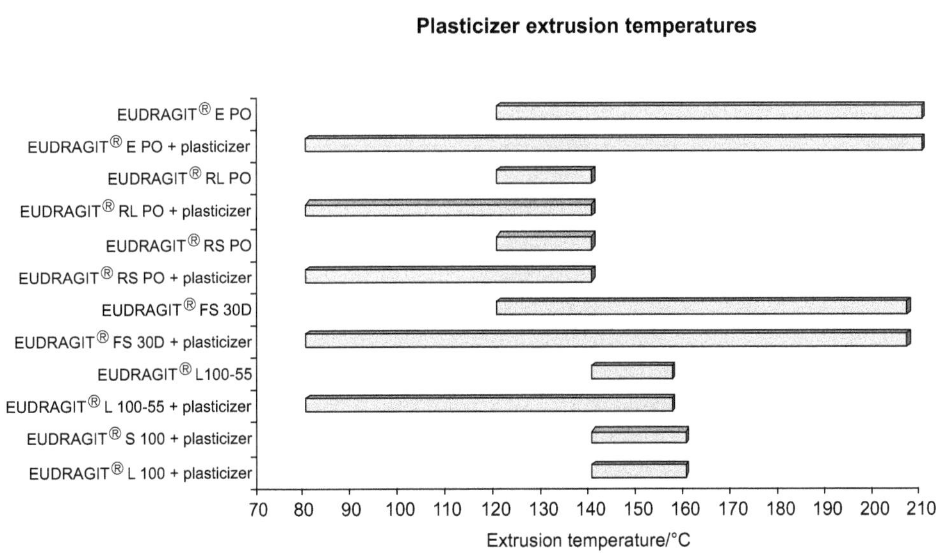

Figure 6.11 *Extrusion temperatures for EUDRAGIT® polymers.*

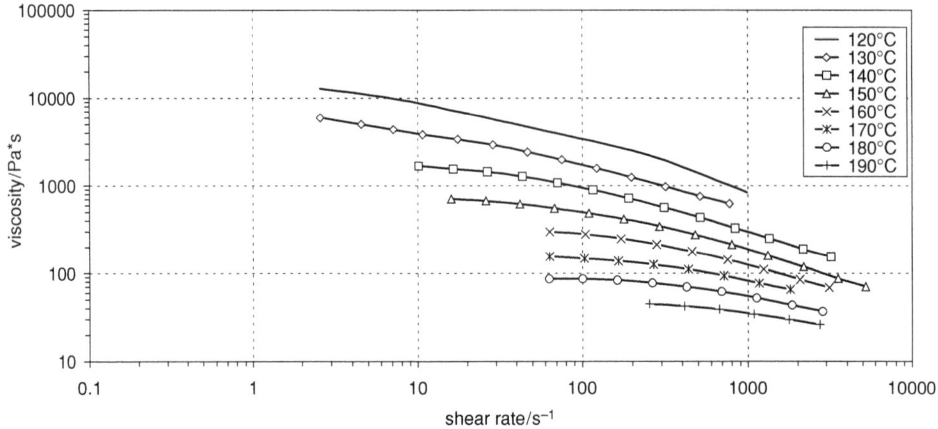

Figure 6.12 Viscosity as function of shear rate for EUDRAGIT® E 100.

viscosity-shear rate data for all pharmaceutical grade poly(meth)acrylates. The use of plasticizer (TEC) reduces the viscosity of high poly(meth)acrylates with higher glass transition temperatures (e.g. EUDRAGIT® L 100-55 and EUDRAGIT® S 100) to acceptable ranges at 150°C [21].

6.2.6 Specific Heat Capacity

The specific heat capacity (heat capacity per unit mass of the polymer) defines how much energy it takes to raise the temperature of a polymer. It is a measure for the drive power of an extruder that is required to process the polymer. The power required to heat the

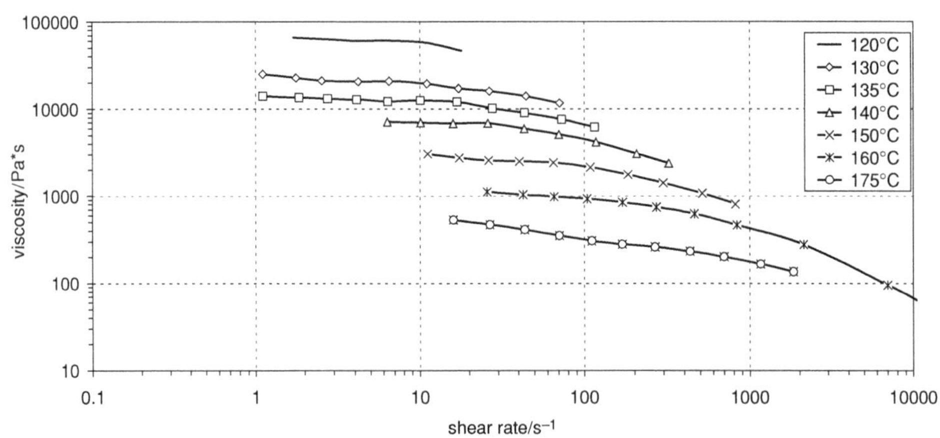

Figure 6.13 Viscosity as function of shear rate for EUDRAGIT® RL 100.

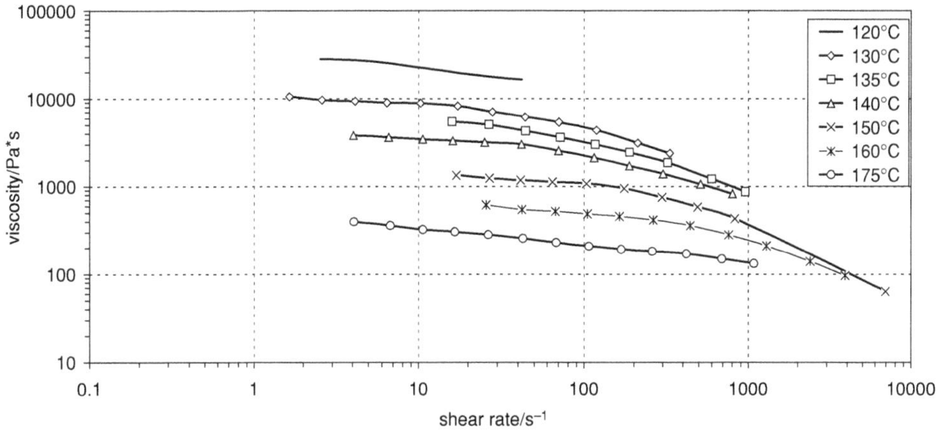

Figure 6.14 Viscosity as function of shear rate for EUDRAGIT® RS 100.

polymer from room temperature to the molten state is defined by the mass flow rate, the heat capacity, the difference in temperature between the feed and melt temperature and the heat of fusion for the polymer. The specific heat capacity of a formulation can help to minimize the power input through an optimized design of the screw.

On the contrary, it also characterizes the cooling capacity to extract that heat from the polymer. This determines the design and length of the cooling unit to downstream the extrudate.

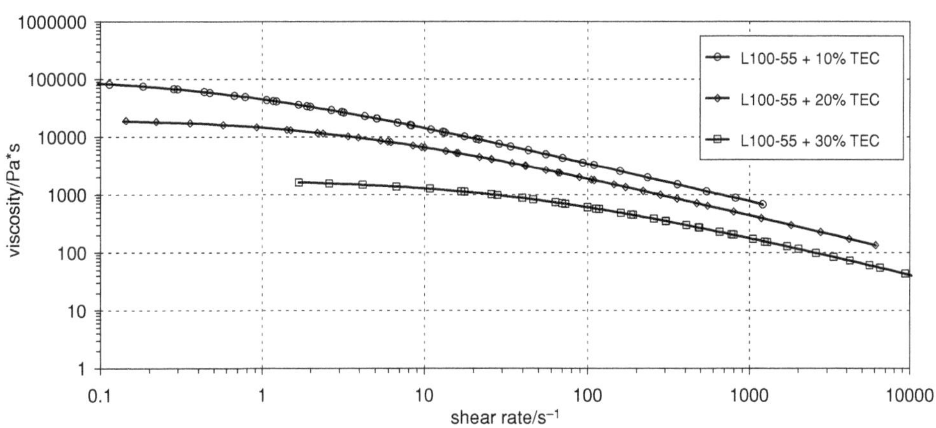

Figure 6.15 Viscosity as function of shear rate for EUDRAGIT® L 100-55 with 30, 40 and 50% TEC at 150°C.

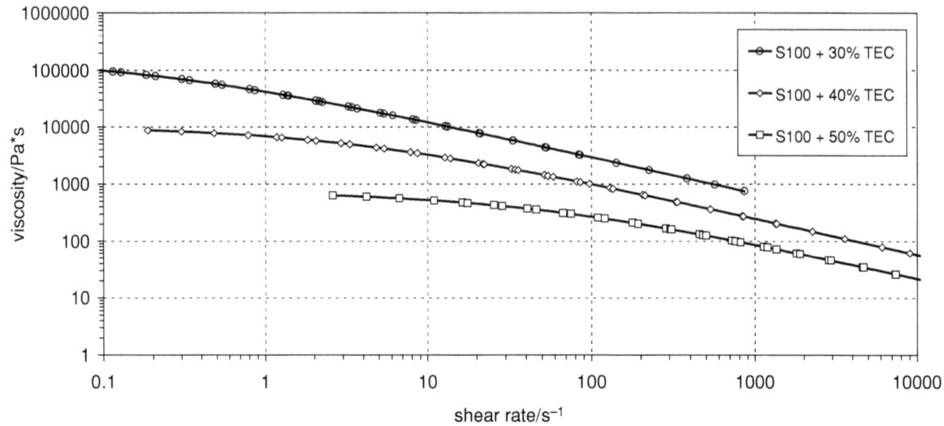

Figure 6.16 *Viscosity as function of shear rate for EUDRAGIT® S 100 with 30, 40 and 50% TEC at 150°C.*

The specific heat capacity can be measured by differential scanning calorimetry (DSC). The specific heat capacity values for all polymers at 120°C are between 1.5 and 2.5 J/(gK) (Figures 6.17 and 6.18).

6.2.7 Hygroscopicity

One challenge in the development of solid solutions is to ensure their physical stability. As only small amounts of crystals in the solid solution can evoke recrystallization of the

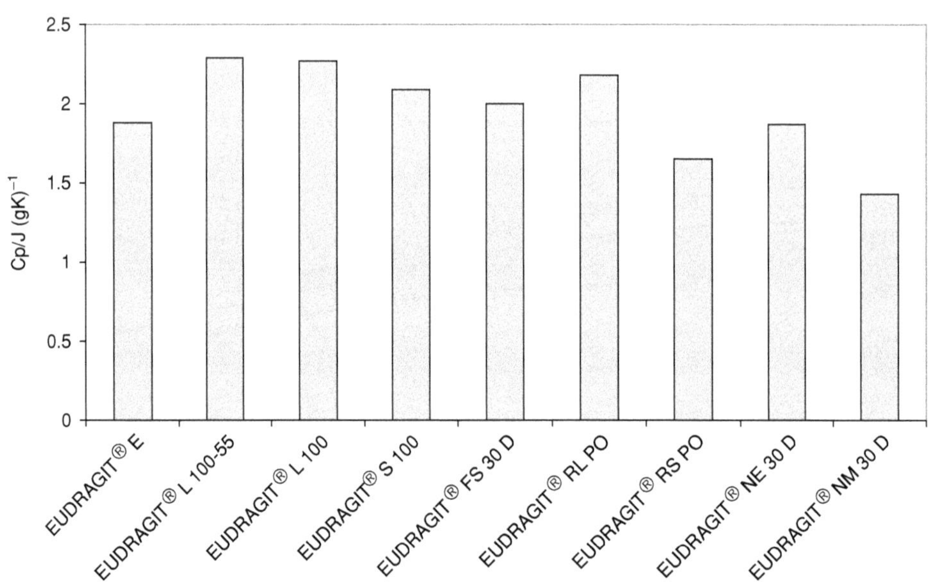

Figure 6.17 *Specific heat capacities of EUDRAGIT® polymers at 120°C (DSC, 10 K/min).*

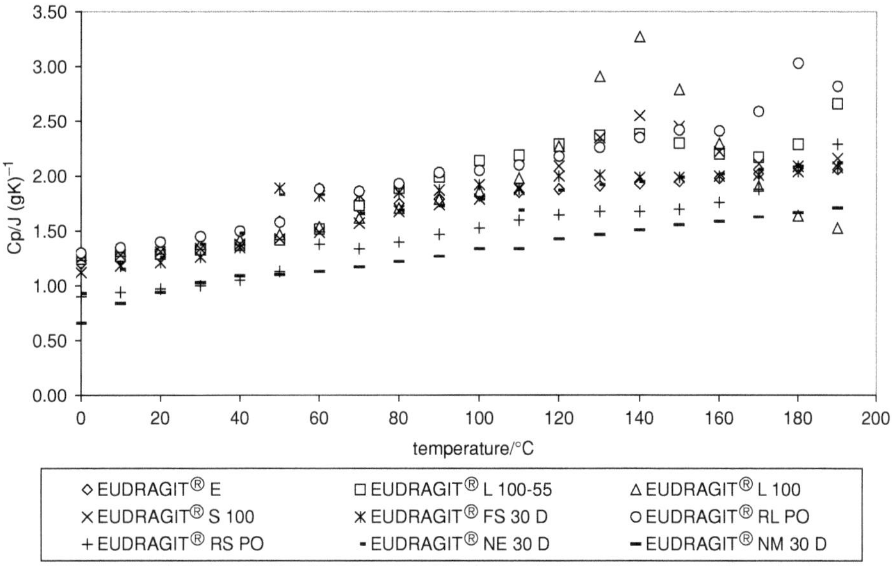

Figure 6.18 *Specific heat capacities of EUDRAGIT® polymers at different temperatures (DSC, 10 K/min).*

molecularly dispersed drug, the stability of the solid state is one major prerequisite for a reproducible drug release. As water acts as plasticizer and increases the mobility of the drug molecules in the carrier, the formulation should not be hygroscopic. Based on their chemical structure, poly(meth)acrylates absorb only small quantities of water and are therefore excellent carriers for stable solid dispersions (Figure 6.19).

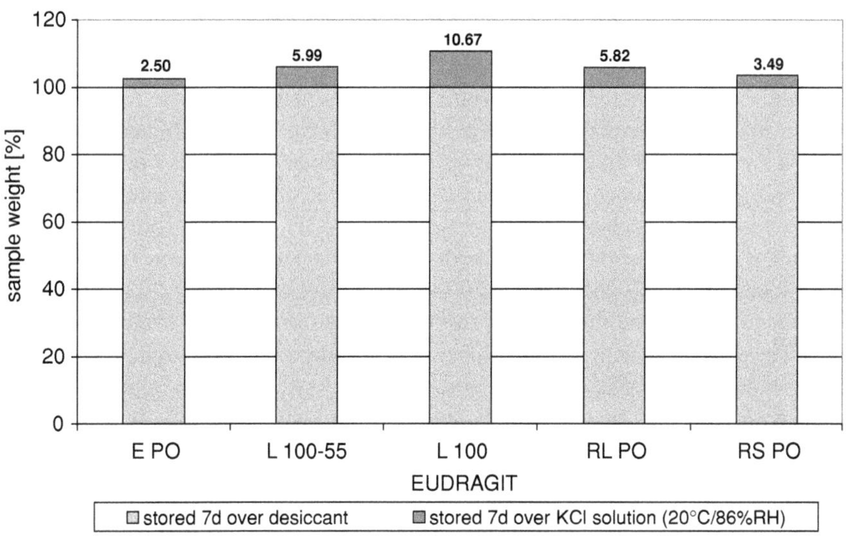

Figure 6.19 *Moisture absorption of different EUDRAGIT® grades at 20°C/86% RH over 7 days.*

6.3 Melt Extrusion of Poly(methacrylates) to Design Pharmaceutical Oral Dosage Forms

Melt extrusion in the pharmaceutical industry is widely used for the preparation of oral drug formulations, transdermal patches and implants.

Melt extrusion with poly(meth)acrylates is mainly used for oral drug delivery and, depending on the different functionality of the polymers, dosage forms are designed: (1) to improve the bioavailability of poorly soluble drugs by increasing the drug solubility and dissolution rate; (2) to mask unpleseant tastes of actives via interactions between drug and polymer; and (3) for sustained- and modified-release applications.

Table 6.4 lists selected examples of drug formulations using poly(meth)acrylates characterized and described in the literature.

6.4 Solubility Enhancement

A prerequisite for the pharmacological action of a drug is its bioavailability, which describes the degree to which a drug becomes available at the site of physiological action after administration. Drug absorption from the gastrointestinal tract can be limited by poor aqueous solubility and poor membrane permeability of the drug molecule [40, 41]. When delivering a drug orally, it first has to dissolve in the gastrointestinal fluid prior to permeating membranes of the gastrointestinal tract to reach the systemic circulation.

For drugs facing dissolution as the rate limiting step to absorption, the poor solubility can be overcome through advanced formulation approaches. One option is the formation of a solid solution of the drug in a polymeric carrier via melt extrusion [42–44]. As the drug is molecularly dispersed in the carrier, no energy is required to break up the crystal lattice during the dissolution process. According to the law of Noyes and Whitney [45], the surface area of the drug particles present as molecules is increased to a maximum which leads to a faster dissolution.

If a drug features different polymorphic structures, another option for solubility enhancement is the transformation to the polymorphic form with the higher solubility. The polymorph has to be stabilized within its form by an embedding into a polymeric matrix. Nollenberger [25] investigated the dissolution behavior of carbamazepine and solid dispersions with carbamazepine and the carrier EUDRAGIT® E. The formation of a solid dispersion by melt extrusion led to a high increase in the dissolution rate of the drug. For drug loads of 10, 30 and 50%, a 100% dissolution of the drug was achieved compared to the pure drug and the physical mixture where only 40–50% were dissolved after 120 min. The extrudates with 30 and 50% carbamazepine were not completely amorphous. A change in the polymorphic form was observed which explained the improved solubility. Carbamazepine was dispersed in modification I in the melt-extruded polymer, whereas in physical mixtures modification III was observed.

Shah et al. [46] describe a concentration of 10 μg/mL as a critical value for poor solubility. With recent advances in molecular screening methods for the identification of potential drugs, an increasing number of such candidates exhibits poor aqueous solubility. Main groups of poorly soluble drugs include antiepileptics, cardiovasculars, antiinfectives, neurologics, oncologics, antidiabetics or antiviral and antibiotic agents [47].

Table 6.4 *Poly(meth)acrylates used in melt extrusion applications.*

Purpose	Drug	Polymer: chemical name	Trade name
Solubility enhancement/bioavailability enhancement	celecoxib [22] nimodipine [23, 24] felodipine [25] itraconazol [26] CB-1 antagonist [27] indomethacin [28] fenofibrate [29] naproxen, furosemide [30]	Poly(butyl methacrylate-co-(2-dimethylaminoethyl) methacrylate-co-methyl methacrylate) 1:2:1	EUDRAGIT® E PO
Solubility enhancement	celecoxib [31] itraconazol [32]	Poly(methyl acrylate-co-methyl methacrylate-co-methacrylic acid) 7:3:1 Poly(methacrylic acid-co-ethyl acrylate) 1:1	EUDRAGIT® FS 30 D EUDRAGIT® L 100-55
Taste masking	ibuprofen [33]	Poly(butyl methacrylate-co-(2-dimethylaminoethyl) methacrylate-co-methyl methacrylate) 1:2:1	EUDRAGIT® E PO
Colon targeting	photosensitizer [34] 5-aminosalicylic acid [17]	Poly(methyl acrylate-co-methyl methacrylate-co-methacrylic acid) 7:3:1 Poly(methacrylic acid-co-methyl methacrylate) 1:2	EUDRAGIT® FS 30 D EUDRAGIT® S 100
Gastro-resistant pellets	theophylline [35]	Poly(methacrylic acid-co-ethyl acrylate) 1:1 Poly(methacrylic acid-co-methyl methacrylate) 1:1 Poly(methacrylic acid-co-methyl methacrylate) 1:2	EUDRAGIT® L100-55, L100 and S100

1

(continued)

Table 6.4 *Poly(meth)acrylates used in melt extrusion applications* (continued).

Purpose	Drug	Polymer: chemical name	Trade name
Enteric tablets	5-aminosalicylic acid [36]	Poly(methacrylic acid-co-ethyl acrylate) 1:1	EUDRAGIT® L100-55
Gastro-retentive controlled drug release	acetohydroxamic acid, chlorpheniramin-maleate [37]	Poly(ethyl acrylate-co-methyl methacrylate-co-trimethylammonioethyl methacrylate chloride) 1:2:0.1 Poly(butyl methacrylate-co-(2-dimethylaminoethyl) methacrylate-co-methyl methacrylate) 1:2:1	EUDRAGIT® RS PO and/or EUDRAGIT® E PO
Controlled release	indomethacin [38]	Poly(methacrylic acid-co-methyl methacrylate) 1:1 Poly(methacrylic acid-co-methyl methacrylate) 1:2 Poly(ethyl acrylate-co-methyl methacrylate-co-trimethylammonioethyl methacrylate chloride) 1:2:0.2	EUDRAGIT® L 100, S100, RD 100[a]
Controlled release and processability	theophylline [16]	Poly(methacrylic acid-co-methyl methacrylate) 1:2	EUDRAGIT® S100
Sustained release	diltiazem hydrochloride or chlorpheniramine maleate [39]	Poly(ethyl acrylate-co-methyl methacrylate-co-trimethylammonioethyl methacrylate chloride) 1:2:0.1 Poly(ethyl acrylate-co-methyl methacrylate-co-trimethylammonioethyl methacrylate chloride) 1:2:0.2	EUDRAGIT® RS PO or EUDRAGIT® RD 100[a]

[1] [a]Contains EUDRAGIT® RL and sodiumcarboxymethylcellulose.

Polymers used as carriers for the formation of solid solutions via melt extrusion have to fulfill several requirements. A high miscibility of the polymer with the drug is one prerequisite for the formation of a solid solution. A low hygroscopicity of the polymer is beneficial for the physical stability of the solid solution, as water can cause the recrystallization of the drug. For an immediate-release formulation, the polymer should have a good solubility in gastrointestinal fluids. The dissolution of a solid solution follows a carrier-controlled dissolution and the drug which is molecularly dispersed in the carrier is released by an erosion process of the solid solution [22, 48].

Polymers that fulfill those criteria and have a high historical prevalence in solubility enhancement by hot-melt extrusion are EUDRAGT® E and EUDRAGIT® L. Several studies with EUDRAGIT® E as carrier illuminate its the high capability for solubility enhancement for different drugs [22, 25]. Depending on the properties of the drug, the polymer can be loaded with up to 50–60% drug. Examples with celecoxib, felodipine and nifedipine (Figure 20) show that solid solutions with EUDRAGIT® E are physically stable [22, 25, 49].

For a targeted release at higher pH values in the intestine the anionic types such as EUDRAGIT® L-55, EUDRAGIT® L, EUDRAGIT® S and EUDRAGIT® FS would be favored.

If the dissolution behavior of the polymers needs to be changed, pH modulators can be incorporated into the extrudates. In case of EUDRAGIT® E which is soluble at low pH values, the dissolution of the extrudate can be shifted to higher pH values by the addition of organic acids such as citric acid, ascorbic acid, fumaric acid and succinic acid. By the incorporation of organic acids into the extrudate in the dissolution, the microenvironmental pH is changed which leads to a faster erosion of the solid solution [25].

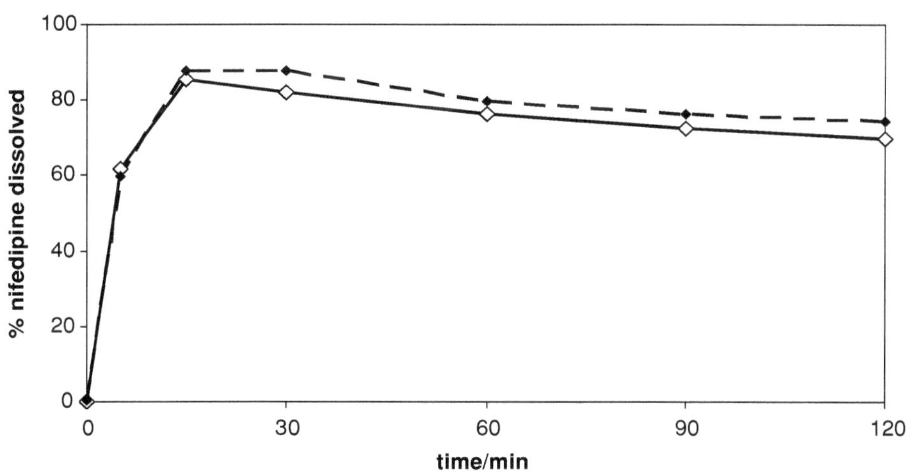

Figure 6.20 *Dissolution profile of extrudates containing 20% nifedipine and 80% EUDRAGIT® E PO (straight line: initial release profile, dashed line: release profile after 6 months storage at 40°C/75% relative humidity), dissolution: USP paddle apparatus, 2 hours in 500 ml SGF pH 1.2.*

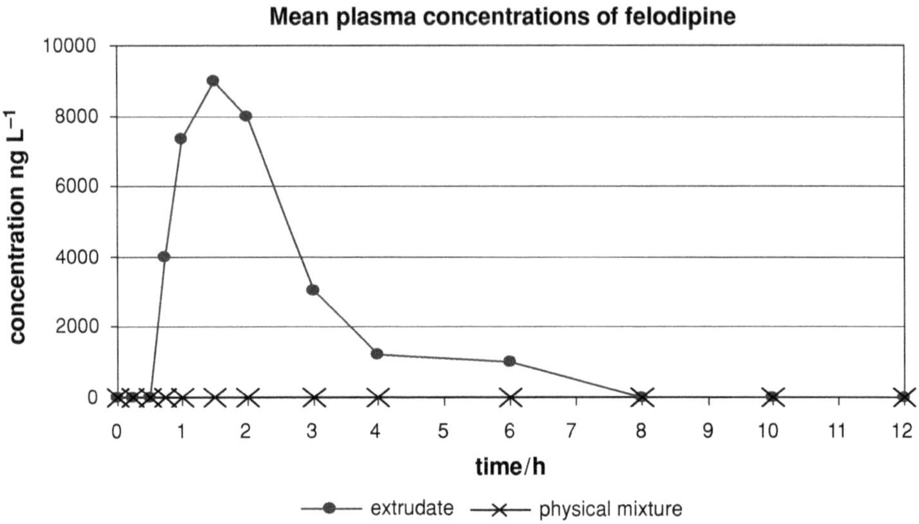

Figure 6.21 *In vivo study in dogs (n = 4) [24].*

Based on the chemical structure, the basic polymer EUDRAGIT® E can build polyelectrolyte complexes with poorly water-soluble acid drugs such as naproxen. Kindermann et al. [30] investigated the release of polyelectrolyte complexes and observed a complex stability in aqueous media. The stability of polyelectrolyte complexes can be influenced by other electrolytes [50]. This concept can be advantageous for triggering the drug release by the addition of pH neutral alkali-halogen electrolytes.

In general, the dissolution of solid solutions happens polymer-controlled as the drug is molecularly dispersed in the polymer [22]. In the dissolution medium, the drug builds a supersaturated solution which becomes obvious when working under non-sink conditions. Based on their high energetic potential, supersaturated solutions tend to recrystallize. In order to avoid the recrystallization in the dissolution medium, the mobility of the drug molecules needs to be reduced. This can either be obtained through a viscosity increase of the microenvironment of the released molecules [22] or by a hindrance through hydrophobic polymer chains. One option to stabilize the supersaturated state is to include hydrophobic polymers such as EUDRAGIT® NE into the extrudate formulation [25].

In vivo studies with the poorly water-soluble drug felodipine (Figure 6.21) showed the poor bioavailability of the drug. If the drug was physically mixed with EUDRAGIT® E, no significant increase in bioavailability could be observed. However, solid solutions with elodipine and EUDRAGIT® E prepared via melt extrusion decisively increased bioavailablity.

6.5 Bioavailability Enhancement of BCS Class IV Drugs

Poor bioavailability is a challenge that is faced by the pharmaceutical industry more and more with new chemical entities. BCS class IV drugs do not only suffer from poor solubility

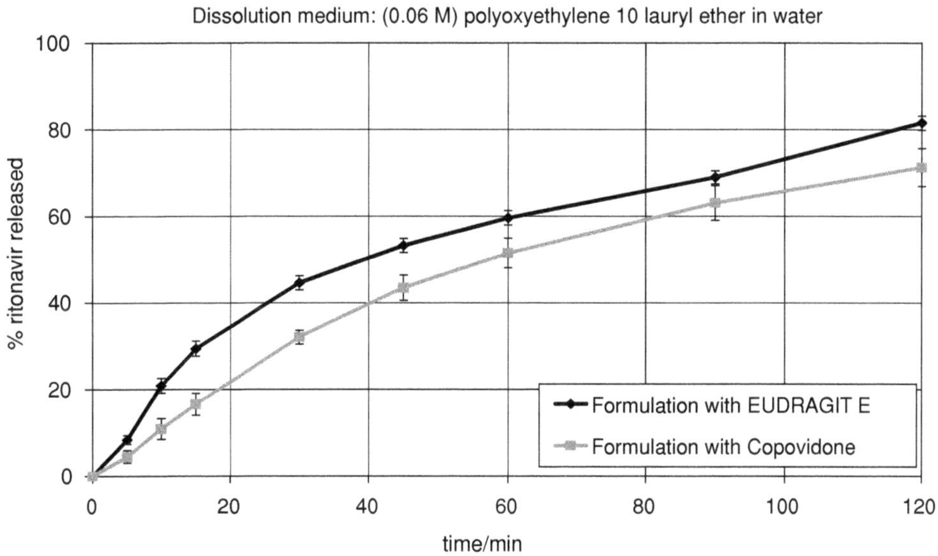

Figure 6.22 *Dissolution of ritonavir from extrudates of EUDRAGIT® E and copovidone; USP Type II paddle apparatus at 75 rpm in 900 mL medium at 37°C ± 0.5°C for 2 hours.*

but also from poor permeability [40]. Embedding those drugs in polymer matrices by melt extrusion can help to increase their bioavailability decisively. Bodinge et al. compared the cationic polymer, EUDRAGIT® E PO, to a neutral polymer (copovidone) for its solubility and permeability-enhancing properties [51]. No excipients commonly known to increase the permeability of poorly permeable drugs such as surfactants were used. Ritonavir (Figure 6.22) and Lopinavir (Figure 6.23) were used as BCS class IV model drugs. The bioavailability values of the formulations were investigated in healthy human volunteers under fasted conditions. Both formulations contained similar drug to polymer ratios with 50 mg drug and 750 mg polymer and were processed using hot-melt extrusion under identical experimental conditions. The in vitro results revealed similar dissolution profiles of both drugs with EUDRAGIT® E and copovidone.

A clinical study was carried out in an open-label, balanced, randomized, two-treatment, two-period, two-sequence, single-dose crossover study in 16 healthy, adult human male volunteers under fasted conditions. Both formulations with EUDRAGIT® E and copovidone were included in the study.

The in vivo data showed a superior bioavailability of the EUDRAGIT® E formulation for both ritonavir and lopinavir (Figures 6.24 and 6.25). Although no in-vitro recrystallization of the drugs could be observed from the copovidone extrudates, an in-vivo recrystallization could be a possible reason for the low plasma concentrations. This shows that EUDRAGIT® E can also provide effective bioavailability enhancement for drugs with poor solubility and low permeability.

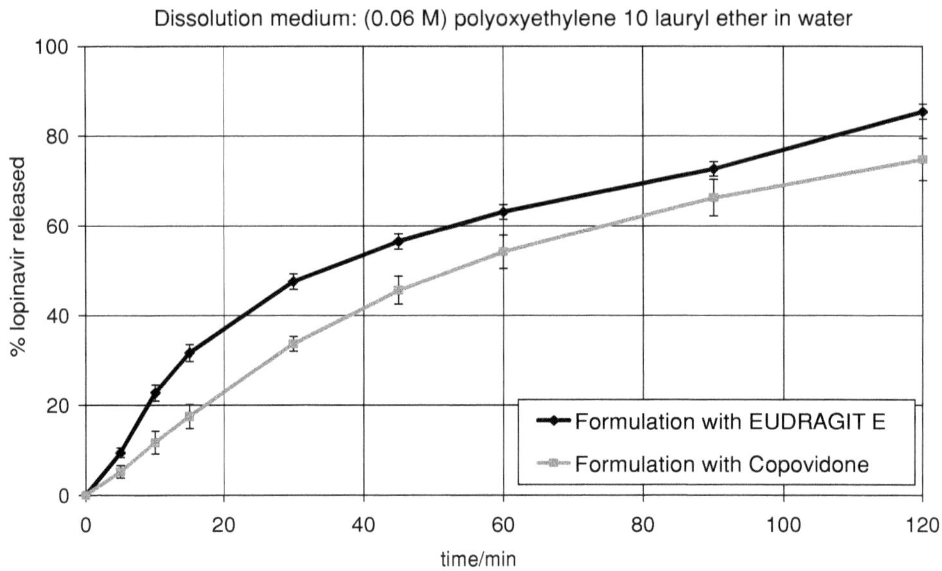

Figure 6.23 *Dissolution of lopinavir from extrudates of EUDRAGIT® E and copovidone; USP Type II paddle apparatus at 75 rpm in 900 mL medium at 37°C ± 0.5°C for 2 hours.*

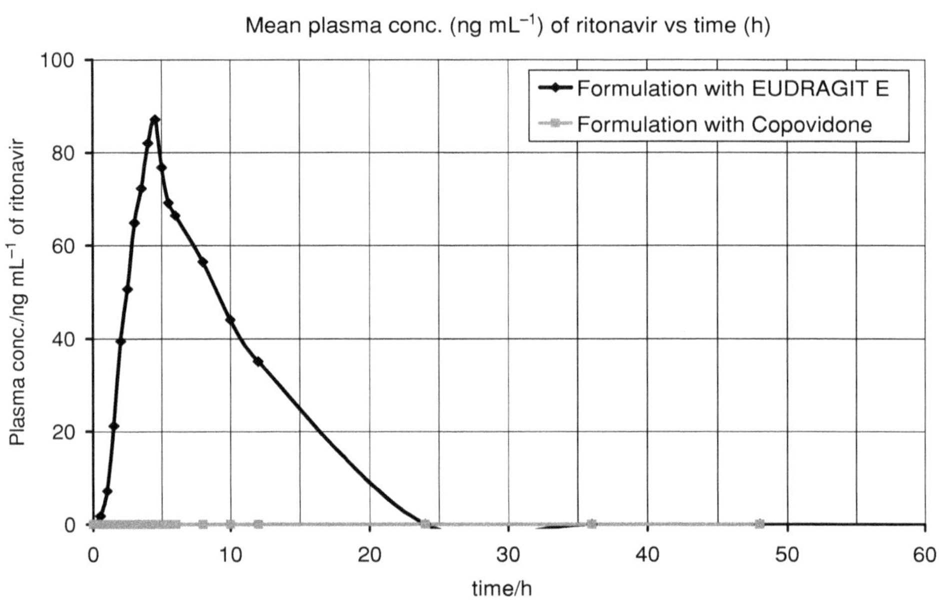

Figure 6.24 *Comparative mean plasma concentration of ritonavir from EUDRAGIT® E PO and copovidone formulations.*

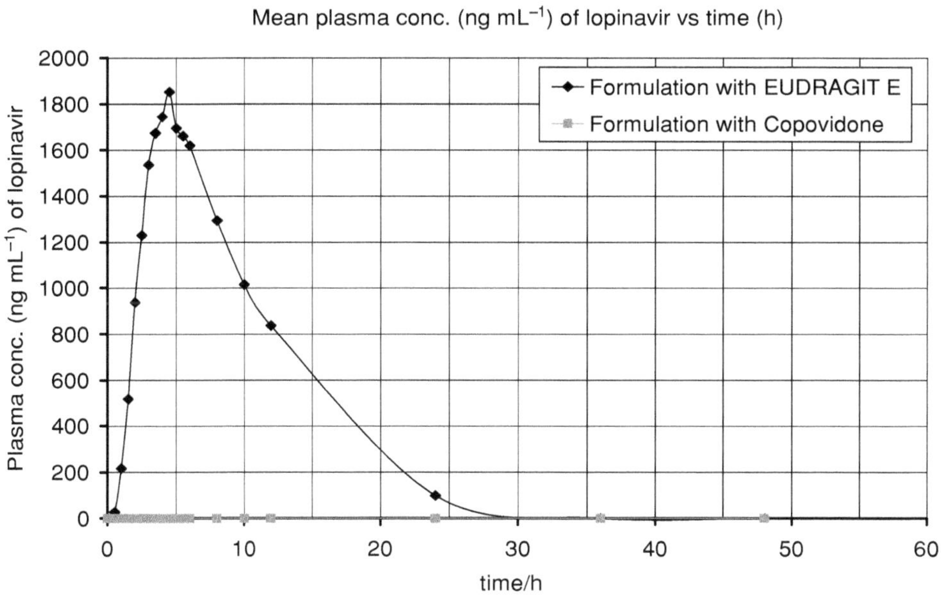

Figure 6.25 *Comparative mean plasma concentration of lopinavir from EUDRAGIT® E PO and copovidone formulations.*

6.5.1 Controlled Release

Controlled-release dosage forms deliver the active ingredient in a controlled way over a certain time up to 24 hours or at a specific point in the gastrointestinal tract. By formulating drugs into controlled-release dosage forms, more constant plasma concentrations of the drug are reached, the dosing intervals are prolonged, toxic plasma concentrations are circumvented reducing side effects and the risk of sub-therapeutic plasma concentrations.

As well as using diffusion-controlled barriers via coating of particles, pellets or tablets, the embedding of a drug in a matrix is an alternative approach in achieving controlled-release properties. In general, controlled-release dosage forms can be divided into monolithic dosage forms such as tablets or multiple unit dosage forms (e.g. pellets or particles).

pH-independent poly(meth)acrylates, such as EUDRAGIT® RL, EUDRAGIT® RS, EUDRAGIT® NE 30 D and EUDRAGIT® NM 30 D, and pH-dependent soluble poly(meth)acrylates, such as EUDRAGIT® FS 30 D or EUDRAGIT® S 100, are used for the preparation of matrix dosage forms as they release the active ingredient mainly by (pore) diffusion. In case of the pH-dependent polymers with dissolution pH values below 7, the drug is also released by erosion when reaching an environment above the dissolving pH. Different manufacturing technologies are used such as direct compression, aqueous and organic granulation, melt granulation and melt extrusion, which all influence the controlled-release behavior of the dosage forms. Melt-extruded matrix systems provide a strong controlled-release effect due to the very dense structure and the molecular dispersion of the active ingredient in the polymer.

6.5.2 Time-controlled-release Dosage Forms

EUDRAGIT® RL and EUDRAGIT® RS are pH-independent insoluble polymers showing swelling characteristics in aqueous media. As functional group both polymers have quaternary ammonium groups with a chloride as counter ion. The difference between the polymers is the percentage of the functional monomer, leading to a difference in permeability.

By melt extruding combinations of these polymers in different ratios, the permeability can be adjusted to achieve a certain dissolution profile. Figure 6.26 depicts the release profile of melt-extruded theopyhlline pellets. The drug was melt extruded with different loadings and weight ratios of EUDRAGIT® RL and EUDRAGIT® RS. Pellets were prepared in a one-step process and the melt was directly cut with a rotating knife at the die face surrounded by a cooled chamber. The melt droplets were cooled rapidly forming round pellets. The pellet size was well controllable and reproducible, leading to a narrow particle size distribution (Figure 6.27). The diameter of the pellets was determined by the die hole and their length was adjusted by the knife speed and the throughput of the extruder. The release profiles showed the influence of the two polymers and their blends. EUDRAGIT® RL, with a higher permeability than EUDRAGIT® RS, shows a faster release as the media penetrates faster into the matrix structure dissolving the drug and releasing it via diffusion. Blends of both polymers showed intermediate-release profiles [52].

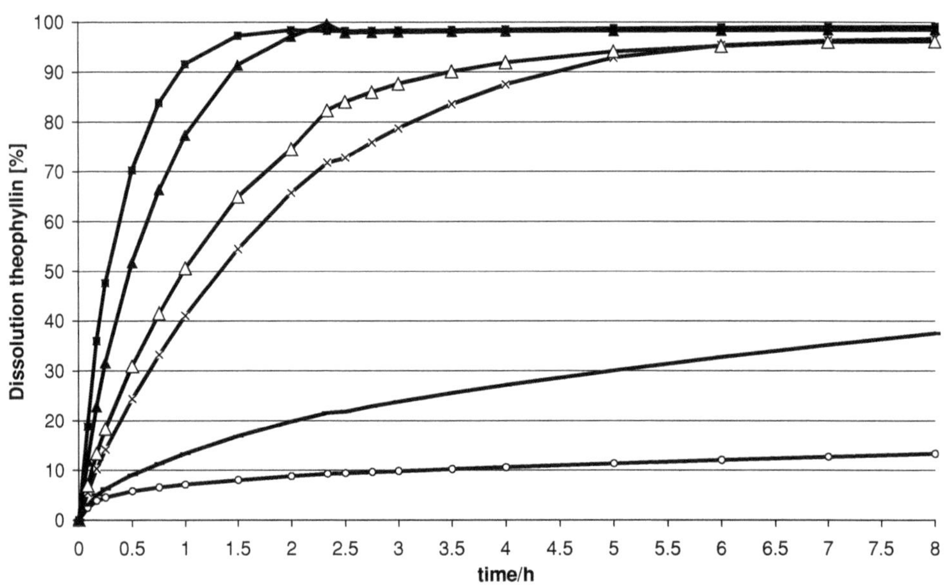

Figure 6.26 Drug release profile of hot-melt-extruded theophylline pellets with different drug loadings and polymer ratios (■ EUDRAGIT® RL:Theopyhlline 65:35%; ▲ EUDRAGIT® RL:Theopyhlline 50:50%; Δ EUDRAGIT® RL:EUDRAGIT® RS: Theopyhlline 25:25:50%; × EUDRAGIT® RL:EUDRAGIT® RS:Theopyhlline 32.5:32.5:35%; – EUDRAGIT® RS:Theopyhlline 50:50%; ○ EUDRAGIT® RS:Theopyhlline 65:35%). Dissolution: USP paddle apparatus, 2 hours in 700 ml pH 1.2; after 2 hours change to pH 6.8 with 214 ml trisodiumphosphate solution.

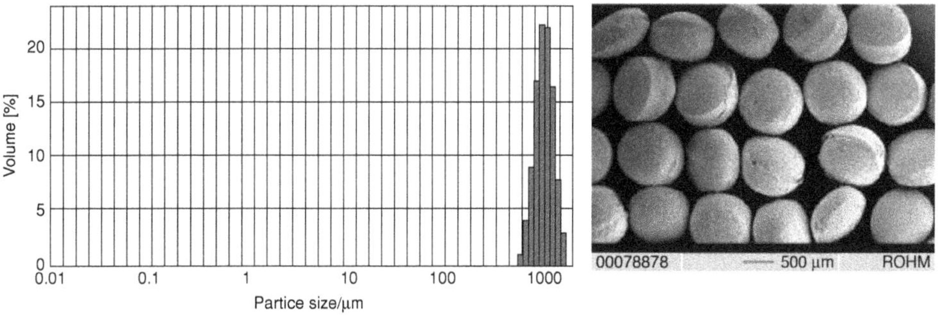

Figure 6.27 *SEM picture and particle size distribution of hot melt extruded EUDRAGIT® RS / theophylline pellets.*

Additionally, other functionality e.g. stronger sustained release, enteric properties or colon targeting can be added to melt-extruded pellets via coating with the respective functional polymers. Melt extruded pellets containing EUDRAGIT® RL and theophylline in a ratio of 70:30 were coated exemplarily with EUDRAGIT® L30D-55 (Figure 6.28) to achieve gastro resistance of the pellets. Due to the density of the melt-extruded pellets, a coating level of 6% EUDRAGIT® L30D-55 was sufficient to achieve enteric properties.

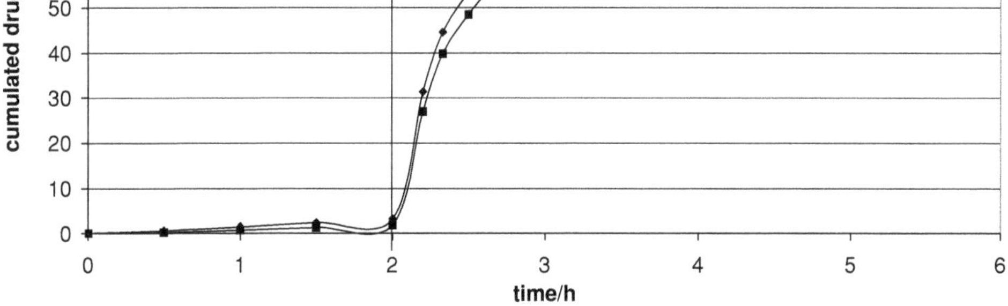

Figure 6.28 *Drug release profile of hot-melt-extruded theophylline pellets, pellet formulation: EUDRAGIT® RL (theopyhlline with EUDRAGIT® L30D-55 coating: ♦ 6% polymer applied; ■ 8% polymer applied). Dissolution: USP paddle apparatus, 2 hours in 700 ml pH 1.2; after 2 hours change to pH 6.8 with 214 ml trisodiumphosphate solution.*

6.5.3 pH-dependent Release

The necessity to formulate drugs into gastro-resistant dosage forms is related to either the properties of the drug such as instability in the acidic media, irritation of the gastric mucosa, for achieving a pH controlled release for optimum resorption, a targeting of certain regions in the GI tract to reach the absorption window or targeting of the colonic region for systemic or localized treatment of diseases such as Crohn's disease or Ulcerative colitis.

Like when applying enteric polymers as coating on hot-melt-extruded pellets, gastro resistance can also be achieved in one step via melt extrusion directly. The anionic poly(meth)acrylates can be melt-extruded alone or in combination with other polymers to achieve gastric resistance. The release profiles affect the choice of the polymer(s) as well as the processability of the polymers and the active.

Figure 6.29 depicts the release profile of melt-extruded pellets. EUDRAGIT® FS 30D, was combined by melt extrusion with different amounts of EUDRAGIT® RS and theophylline as model drug. After two hours, all formulations show less than 10% drug release in pH 1.2 and in pH 6.8. Depending on the amount of EUDRAGIT® RS, the release in pH 7.2 can be varied from a fast to a very slow drug release.

Schilling et al. fomulated gastro-resistant pellets using EUDRAGIT® S 100, theophylline (30% drug loading) and various plasticizers via melt extrusion. As described in chapter 6.2.2., plasticizers are required to enable melt extrusion below the degradation temperature

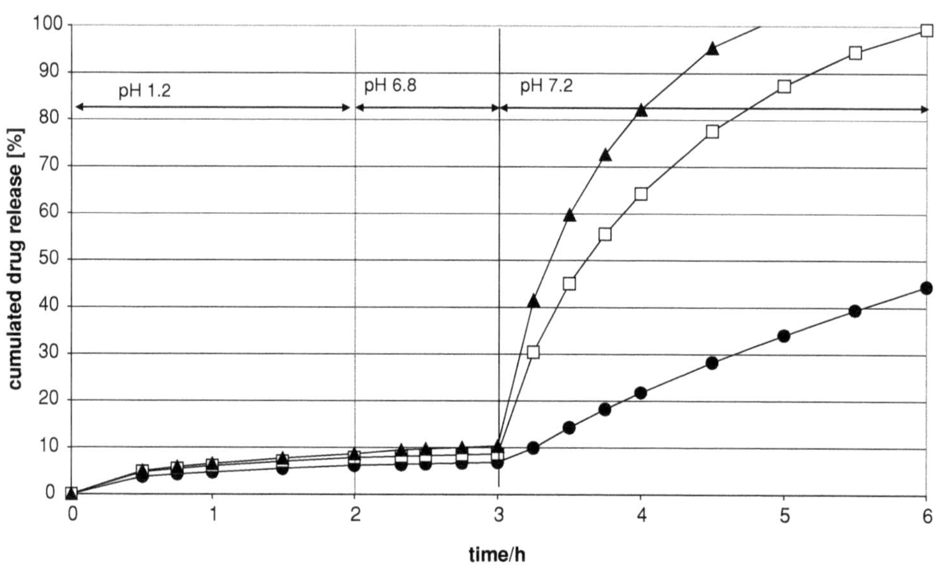

Figure 6.29 Drug release profile of hot-melt-extruded theophylline pellets, drug loading 50%, • 30% EUDRAGIT® RS and 20% EUDRAGIT® FS 30 D; □ 20% EUDRAGIT® RS and 30% EUDRAGIT® FS 30 D; ▲ 30% EUDRAGIT® RS and 20% EUDRAGIT® FS 30 D). Dissolution: USP paddle apparatus, 2 hours in 700 ml pH 1.2; after 2 hours change to pH 6.8 with 214 ml trisodium phosphate solution and after 1 hour change to pH 7.5 with 50 ml trisodium phosphate solution.

of EUDRAGIT® S 100 and EUDRAGIT® L 100. Amongst all tested plasticizers, triethylcitrate showed no adverse effect on gastric resistance whereas pellets with PEG 8000, due to its high water solubility, failed to stay beneath the 10% specification for enteric dosage forms. In pH 7.2, higher plasticizer quantities (TEC) increased the release from the pellets [16].

Andrews et al. manufactured melt extruded enteric tablets, using -5-Aminosalicylic acid as model drug and EUDRAGIT® L100-55 as gastro-resistant polymer. The melt was either directly cut into small tablets or milled and subsequently directly compressed without other excipients into matrix tablets. The extruded tablets showed an effective acid protection, whereas milled extrudates compressed into tablets failed to be gastro resistant. As the polymer is melted and compressed with high pressure in the extruder, the directly cutted tablets show low porosity. Tablets produced from milled extrudates showed reduced tablet hardness and faster disintegration times due to the poor plastic deformation of the powder. Furthermore, the lower porosity of the compressed tablets allows a faster penetration of the dissolution media into the tablet, leading to a faster release in the gastric media [36].

6.5.4 Taste Masking

Masking the bitter or unpleasant taste of active ingredients is an important approach to increase patient compliance, especially in pediatric and geriatric treatment. Creating a coating barrier on the dosage form to protect the taste receptors from the active ingredient is a well-known approach. As solvent- and aqueous free one step process, melt extrusion has proved to be a fast and easy alternative to achieve taste masked granules by utilizing ionic interactions between the active and a functional polymer, e.g. a basic drug with an anionic polymer or an acidic drug with a cationic polymer. The melting of the active and the softening of the polymers in the extruder provides a sufficient mobility of the drug in the polymer, triggering the interactions between the functional groups.

Having a pH-dependent release in acidic media but not in higher pH values, EUDRAGIT® E PO has traditionally been used for taste masking via coating applications. The tertiary aminofunction of the polymer can also be utilized to interact with functional groups of acidic drugs masking the bitter taste. An effective taste masking of ibuprofen was achieved by melt-extruding ibuprofen with EUDRAGIT® E PO, using different molar ratios of the functional groups in both substances. The melt-extrusion temperature was well above the melting temperature to support the dissolution of the drug in the polymer and to ease the formation of the interaction. Taste masking of the solid dispersion was either tested with the directly obtained granules (2 mm) or after milling the extrudates to a particle size $< 250\ \mu m$. In case of the granules, all tested ratios of polymer to ibuprofen (1:0.5, 1:1 or 1:1.5) showed excellent taste masking above 60 s. The milled particles however showed that taste masking at higher ratios of ibuprofen (1:1.5) lead to a slightly bitter taste of ibuprofen after one minute, as free ibuprofen groups were present at the surface of the milled extrudate. The interactions between ibuprofen and EUDRAGIT® E PO could be verified by IR-spectroscopy, where a carboxylate formation in the ibuprofen spectra could be identified. As an acidic drug, ibuprofen shows poor solubility in acidic media. In addition to the taste-masking effect, melt extrusion of ibuprofen and EUDRAGIT® E also showed a solubility-enhancing effect of ibuprofen in acidic media [53].

Gryczke et al. formulated tasted-masked ibuprofen granules prepared by melt extrusion with EUDRAGIT® E in oral disintegrating tablets using different super-disintegrants. Both taste masking and good palatability was achieved at well as the same disintegration time and crushing resistance as the reference product containing ibuprofen [33].

Interactions that provide this special taste-masking effect may lead to very stable drug-polymer complexes causing slow release or even failure in the release of the drug. For melt-extruded solid dispersions of naproxen and EUDRAGIT® E or furosemide and EUDRAGIT® E, Kindermann et al. described the formation of very stable complexes showing poor release of the drug from the extrudates. However, by adding electrolytes to the dissolution media, a destabilization of the complex could be achieved leading to a strong increase in dissolution [30].

The same principle can be used for basic drugs or salts with anionic polymers. Anionic polymers such as EUDRAGIT® L 100, EUDRAGIT® L 100-55, EUDRAGIT® S 100 and EUDRAGIT® FS 30 D are suitable to be used in melt extrusion and taste-masking properties of melt-extruded granules were shown for verapamil HCl using EUDRAGIT® L 100-55 or EUDRAGIT® FS 30 D [54].

6.6 Summary

The interest in melt extrusion in the development of oral dosage forms has been increasing over the past years, and this trend is expected to continue in future. In this chapter poly(meth)acrylates as polymeric carrier for the preparation of melt extruded dosage forms are described. Their thermoplastic behavior, viscosity and thermostability allow to process most of the poly(meth)acrylates without the necessity of adding plasticizers in a wide temperature range up to 200°C. Effective plasticizers have been investigated in literature to lower the extrusion temperature when required. Poly(meth)acrylates are divided into pH-dependent soluble and pH-independent insoluble polymers and can be melt extruded with drug molecules and excipients to achieve various release profiles. pH-dependent soluble polymers are mainly used for solubility enhancement and utilizing their functional groups for taste masking via interactions. Anionic poly(meth)acrylates are used alone or in combination for enteric protection, gastrointestinal and colon targeting. The insoluble poly(meth)acrylates provide sustained release matrix systems and can be used to prevent recrystallization of a poorly soluble drugs from an oversaturated solution. Combining poly(meth)acrylates in melt extrusion with other polymers or excipients and shaping the melt into different dosage forms allows to achieve specific release profiles and functionalities.

References

(1) Skalsky, B. and Petereit, H.-U. (2008) Chemistry and application properties of polymethacrylate systems. In: Aqueous polymeric coatings for pharmaceutical dosage forms, J.W. McGinity, and L.A. Felton (ed.). Informa Healthcare: New York, London.

(2) Crowley, M.M., Zhang, F., Repka, M.A., Thumma, S., Upadhye, S.B., Battu, S.K., McGinity, J.W. and Martin, C. (2007) Pharmaceutical applications of hot-melt extrusion: Part I. *Drug Development & Industrial Pharmacy*, **33**, 909–926.

(3) Follonier, N., Doelker, E. and Cole, E.T. (1994) Evaluation of hot-melt extrusion as a new technique for the production of polymer-based pellets for sustained release capsules containing high loadings of freely soluble drugs. *Drug Development & Industrial Pharmacy*, **20**(8), 1323–1339.
(4) Kidokoro, M., Shah, N.H., Malick, A.W., Infeld, M.H. and McGinity, J.W. (2001) Properties of tablets containing granulations of ibuprofen and an acrylic copolymer prepared by thermal processes. *Pharmaceutical Development & Technology*, **6**(2), 263–275.
(5) Zhu, Y.C., Shah, N.H., Malick, A.W., Infeld, M.H. and McGinity, J.W. (2002) Solid-state plasticization of an acrylic polymer with chlorpheniramine maleate and triethyl citrate. *International Journal of Pharmaceutics*, **24**(2), 301–310.
(6) Aitken-Nichol, C., Zhang, F. and McGinity, J.W. (1996) Hot melt extrusion of acrylic films. *Pharmaceutical Research*, **13**(5), 804–808.
(7) Leuner, C. and Dressman, J.B. (2000) Improving drug solubility for oral delivery using solid dispersions. *European Journal of Pharmaceutics & Biopharmaceutics*, **50**, 47–60.
(8) European Pharmacopoeia (2011), 7th Edition, Deutscher Apothekerverlag Stuttgart, GOVI-Verlag-Pharmazeutischer Verlag GmbH Eschborn.
(9) United States Pharmacopeia (USP) 34—National Formulary (NF) 29 (2011), US Pharmaceutical Convention Inc., Rockville.
(10) Japanese Pharmaceutical Excipients (2004), The Japan Pharmaceutical Excipients Council, YAKUJI NIPPO, Ltd.
(11) EUDRAGIT° Application Guidelines (2011), 12th Edition, Evonik Industries AG.
(12) Jones, D.S. (2007) Thermorheological characterization. In *Thermal Analysis of Pharmaceuticals*, CRC Press, Boca Raton.
(13) Wu, C. and McGinity, J.W. (2003) Influence of methylparaben as a solid-state plasticizer on the physicochemical properties of Eudragit RS PO hot-melt extrudates. *European Journal of Pharmaceutics & Biopharmaceutics*, **56**(1), 95–100.
(14) Zhu, Y., Mehta, K.A. and McGinity, J.W. (2006) Influence of plasticizer level on the drug release from sustained release film coated and hot-melt extruded dosage forms. *Pharmaceutical Development & Technology*, **11**(3), 285–294.
(15) Ghebremeskel, A.N., Vemavarapu, C. and Lodaya, M. (2007) Use of surfactants as plasticizers in preparing solid dispersions of poorly soluble API: Selection of polymer–surfactant combinations using solubility parameters and testing the processability. *International Journal of Pharmaceutics*, **328**, 119–129.
(16) Schilling, S.U., Lirola, H.L., Shah, N.H., Waseem Malick, A. and McGinity, J.W. (2010) Influence of plasticizer type and level on the properties of Eudragit S100 matrix pellets prepared by hot-melt extrusion. *Journal of Microencapsulation*, **27**(6), 521–532.
(17) Bruce, L.D., Shah, N.H., Malick, A.W., Infeld, M.H. and McGinity, J.W. (2005) Properties of hot-melt extruded tablet formulations for the colonic delivery of 5-aminosalicylic acid. *European Journal of Pharmaceutics & Biopharmaceutics*, **59**(1), 85–97.
(18) Schilling, S.U., Shah, N.H., Malick, A.W., Infeld, M.H. and McGinity, J.W. (2007) Citric acid as a solid-state plasticizer for Eudragit RS PO. *Journal of Pharmacy & Pharmacology*, **59**(11), 1493–1500.

(19) Zhu, Y., Shah, N.H., Malick, A.W., Infeld, M.H. and McGinity, J.W. (2002) Solid-state plasticization of an acrylic polymer with chlorpheniramine maleate and triethyl citrate. *International Journal of Pharmaceutics*, **241**(2), 301–310.
(20) Wu, C. and McGinity, J.W. (2001) Influence of ibuprofen as a solid-state plasticizer in Eudragit RS30D on the physico-chemical properties of coated beads. *AAPS Pharmaceutical Science & Technology*, **2**(4), 24.
(21) Asgarzadeh, F., Rambaud, E., Li, J.-X., Moß, J. and Nollenberger, K. (2011) Rheological characterization of (meth)acrylic copolymers suitable for melt extrusion applications. Poster, CRS Annual Meeting.
(22) Albers, J. (2008) Hot-melt extrusion with poorly soluble drugs. Thesis, Universität Düsseldorf.
(23) Jijun, F., Lishuang, X., Xiaoli, W., Shu, Z., Xiaoguang, T., Xingna, Z., Haibing, H. and Xing, T. (2011) Nimodipine (NM) tablets with high dissolution containing NM solid dispersions prepared by hot-melt extrusion. *Drug Development & Industrial Pharmacy*, **37**(8), 934–944.
(24) Sun, Y., Rui, Y., Wenliang, Z. and Tang, X. (2008) Nimodipine semi-solid capsules containing solid dispersion for improving dissolution. *International Journal of Pharmaceutics*, **359**(1–2), 144–149.
(25) Nollenberger, K. (2009) Löslichkeitsverbesserung schwerlöslicher Arzneistoffe durch Schmelzextrusion mit Polymethacrylaten. Thesis, University of Frankfurt.
(26) Six, K., Verreck, G., Peeters, J., Brewster, M. and Van Den Mooter, G. (2004) Increased physical stability and improved dissolution properties of itraconazole, a class II drug, by solid dispersions that combine fast- and slow-dissolving polymers. *Journal of Pharmaceutical Sciences*, **93**(1), 124–131.
(27) Ranzani, L.S., Font, J., Galimany, F., Santanach, A., Gomez-Gomar, A.M., Casadevall, G. and Gryczke, A. (2011) Enhanced in vivo absorption of CB-1 antagonist in rats via solid solutions prepared by hot-melt extrusion. *Drug Development & Industrial Pharmacy*, **37**(6), 694–701.
(28) Liu, H., Wang, P., Zhang, X., Shen, F. and Gogos, C.G. (2010) Effects of extrusion process parameters on the dissolution behavior of indomethacin in Eudragit E PO solid dispersions. *International Journal of Pharmaceutics*, **383**(1–2), 161–169.
(29) He, H., Yang, R. and Tang, X. (2010) In vitro and in vivo evaluation of fenofibrate solid dispersions prepared by hot-melt extrusion. *Drug Development & Industrial Pharmacy*, **36**(6), 681–687.
(30) Kindermann, C., Matthée, K., Strohmeyer, J., Sievert, F. and Breitkreutz, J. (2011) Tailor-made release triggering from hot-melt extruded complexes of basic polyelectrolyte and poorly water-soluble drugs. *European Journal of Pharmaceutics & Biopharmaceutics* **79**(2), 372–381.
(31) Abu-Diak, O.A., Jones, D.S. and Andrews, G.P. (2011) An investigation into the dissolution properties of celecoxib melt extrudates: understanding the role of polymer type and concentration in stabilizing supersaturated drug concentrations. *Molecular Pharmacology*, **8**(4), 1362–1371.
(32) Miller, D.A., DiNunzio, J.C., Yang, W., McGinity, J.W. and Williams, R.O. 3rd (2008) Enhanced in vivo absorption of itraconazole via stabilization of supersaturation following acidic-to-neutral pH transition. *Drug Development & Industrial Pharmacy*, **34**(8), 890–902.

(33) Gryczke, A., Schminke, S., Maniruzzaman, M., Beck, J. and Douroumis, D. (2011) Development and evaluation of orally disintegrating tablets (ODTs) containing Ibuprofen granules prepared by hot melt extrusion Colloids Surf B. *Biointerfaces*, **86**(2), 275–284.
(34) Cassidy, C.M., Tunney, M.M., Caldwell, D.L., Andrews, G.P. and Donnelly, R.F. (2011) Development of novel oral formulations prepared via hot melt extrusion for targeted delivery of photosensitizer to the colon. *Photochemistry & Photobiology*, **87**(4), 867–876.
(35) Schilling, S.U., Shah, N.H., Waseem Malick, A. and McGinity, J.W. (2010) Properties of melt extruded enteric matrix pellets. *European Journal of Pharmaceutics & Biopharmaceutics*, **74**(2), 352–361.
(36) Andrews, G.P., Jones, D.S., Diak, O.A., McCoy, C.P., Watts, A.B. and McGinity, J.W. (2008) The manufacture and characterisation of hot-melt extruded enteric tablets, *European Journal of Pharmaceutics & Biopharmaceutics*, **69**(1), 264–273.
(37) Fukuda, M., Peppas, N.A. and McGinity, J.W. (2006) Floating hot-melt extruded tablets for gastroretentive controlled drug release system. *Controlled Release*, **115**(2), 121–129.
(38) Zhu, Y., Shah, N.H., Malick, A.W., Infeld, M.H. and McGinity, J.W. (2006) Controlled release of a poorly water-soluble drug form hot-melt extrudates containing acrylic polymers. *Drug Development & Industrial Pharmacy*, **32**, 569–583.
(39) Zhu, Y., Mehta, K.A. and McGinity, J.W. (2006) Influence of plasticizer level on the drug release from sustained release film coated and hot-melt extruded dosage forms. *Pharmaceutical Development & Technology*, **11**(3), 285–294.
(40) Dressman, J.B., Butler, J., Hempenstall, J. and Reppas, C. (2001) The BCS: Where do we go from here? *Pharmacy Technology*, **25**, 68–76.
(41) Amidon, G.L., Lennernäs, H., Shah, V.P. and Crison, J.R. (1995) A hheoretical basis for a biopharmaceutic drug classification - the correlation of in-vitro drug product dissolution and in-vivo bioavailability. *Pharmaceutical Research*, **12**, 413–420.
(42) Breitenbach, J. (2002) Melt extrusion: from process to drug delivery technology. *European Journal of Pharmaceutics & Biopharmaceutics*, **54**, 107–117.
(43) Forster, A., Hempenstall, J. and Rades, T. (2001) Characterization of glass solutions of poorly water-soluble drugs produced by melt extrusion with hydrophilic amorphous polymers. *Journal of Pharmacy & Pharmacology*, **53**, 303–315.
(44) Forster, A., Hempenstall, J., Tucker, I. and Rades, T. (2001) Selection of excipients for melt extrusion with two poorly water-soluble drugs by solubility parameter calculation and thermal analysis. *International Journal of Pharmaceutics*, **226**, 147–161.
(45) Noyes, A.A. and Whitney, W.R. (1897) The rate of solution of solid substances in their own solutions. *Journal of American Chemical Society*, **19**, 930–934.
(46) Shah, V.P., Konecny, J.J., Everett, R.L., McCullough, B., Noorizadeh, A.C. and Selly, J.P. (1989) In-vitro dissolution profile of water-insoluble drug dosage forms in the presence of surfactants. *Pharmaceutical Research*, **6**, 612–618.
(47) Lindenberg, M., Knopp, S. and Dressman, J.B. (2004) Classification of orally administered drugs on the World Health Organization Model list of Essential Medicines according to the biopharmaceutics classification system. *European Journal of Pharmaceutics & Biopharmaceutics*, **58**, 265–278.

(48) Craig, D.Q.M. (2002) The mechanisms of drug release from solid dispersions in water-soluble polymers. *International Journal of Pharmaceutics*, **231**, 131–144.
(49) Nollenberger, K., Listro, T., Gryczke, A., Dorairaju, G. and Brunnengräber, Ch. (2010) Scale-up of the melt extrusion process using EUDRAGIT® polymers containing nifedipine as a model drug for solubility enhancement. Poster AAPS Annual Meeting.
(50) Quinteros, D.A., Rigo, V.R., Kairuz, A.F.J., Olivera, M.E., Manzo, R.H. and Allemandi, D.A. (2008) Interaction between a cationic polymethacrylate (Eudragit E) and anionic drugs. *European Journal of Pharmaceutical Sciences*, **33**, 72–79.
(51) Bodinge, S. et al. (2011) Bioavailability enhancement of antiretrovirals via melt extrusion using EUDRAGIT® polymers. Poster CRS Annual Meeting & Exposition.
(52) Gryczke, A. and Nollenberger, K. (2006) Melt extruded theophyllin micro spheres for better controlled release. Poster CRS, Vienna.
(53) Gryczke, A., Meier, Ch. and Petereit, H.-U. (2003) Basic investigations on taste masking of anionic drugs by melt extrusion. Poster AAPS Annual Meeting.
(54) Petereit, H.-U., Gryczke, A. and Meier, Ch. (2002) Basic investigations on taste masking of drugs by melt extrusion. Poster AAPS Annual Meeting.

7

Hot-melt Extrusion of Ethylcellulose, Hypromellose and Polyethylene Oxide

Mark Hall and Michael Read
The Dow Chemical Company, Midland Michigan

7.1 Introduction

Ethylcellulose, hypromellose, methylcellulose and polyethylene oxide have long histories of use in the pharmaceutical industry. Typical applications for ethylcellulose include multiparticulate or matrix tablet coatings to provide controlled release, taste masking or moisture barrier protection; controlled-release excipients in matrices; and binders in roller compaction or direct compression [1]. Hypromellose and methylcellulose applications include controlled-release excipients in matrix systems [2], granulation binders [3], tablet coatings [4] and hard shell capsules [5]. Polyethylene oxide applications include controlled-release excipients in matrix tablets and osmotic systems, binders in direct compression processing and applications requiring good mucoadhesion [6].

The use of these excipients in pharmaceutical hot-melt extrusion (HME) is relatively new and is rapidly evolving. However, each of these products has historical applications utilizing extrusion processing. Ethylcellulose was extruded as an insulation coating on wire as early as 1949 [7]. Hypromellose has been used as a binder in extruded formulations. Chalasani and Johnson patented the use of hypromellose in honeycomb forms that are used in applications such as catalysts, adsorption and filters [8]. Miller et al. previously taught the use of water as a processing aid for polyethylene oxide [9]. This document reviews the use of these excipients and summarizes the key properties of these polymers in terms of hot-melt extrusion.

Hot-melt Extrusion: Pharmaceutical Applications, First Edition. Edited by Dennis Douroumis.

7.2 Background

The structures of ethylcellulose, hypromellose, methylcellulose and polyethylene oxide are depicted in Table 7.1.

Ethylcellulose (EC), hypromellose (HPMC) and methylcellulose (MC) are all cellulosic polymers. EC is an ethyl ether of cellulose. This polymer is hydrophobic and thus water insoluble. EC is available in various molecular weight grades. HPMC is water-soluble cellulose ether produced via addition of methyl and hydroxypropyl groups to the cellulose backbone. Numerous products are commercially available which encompass a range of methyl and hydroxypropyl substitution levels and polymer molecular weight (viscosity grades). MC is another water-soluble cellulose ether, this time produced via addition of methyl groups to the cellulose backbone. Several grades are commercially available,

Table 7.1 Structures of ethylcellulose, hypromellose, methylcellulose and polyethylene oxide.

Polymer	Nomenclature	Structure
Ethylcellulose	EC	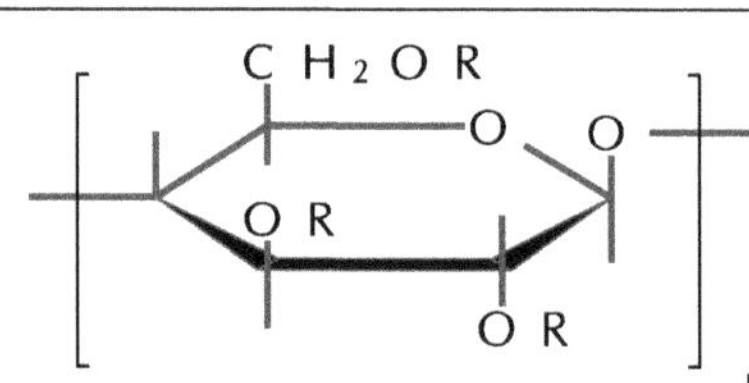 R = H, CH2CH3
Hypromellose	HPMC	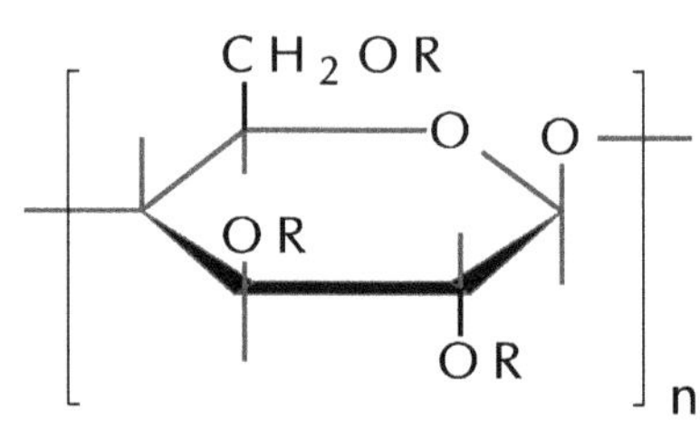 R = H, CH_3, $CH_2CHOHCH_3$
Methylcellulose	MC	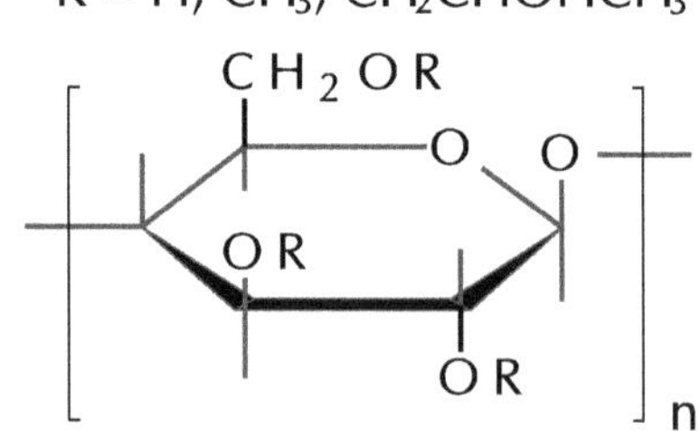 R = H, CH3
Polyethylene oxide	PEO	-(-CH2CH2O-)n-

encompassing a range of molecular weights (viscosity grades). Polyethylene oxide (PEO) is a highly crystalline, hydrophilic polymer. It is available in molecular weights ranging 100,000–7,000,000 Daltons.

7.3 Thermal Properties

The thermal properties of representative grades of EC, HPMC, MC and PEO were previously reported [10]; key properties are summarized in Table 7.2.

EC polymers have glass transition temperatures ranging over 131–137°C and crystalline melting points ranging over 172–190°C. The decomposition temperatures are all above 215°C in air. Interestingly, the decomposition temperatures are significantly increased in a nitrogen environment, suggesting some sort of oxidative degradation mechanism. There is a general trend that the higher viscosity (molecular weight) grades tend to have higher glass transition and melting temperatures. The inverse trend was observed for decomposition temperature. Lower viscosity grades have higher decomposition temperatures. EC is a good candidate for HME since it exhibits thermoplastic behavior at temperatures above its glass transition temperature.

HPMC is an amorphous polymer, as noted by the absence of melting events. The glass transition temperatures encompass the broad range of 168–209°C. The decomposition temperatures in air and nitrogen are very similar, and all are in excess of 200°C. There is a general trend that higher viscosity (higher molecular weight) grades tend to have higher glass transition temperatures. HPMC has proven to be challenging to extrude due to the combination of high glass transition temperatures with relatively low decomposition temperatures. The addition of processing aids such as plasticizers was recommended as a means to broaden the processing window [11].

PEO has glass transition temperatures well below 0°C and accordingly this thermal property has no impact on HME. As mentioned above, PEO is a highly crystalline polymer with melting temperatures of ~70°C. Decomposition temperatures in air are 190–200°C, but in general increase by approximately 20°C in a nitrogen environment. PEO is an ideal candidate for HME due to its broad processing window. The neat polymer can be extruded at temperatures modestly above its crystalline melting point up to their decomposition temperatures.

7.4 Processing Aids/Additives

The addition of processing aids to is a common practice to broaden the extrusion processing window [12, 13]. Strictly speaking, plasticizers lower the glass transition temperature of a polymer. However, they typically also have a positive impact on extrusion processing, reducing the required processing temperature of the composition. Plasticizers may impact drug release rate and shelf life stability of the formulation, so they should be used with caution. Plasticizer and antioxidants are the most common types of processing aids, although many varieties are available. Unfortunately, there are a limited number of plasticizers and antioxidants that satisfy the regulatory requirements for pharmaceutical use.

Table 7.2 *Summary of thermal data for ethylcellulose, hypromellose, methylcellulose and polyethylene oxide.*

Material	Glass transition temperature, first heat (°C)	Glass transition temperature, Second heat (°C)	Crystalline melting temperature, first heat (°C)	Crystalline melting temperature, second heat (°C)	Decomposition temperature, in air (°C)	Decomposition temperature, in nitrogen (°C)
ETHOCEL STD 4 PRM	131	122	172	169	244	>300
ETHOCEL STD 10 PRM	136	128	181	177	216	276
ETHOCEL STD 100 PRM	137	133	190	188	218	225
METHOCEL E5 LV PRM	173	118, 139	N/A	N/A	228	267
METHOCEL E50 PRM	168	158	N/A	N/A	251	260
METHOCEL E4M PRM	189	159	N/A	N/A	257	260
METHOCEL F4 LV PRM	170	142	N/A	N/A	241	245
METHOCEL F50 PRM	175	140, 165	N/A	N/A	265	266
METHOCEL F4M PRM	189	171	N/A	N/A	276	276
METHOCEL K3 LV PRM	174	143	N/A	N/A	243	249
METHOCEL K100 PRM	196	186	N/A	N/A	261	257
METHOCEL K4M PRM	209	191	N/A	N/A	271	275
POLYOX N-10 WSR NF	<0	<0	69	63	190	198
POLYOX N-80 WSR NF	<0	<0	68	63	196	205
POLYOX N-750 WSR NF	<0	<0	70	67	201	260
POLYOX 250 WSR NF	<0	<0	71	66	196	222
POLYOX 1105 WSR NF	<0	<0	72	67	200	227
POLYOX N12K WSR NF	<0	<0	71	67	189	218
POLYOX N60 WSR NF	<0	<0	71	66	189	235
POLYOX 301 WSR NF	<0	<0	72	67	190	216
POLYOX 303 WSR NF	<0	<0	72	67	195	199

The extrusion of HPMC has proven challenging, due to the relatively high glass transition temperatures of these materials combined with a relatively low degradation temperature. The net result is a narrow processing window. Alderman and Wolford suggested including high levels of plasticizer in the formulation, at least 30% by weight [11]. The authors noted that when higher levels of plasticizer (50–90 wt%) are used, the resulting extrudate is very flexible. They recommended the use of propylene glycol and polyethylene glycols with molecular weight less than 1000. No limits on the amount of active pharmaceutical ingredient (API) that can be used were taught.

The addition of Vitamin E D-α-tocopherol polyethylene glycol succinate (TPGS) to PEO was studied by Repka and McGinity [14]. Vitamin E TPGS was found to function as a plasticizer. Its addition was observed to decrease the melt viscosity of the formulation during extrusion processing, as evidenced by reductions in barrel melt pressure, drive-motor amperage and torque. Further, the addition of Vitamin E TPGS was found to decrease the tensile strength and increase percent elongation of the films produced. Both of these observations are indicative of vitamin E TPGS functioning as a plasticizer for PEO. Further, vitamin E TPGS is known to function as a PEO antioxidant, which reduces molecular weight loss during extrusion [15]. Other effective antioxidants noted include vitamin E succinate and vitamin E. This additional property makes it a particularly useful processing aid for PEO [16].

Zhang and McGinity studied the use of PEG 3350 as a processing aid for PEO [17]. In this study, chlorpheniramine maleate was extruded with PEO to produce matrix tablets. PEG 3350 was added in an effort to reduce molecular weight loss during extrusion. In addition to reducing molecular weight loss, PEG 3350 addition was also found to increase the rate of chlorpheniramine maleate release.

7.5 Unconventional Processing Aids: Drugs, Blends

Several authors have reported the use of APIs as non-traditional processing aids for EC, HMPC, MC and PEO. For example, deBrabander et al. found that ibuprofen effectively functions as a plasticizer for EC during HME [18]. The plasticizing effect was noted during extrusion processing. A single T_g was observed in the resulting film, indicating the formation of a solid dispersion. Further, the T_g was observed to decrease as the amount of ibuprofen in the formulation increased.

Similarly, Rambali et al. studied HME of itraconazole with HPMC and hydroxypropyl-ß-cyclodextrin [19]. They noted that formulations containing higher levels of itraconazole exhibited lower torque during the extrusion process.

Combination of excipients is another approach to improve extrusion processing. For example, Crowley et al. combined high and low molecular weight PEO grades. In this study, chlorpheniramine maleate was extruded with 100,000 and 1,000,000 Da molecular weight grades [15]. The chlorpheniramine maleate concentration was held at 20% and PEO 100,000 was added at 10, 20 and 40 wt% of the total formulation. Decreases in extruder torque were observed as the amount of PEO 100,000 increased. Interestingly, the dissolution rate was not appreciably changed compared to that of 80/20 PEO 1,000,000/chlorpheniramine maleate.

Coppens et al. [20, 21] discussed the use of combinations of PEO with HPMC, PEO with EC and combinations of PEO, EC and HPMC. They noted that combinations such as these could be used to broaden the extrusion processing window. For example, the combination of 32 wt% PEO N-10 with 48 wt% HPMC E4M was easily processed with 20 wt% nifedipine or ketoprofen, thus avoiding the addition of a conventional HPMC plasticizer. They also observed that the combinations could provide dissolution profiles that could not be achieved via the use of any single excipient. They also found that the composition of the blend also impacted the morphology of the API in the extrudate. This morphology will have a direct impact on the aqueous solubility of the API.

Verreck et al. studied the use of supercritical CO_2 as a processing aid for EC. The initial studies were performed without inclusion of an API [22]. They found that supercritical CO_2 was an effective temporary plasticizer for EC. A decrease in melt viscosity was observed during extrusion, as measured via extruder torque. Further, a 65°C lower processing temperature was possible. A highly porous extrudate was produced. The morphology ranged from foam to fibrous, depending upon the amount of CO_2 added. No indication of degradation was detected via size exclusion chromatography (SEC). The glass transition temperature of the extrudate was unchanged, but the amount of crystallinity present was found to increase as the amount of CO_2 added was increased. The extrudate was easier to mill than conventional extrudate. A second study incorporated *p*-amino salicylic acid [23]. They found that supercritical CO_2 addition again permitted reduced processing temperature, this time approximately 30°C. Analysis of the extrudate showed that the amount of *p*-amino salicylic acid degradation was significantly reduced (5% compared to 17% for the extrudate produced without CO_2 injection).

Supercritical CO_2 was also investigated as a processing aid for PEO [24]. In this paper, PEO (MW = 200,000) was extruded with carvedilol (7.5–14 wt%). Eudragit EPO (21–46 wt%) was included in some of the formulations. Both control formulations (without CO_2) and experimental formulations (with CO_2 injected at 1200 psi) were generated. Extrusion was performed using a bench-top 16 mm twin-screw extruder with a 25/1 length/diameter ratio. The extrudate was cooled and then granulated. The CO_2 was injected into the extruder barrel via a port at approximately 2/3 the length of the barrel. A significant decrease in extruder drive motor torque and die pressure was observed upon addition of the supercritical CO_2 with all process conditions unchanged. This indicated that the CO_2 was reducing the melt viscosity of the extrudate. There was also strong evidence to suggest that carvedilol was acting as a plasticizer. The drive motor torque and die pressure decreased as the amount of carvedilol increased. Parallel plate rheology evaluations of the extrudate confirmed that the effect of supercritical CO_2 only occurred in the extruder. No difference in rheology was observed between equivalent compositions extruded with and without CO_2 addition. Thermal analysis of the extrudate via micro-thermal analysis found that supercritical CO_2 addition led to a higher level of PEO crystallinity in the extrudate.

The authors [24] proposed that this resulted from increased polymer chain mobility which permitted the PEO chains to reconfigure into the thermodynamically preferred crystalline form. Dissolution testing was performed with 0.2 M HCl and pH 7.2 buffers. Faster dissolution was observed for the extrudates produced with supercritical CO_2. It was felt that this resulted from an increased internal surface area which resulted from the foam-like extrudate. Overall, the authors concluded that supercritical CO_2 is a viable approach during HME of PEO formulations. The advantage of this technology is that higher output rates may be reached with minimal negative impact on the properties of the extrudate.

7.6 Case Studies

7.6.1 Ethylcellulose

Many authors have utilized ethylcellulose in hot-melt extrusion. For example, Crowley et al. [25] studied formulations containing 30% guaifenesin and 70% EC. The performance attributes of matrix tablets produced via HME and direct compression were compared. The HME tablets were produced using a single-screw extruder equipped with a 6 mm diameter rod-shaped die. The extrudate was manually cut to produce 250 mg tablets. The authors observed that tablets produced with smaller particle size EC exhibited slower guaifenesin release for tablets produced by both HME and direct compression. They proposed that the smaller particle size EC produced tablets with lower porosity than tablets produced with larger particle size EC. They also observed that guaifenesin release rates from HME tablets were slower than those produced via direct compression. The effect of extrusion temperature on guaifenesin release rate was also studied. They found that extruding at higher temperatures (90–100°C) produced tablets that had a slower guaifenesin release rate. They reasoned this was because these tablets were less porous and that the diffusion pathway was more tortuous.

7.6.2 Combinations of Excipients

As noted above, combinations of EC, HPMC/MC and/or PEO excipients can be used as an alternative to conventional plasticizers as an approach to broaden the HME processing window. Many authors have performed similar investigations aimed at understanding the impact of such compositional changes on API dissolution behavior [26–38].

DeBrabander et al. have performed multiple studies combining EC with various hydrophilic polymers. They studied combinations of EC with HPMC as an approach to alter the release rate of ibuprofen [26]. A co-rotating twin-screw extruder equipped with a 3 mm rod die was used to produce an extruded strand, which was manually cut to produce the 2 mm long mini-matrices. Increasing the ratio of HPMC:EC in the formulation resulted in more rapid ibuprofen release. The ibuprofen content of the mini-matrices was assayed via high-performance liquid chromatography (HPLC). The authors found that 98.9% of the ibuprofen remained after extrusion.

In a follow-up study, DeBrabander et al. investigated the impact on ibuprofen bioavailability with the addition of xanthan gum to EC and HPMC containing mini-matrices [27]. The mini-matrices were once again produced using a co-rotating twin-screw extruder equipped with a 3 mm rod die, followed by manually cutting the extruded strand into 2 mm length mini-matrices. In vivo testing indicated that the HME-produced mini-matrices provided a more constant drug absorption pattern than observed with a commercially available product.

A subsequent paper documented the impact of xantham gum level on in vitro and in vivo results [28]. The authors reported that increasing xantham gum level accelerated in vitro ibuprofen release. They believe this was due to faster swelling of the mini-matrices followed by more rapid erosion. However, ibuprofen release was primarily diffusion controlled, but this swelling was important to permit complete ibuprofen release. Significant differences in vitro dissolution were observed for several of the compositions studied. Interestingly, in vivo differences were much smaller. Overall, the authors concluded that the combination of EC with a hydrophilic additive is a useful approach to generated tailored drug-release profiles.

The EC/xanthan gum studies were continued using metoprolol tartrate as the model drug [29]. This study included dibutyl sebacate (DBS), diethyl phthalate (DEP), triethyl citrate (TEC) and triacetin as EC plasticizers. Plasticizer screening via differential scanning calorimetry (DSC) indicated that all were equally effective in terms of reducing EC glass transition temperature. However, TEC and triacetin containing formulations exhibited a burst effect in drug release testing, and were subsequently removed from the program. DBS was chosen as the plasticizer for continued study. The authors found that utilizing an EC/DBS ratio of at least 2/1 was required to produce physical defect-free mini-matrices. Increasing the level of xanthan gum in the formulations accelerated metropolol tartrate release. Raman spectroscopy was used to study drug content uniformity in the mini-matrices. The results indicated that the metropolol tartrate was homogeneously distributed. A process reproducibility study was performed using 2/1 EC/DBS, 5% xanthan gum and 30% metropolol tartrate. Excellent reproducibility was observed. The authors concluded that HME processing of this formulation was very robust.

Verhoeven et al. continued these studies utilizing polyethylene glycol (PEG) and PEO [30]. Mini-matrices containing EC, DBS (2/1 EC/DBS ratio), metoprolol tartrate (30 wt%) and various PEG or PEO products were produced via HME. Similar to previously reported results, the authors found that increasing the amount of hydrophilic polymer increased drug release rate. They noted an interesting effect of PEO molecular weight over the range studied (MW 6000, 100M, 1MM, 7 MM). At low PEG/PEO concentrations, no effect of molecular weight was noted. However, at 5 and 10%, drug release was slower for formulations containing low molecular weight hydrophilic polymer. At higher concentrations (20 and 70%), the expected slowing of drug release as molecular weight increased was observed. In vitro studies were found to be consistent with in vivo results. The authors felt that the addition of PEG or PEO of various molecular weights to these EC-based mini-matrices could be used to produce a wide variety of dissolution profiles, affecting both the initial release rates but also the overall shape of the profile. A similar result was noted by Hall et al. [31].

These PEG/PEO studies were completed utilizing mathematical modeling to understand the mechanism of drug release from these mini-matrices [32]. A model based on Fick's law indicated that mass transport in mini-matrices was primarily diffusion controlled for formulations incorporating high PEG/PEO content ($\geq$ 20%) and for formulations with intermediate PEG/PEO content (2.5–10%) incorporating low molecular weight PEG/PEO (molecular weight $\leq$ 100,000 Da). A more complex model was required to describe the observed metoprolol tartrate dissolution from compositions with low PEG/PEO content ($<$ 2.5%) and those with intermediate concentration (2.5–10%) of high molecular weight PEO (MW $>$ 100,000). In these formulations, it was felt that changes in the porosity as drug and water-soluble polymer diffused out of the mini-matrices were a primary factor affecting drug dissolution.

PEO extrusion in combination with chitosan and/or xanthan gum was reported by Fukuda et al. [33]. The objective of this study was to produce matrix tablets via HME and then study the influence of pH, buffer species and ionic strength on chlorpheniramine maleate dissolution. The formulations were composed of 27 wt% PEO N-80 (MW = 200,000), 3 wt% glycerol monostearate (plasticizer) and 10 wt% chlorpheniramine maleate, with the remainder being combinations of chitosan (Daichitosan H or M) and/or microcrystalline cellulose and/or xanthan gum. Chitosan and xanthan gum were being evaluated as controlled-release polymers. These materials were processed on a single-screw extruder to produce a cylindrical strand which was manually cut to produce 300 mg tablets. Differential scanning calorimetry (DSC) analysis of the extrudate indicated the absence of crystalline chlorpheniramine maleate. Accelerated stability testing was performed at 40°C for 1 month, 40°C/75% RH for 1 month and 50°C for 15 days. No recrystallization of chlorpheniramine maleate was observed via DSC and no change in dissolution profiles was observed, indicating that these compositions were stable under these storage conditions.

Dissolution studies demonstrated that tablets produced via HME had chlorpheniramine maleate release that was unaffected by changes in pH and buffer type. An interesting observation was that tablets of equivalent compositions produced via direct compression exhibited chlorpheniramine maleate that was strongly influenced by pH and buffer changes. The authors suggested that this difference in performance was inherent to the HME-produced tablets, as they effectively have a higher density (lower porosity).

The addition of PEO (MW = 7 million) to function as an API crystal growth inhibitor was studied by Bruce et al. [34]. The primary matrix-forming polymer was Eudragit L100-55. The API was guaifenesin (25 wt%). Triethyl citrate (3 wt%) was also included. PEO was 2.5 wt% of the formulation. Cylindrical extrudates 6 mm in diameter were produced via a single-screw extruder and were manually cut to generate 250 mg tablets. The extrudate was evaluated via scanning electron microscopy (SEM) immediately and after 4 weeks of storage at 25°C/60% RH. No crystalline guaifenesin was observed on the initial samples, but was observed on the aged sample. However, the amount of crystalline guaifenesin was markedly reduced for the sample containing the PEO compared to a control sample. Dissolution testing in pH 6.8 phosphate buffer showed that PEO addition had no impact on the dissolution profile. Aging also had no impact on dissolution profile, indicating the formulation was stable.

Young et al. evaluated the addition of HPMC to melt-extruded dosage forms based on a methacrylic acid copolymer and containing theophylline [35]. The primary excipient in the formulation was Eudragit L 100-55, although it was incorporated as Acryl-EZE. HPMC (METHOCEL K4M Premium) was added at 2.5 and 5 wt% to study its impact on the mechanism and kinetics of theophylline release. Triethyl citrate was also included in the formulation. The authors noted that the addition of HPMC resulted in an increase in the extruder drive motor amps and higher die pressures. They found that HPMC addition increased theophylline release in 0.1N HCl medium. They proposed that this resulted from the HPMC functioning as a porosity modifier in the matrix as it dissolved, since the Eudragit polymer was insoluble at this pH. However, when the pH of the dissolution media was adjusted to 6.8 via addition of a phosphate buffer solution, they observed that the HPMC decreased the rate of theophylline release. The formation of an entangled gel network that reduced the erosion rate of the tablet was proposed as an explanation. At low pH, dissolution from the HPMC containing matrices was diffusion controlled. As the amount of HPMC increased, dissolution became more erosion controlled. At pH = 6.8, erosion became the primary mechanism for dissolution of the HPMC containing matrices.

Combinations of HPMC and PEO were studied in terms of the impact on dissolution profile and crystalline state of an API in the extrudate [36]. Acetaminophen (APAP) was the model drug included in this study. The authors found that combining low molecular

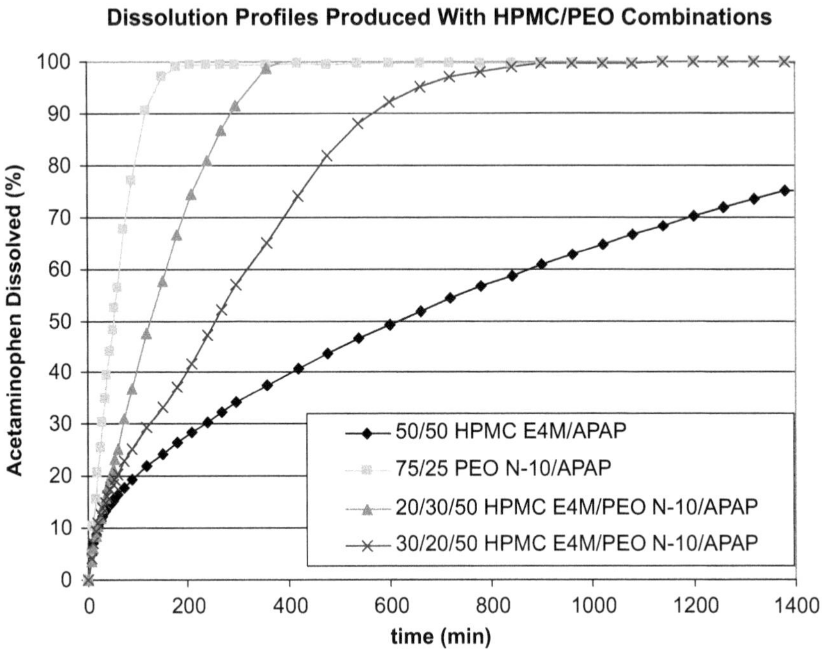

Figure 7.1 *APAP dissolution profiles produced via combinations of HPMC and PEO.*

weight PEO with high viscosity grades of HPMC resulted in dissolution profiles that fell between those obtained with either excipient by itself (Figure 7.1).

DSC analyses were used to evaluate the impact of HPMC/PEO combinations on the level of APAP crystallinity in the extrudate. The observed APAP melting temperature and heat of fusion decreased as the amount of excipient in the formulation increased. This indicated a molecular level interaction between APAP and the excipients resulting in disruption of the APAP crystal. The authors felt that the HPMC/APAP interactions were more significant than those of PEO/APAP.

The addition of nanoclay particles to PEO extrusion was studied by Lyons et al. [37]. The formulations evaluated incorporated 14 wt% carvedilol, an organically modified layered silicate at 2, 4 and 6 wt%, with the remainder of the formulation being PEO (MW = 5 million). Extrusion was performed using a bench-top 16 mm diameter twin-screw extruder in co-rotating mode. The cylindrical extrudate was cooled and granulated. The granules were then injection molded to produce ISO 294-1 and ISO 6239 type A test specimens. These test specimens were used for subsequent testing. The addition of the nanoclay had minimal impact on the extrusion parameters monitored (die pressure, drive motor torque). The authors felt this was as expected, due to the small amount of filler added. Parallel plate rheometry of these test specimens also showed a small impact on viscosity. Dissolution testing in a pH 1.2 buffer showed that carvedilol release rate decreased as the amount of nanoclay increased. The authors proposed that the addition of the nanoclay resulted in the generation of a tortuous path that slowed the ingress of solvent into the matrix. They concluded that the addition of nanoclay particles to dosage forms produced via HME may be a useful approach to modulate drug release.

Combinations of EC and PEO via HME to produce formulations exhibiting a broad spectrum of dissolution profiles was claimed by Hall et al. [38]. EC must be at least 15 wt% of the composition. The authors report that the dissolution profiles obtained can be systematically altered by varying the relative concentration of EC and PEO as well as the molecular weight (viscosity grade) of the excipient.

7.6.3 Solubilization

Improved bioavailability of poorly water-soluble drugs is a significant area of study for HME. The desired result is the preparation of an amorphous solid dispersion or solid solution, wherein the amorphous form of the API is intimately mixed with the polymeric matrix. Solid dispersions can also result. In this case, small domains of crystalline drug are present in the extrudate. Both of these forms tend to provide improved water solubility [39].

Solid dispersions of ketoprofen in PEO of MW 100,000 were studied by Schachter et al. [40]. The solid dispersions were produced by mixing PEO and ketoprofen, granulating the mixture with water, melt blending the granules in a Brabender Plasticorder and then pressing the melted mass into plaques. DSC and x-ray diffraction (XRD) analyses of the plaques indicated that the ketoprofen dissolved into the PEO amorphous phase. The solid dispersion was found to be stable after one month of accelerated aging conditions (40°C, 75% RH). Ibuprofen, tolbutamide, sulfathiazole and hydroflumethazide were evaluated similarly. The initial results reported suggest that solid dispersions were successfully produced with these APIs also.

Follow-up characterization of the PEO/ketoprofen solid dispersion was performed via solid-state nuclear magnetic resonance (SSNMR) [41]. The solid dispersions produced above were analyzed. The results indicated that the PEO/ketoprofen interactions were very strong, enough so that the ketoprofen crystalline lattice was thoroughly disrupted, even at temperatures below its melting point. An increase in ketoprofen mobility in the composition, compared to neat ketoprofen, was reported. These results indicated that melt mixing of ketoprofen PEO resulted in the formation of a solid dispersion.

Nifedipine was studied as a poorly water-soluble model drug with PEO alone and combined with HPMC and/or EC [42]. Various formulations were extruded on a laboratory-scale single-screw extruder to produce a 0.325 inch diameter rod. This rod was manually cut to generate tablets of approximately 300 mg in size. A formulation comprising 20 wt% nifedipine and 36 wt% PEO N-10, 28 wt% HPMC E4M and 16 wt% EC Std 10 was found to have 100% nifedipine release at 600 min. Direct compression tablets of the same composition exhibited only 20% release at 600 minutes, indicating a significant improvement in nifedipine solubility in this extruded formulation. Further, the formulation exhibited acceptable dissolution stability when subjected to a 6 month accelerated stability test (6 months @ 40°C/75% RH). An 80/20 PEO N-10/nifedipine formulation was similarly produced. As shown in Figure 7.2, complete nifedipine dissolution was observed at 300 minutes from these extruded tablets compared to 15% for direct compression tablets of the same composition. The composition was also observed to have acceptable dissolution stability when subjected to accelerated stability testing. In both of these extruded formulations, the nifedipine was found to exist in a crystalline form.

HME of PEO with nifedipine was also studied by Li et al. [43]. PEO N-80 (molecular weight 200,000 Da) was used in this study. Nifedipine comprised 20 wt% of the formulation. Extrusion was performed using a DACA microcompounder bench-top twin-screw

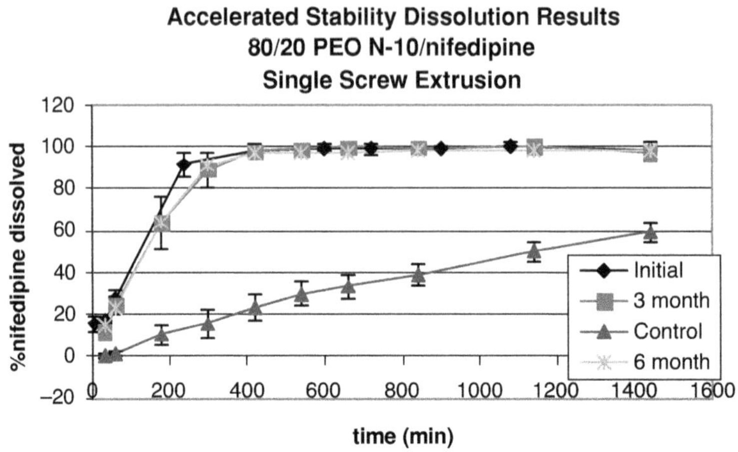

Figure 7.2 *Dissolution results for accelerated stability testing of 80/20 PEO N-10/nifedipine tablets produced via HME and direct compression.*

extruder at processing temperatures ranging from 70 to 200°C. The extrudate was cryogenically milled to produce powders. The authors found that extrusion below the nifedipine melting temperature (172°C) was easily accomplished. Extrudates prepared at processing temperatures below 120°C were opaque and light yellow. Extrudates produced at and above 120°C were transparent and bright yellow (the native color of nifedipine). They surmised that, once PEO melted, it became a nifedipine solvent. This was confirmed via DSC and hot stage microscopy. DSC and XRD analyses indicated the absence of crystalline nifedipine in the extrudate produced at any temperature. Raman spectroscopy was used to evaluate the content uniformity of the extrudate. They found that extrudate produced below 120°C tended to be non-uniform. Distinct domains of PEO and nifedipine and crystalline nifedipine were observed. However, extrudate produced at and above 120°C exhibited excellent content uniformity.

The authors proposed that these differences are caused by the melt viscosity of the PEO at the respective processing temperatures. At temperatures below 120°C, PEO viscosity is too high to permit good mixing in this extrusion equipment whereas good mixing was achieved at higher temperatures. Dissolution testing of the extrudate in pH 6.8 phosphate buffer showed a significant improvement in nifedipine solubility compared to pure nifedipine or a physical mixture of nifedipine with PEO N-80. The extrudates reached a plateau concentration of approximately 8 μg/ml in 10 minutes and maintained this level for the 120 minute duration of the test. The un-extruded materials had a concentration of only approximately 2 μg/ml at 10 minutes and slowly increased to nearly 5 μg/ml at the conclusion of the test. Stability testing was performed at 25°C/60% RH and 40°C/75% RH for 6 months. Less than 1% change in potency was observed. However, some recrystallization of the amorphous nifedipine was observed.

Miller et al. studied PEO extrusion with micronized HPMC/itraconazole particles as a means of improving itraconazole bioavailability [44]. In this work, micronized particles composed of 1:1 itraconazole:PVP and itraconazole:HPMC were produced via flash evaporation followed by milling. The micronized powders were then extruded utilizing a 70:30 poloxamer 407: PEO 200M excipient combination. The extrudate was subsequently triturated to produce a powder for testing. Analyses of the extrudate indicated that amorphous itraconazole was still present after extrusion. In vitro dissolution testing indicated that itraconazole dissolution was improved for the extruded formulation compared to that of the micronized particles alone. The authors also observed that the HPMC/itraconazole maintained supersaturation conditions much better than the PVP/itraconazole formulation. This was not unexpected, due to the ability of HPMC to do so better than PVP [45, 46]. In vivo testing was also performed and a similar bioavailability was observed for both formulations.

HPMC has also been used to improve the aqueous solubility of poorly water-soluble drugs. Verreck et al. [47] and Six et al. [48] studied solid dispersions of HPMC and itraconazole. They began by producing solvent cast films at various itraconazole/HPMC ratios to determine the optimal ratio for further study. Itraconazole dissolution and DSC analyses were used to select 60/40 HPMC/itraconazole. Extrusion was performed with a co-rotating twin-screw extruder. The extrudate was milled to produce a powder. Dissolution studies indicated that 90% of the itraconazole was released in 120 min, whereas only 2% itraconazole release was observed using a physical mixture of itraconazole and HPMC. Follow-on work attempted formulation optimization to improve itraconazole dissolution. Here the formulation was 75/25 HPMC/itraconazole. The extrudate was milled and then sieved such that particles less than 355 micron were studied. In this case, 80% of the itraconazole was released within 30 minutes. This was much improved over the dissolution of crystalline itraconazole (0% @ 30 minutes) and glassy itraconazole (5% @ 30 minutes).

Miller et al. investigated the impact of HPMC molecular weight on the ability to improve the solubility of itraconazole from a solid dispersion [49]. Solid dispersions of itraconazole and HPMC E5 and E50 (1:2 itraconazole:HPMC) were produced using a Haake MiniLab Micro Compounder. DSC analyses of the extrudate confirmed itraconazole was amorphous. Dissolution testing was performed using a technique where the first 2 hours were conducted in 0.1N HCl, followed by a pH adjustment to 6.8 via the addition of tribasic sodium phosphate. The formulation containing E50 had the most rapid dissolution rate, with approximately 68% of the itraconazole released at 30 minutes compared to 47% for the E5 formulation. HPMC E50 also had better itraconazole stabilization after the shift to neutral pH. The authors felt that the interaction between HPMC and itraconazole was responsible for this improvement, specifically the ability of HPMC to provide hydrogen-bonding donor sites. The relative outperformance of the HPMC E50 was unexpected. This observation was attributed to an increase in localized viscosity surrounding the dissolved drug molecule, which effectively retards recrystallization. In vivo testing showed a good correlation with the in vitro results generated using the modified method as described above. The mean $C_{\max}$ value was 732 ng/ml and AUC was 6195 ng h/mL. In vivo testing also included Sporanox® as a control. The mean $C_{\max}$ value was 179.2 ng/ml and AUC was 2186 ng h/mL. The results indicated that the HPMC E50-based formulation provided a significant improvement in bioavailability.

Recently, DiNunzio et al. compared itraconazole/HPMC solid dispersions produced via kinetisol processing with those produced via HME [50]. Solid dispersions containing 1:2 itraconazole:HPMC E5 were produced by both techniques. Milled extrudate was used for subsequent testing. The resulting solid dispersion contained amorphous itraconazole, as determined via DSC and XRD. In vitro dissolution testing under supersaturated conditions showed that both processing techniques yielded materials with improved itraconazole solubility compared to the unprocessed API. The authors noted that the solid dispersion produced via kinetisol processing has faster dissolution rates compared to the HME product. In vivo testing showed no statistically significant difference in performance. The authors concluded that kinetisol processing is a viable alternative melt processing technique to hot-melt extrusion.

Kinetisol processing was also studied as a means to produce solid dispersions of hydrocortisone and HPMC E3 [51]. Hydrocortisone is a thermally sensitive API, which decomposes at temperatures above 180°C. Extrusion processing of 9/1 HPMC/hydrocortisone compositions was performed at180°C and kinetisol processing was performed at 160 and 180°C. An assay of the product indicated that the material produced via kinetisol processing retained higher potency. Amorphous solid dispersions were generated in all cases, although the XRD profile suggested that phase separation was present. Improved cortisone solubility was observed.

Zheng et al. investigated solubility improvement of nimodipine by producing solid dispersions with HPMC via HME [52]. They utilized HPMC type 2910, 5 cp viscosity in their experiments. Simple binary formulations were studied, with nimodipine content levels of 10, 30 and 50 wt%. A single glass transition temperature T_g was observed with 10 wt% nimodipine. This T_g was intermediate between the T_gs of nimodipine and the HPMC, indicating that there was good miscibility between the two components. At the higher nimodipine concentrations, two T_gs were observed which indicated phase separation.

DSC results of the resulting extrudate showed no evidence of nimodipine crystallinity. XRD confirmed the absence of crystalline nimodipine in all compositions. FTIR (Fourier transform infrared) analyses suggested that hydrogen bonding was occurring between the secondary amine functionality of nimodipine and the hydroxyl groups of the HPMC. This likely contributed to the formation of an amorphous solid dispersion. Scanning electron microscopy (SEM) of the extrudate surfaces identified the presence of crystalline nimodipine for all compositions. The authors concluded that this analysis method may be more sensitive than DSC and XRD, as noted above. In vitro dissolution testing (pH 4.5 acetate buffer with 0.5% (w/v) sodium dodecyl sulfate or SDS) showed at least a threefold improvement in nimodipine solubility compared to the unaltered API. Nimodipine bioavailability of these solid dispersions was subsequently evaluated [53]. The in vivo results confirmed that the solid dispersions increased nimodipine bioavailability with approximately a 2.5× increase in C_{max} and 1.75× increase in AUC(0-12) compared to a simple blend of the raw materials.

The combination of surfactants with HPMC for preparing solid dispersions of poorly soluble drugs was studied by Ghebremeskel et al. [54, 55]. The authors utilized HPMC type 2910, 5 cp viscosity (HPMC E5) in their program with a proprietary poorly water-soluble drug. The formulations were based on 1/1 HPMC/API to which various levels of surfactants was added (2, 5, 10 wt%). The desired result was that the surfactant acted as a plasticizer in terms of HME processing, but also improved API dissolution. DSC was used initially to evaluate the efficacy of surfactants being evaluated. The desired result was a decrease in polymer T_g as the amount of surfactant increased. Tween 80, docusate sodium and sodium lauryl sulfate all showed significant reductions. For example, the incorporation of 10% Tween 80 reduced the T_g by 27°C. Formulations composed of 45:45:10 HPMC E5: API:

plasticizer were used for HME evaluation. The desired impact on HME processing was observed (e.g. the amount of torque decreased).

Characterization of the HPMC/API/surfactant solid dispersions showed that the API was present in its amorphous form. Both DSC and XRD showed no sign of crystalline API. The inclusion of the surfactant in the formulation successfully improved API solubility. The addition of Tween 80 increased API solubility to 100% in less than 15 minutes compared to 100 minutes for the HPMC/API formulation.

The authors speculated that the observed improvement could be due to micellar solubilization via the surfactant combined with a reduction in interfacial tension between the samples and dissolution media, thereby improving the wetting of the sample. The stability of an API/HPMC E5/Tween 80 formulation was subsequently studied and compared to an API/HPME E5 control. Two sets of accelerated stability conditions were used (30°C/60% RH, 6 months and 60°C/85% RH, 2 months). XRD and optical microscopy showed minor conversion of the API from its amorphous to crystalline form. This was most pronounced at the 60°C/85% RH stability conditions where conversion was observed to begin within the first 4 hours. API percent crystallinity was 5.7% for the API/HPMC E5 formulation and 6.2% for the API/HPMC E5/Tween 80 formulation at the conclusion of the 2 month stability testing interval at these conditions. API percent crystallinity of 1.2% and 4.4% were measured at the conclusion of the 6-month 30°C/60% RH stability tests for the API/HPMC E5 and API/HPMC E5/Tween 80 samples, respectively. The inclusion of Tween 80 in the formulation yielded a slight increase in the amount of crystalline API. Dissolution testing of these materials at the conclusion of the stability test showed no change in API solubility.

The efficacy of HPMC for drug solubilization is largely due to its ability to prevent API recrystallization in an aqueous environment at concentrations above the solubility limit of the API (e.g. supersaturated conditions). This performance attribute has been studied by several investigators [56–58]. The mechanism for this effect has not yet been determined. However, it is acknowledged that it likely involves an intermolecular attraction in solution between the API and HPMC, which ultimately results in the formation of a physical barrier which prevents recrystallization and precipitation of the API from solution.

7.6.4 Film

The use of PEO and HPMC in oral films has been reported by many investigators. Several potential advantages of oral films as a dosage form have been cited. These include improved patient compliance and improved efficacy thereby decreasing frequency of dosing (convenient for infant and geriatric medications).

Repka et al. investigated PEO-based films for clotrimazole delivery [59]. The authors utilized PEO MW 100,000 in combination with HPC, polycarbophil, PEG 3350 (plasticizer), butylated hydroxytoluene (BHT) and propylgallate as antioxidants, and 10 wt% clotrimazole. A single-screw extruder was used to produce films of 0.34–0.356 mm in thickness. Excellent content uniformity of the films was reported. Wide angle X-ray diffraction showed that the clotrimazole was molecularly dispersed within the HME films. Zero-order clotrimazole release was observed over 6 hours.

Follow-on studies by Prodduturi et al. found that clotrimazole was initially present in its amorphous form within the PEO-based films, but crystallized during storage for 3 months at 25°C/60%RH [60]. These films were produced with PEO MW 200,000 and a combination of PEO MW 300,000/PEO MW 200,000 (5/1) and clotrimazole levels of 5, 10, 15 and 20 wt%. The films were reported to have equivalent or superior bioadhesion compared to other literature values. The films containing PEO MW 300,000 were observed to have stronger bioadhesive properties and also exhibited much higher elongation than films produced with PEO MW 200,000. The combination of HPC with PEO was then evaluated as a means to reduce the tendency of clotrimazole to recrystallize in PEO films [61].

A 1-inch diameter single-screw extruder was again used to produce films approximately 0.54 mm in thickness. The HPC/PEO ratios studied were 90/10, 75/15, 55/35, 35/55 and 0/90. PEO MW 200,000 was used in all cases, except the 0/90 composition where the combination of PEO MW 300,000 and PEO MW 200,000 was used (75% MW 300,000, 15% MW 200,000). Clotrimazole was 10 wt% of the formulation. The authors found that water absorption of the film decreased with increasing PEO content. They also found that PEO presence was required for bioadhesion, with levels less than 35 wt% having unacceptable performance. Sustained clotrimazole release was observed for all compositions, with the release rate slowing with increasing PEO content. Zero-order kinetics were observed, with the mechanism of release believed to be exclusively by erosion. The films contained amorphous clotrimazole immediately after extrusion, but the films incorporating only PEO exhibited clotrimazole recrystallization within 3 months of storage at 25°C/Drierite or 25°C/60% RH. The authors concluded that the optimal film composition was 55:35:10 HPC:PEO:clotrimazole.

Munjal et al. studied PEO-based intra-oral transmucosal patches incorporating Δ^9-tetrahydrocannabinol produced via a hot-melt fabrication process [62, 63]. The primary advantages cited for this delivery approach were avoidance of first-pass metabolism of the API and improved patient compliance. A hot-melt casting method was used as an alternative to extrusion. This was a batch method wherein PEO (MW = 10,000) and processing aids were heated and manually mixed until a homogeneous molten mass was produced. A solution of the API in ethanol was then added to the mixture and manually stirred until homogeneous. The process yielded a flexible patch. Vitamin E succinate, PEG 400, isopropyl myristate and Capmul® PG-12 were found to be the preferred processing aids, as determined by their ability to reduce the required fabrication temperature. Vitamin E succinate was found to negatively interact with the API. Follow-up studies found ascorbic acid to be the preferred anti-oxidant to prevent API degradation in this system.

Crowley et al. utilized PEO as the primary polymeric excipient for the preparation of testosterone containing transdermal films [64]. The bi-layer film consisted of a drug reservoir layer and an inert backing layer. The film was designed to be administered buccally and provide controlled drug release. The backing layer was designed to have no bioadhesive characteristics and was typically based on a hydrophobic polymer (e.g. ethylcellulose or water-insoluble polymethacrylic acid copolymer). The backing layer could also incorporate PEO to improve adhesion to the drug reservoir layer. The drug reservoir layer must be bioadhesive and combinations of PEO grades were typically used. In one example, the PEO portion consisted of PEO MW = 200,000, PEO MW = 1,000,000 and PEO MW = 4,000,000 in a ratio of 2:1.3:1. The bi-layer film could be produced via multilayer coextrusion or by extruding the reservoir layer and backing layer individually, followed by a lamination step. Alternately, the bi-layer film could be produced by extrusion laminating either the drug reservoir layer or backing layer onto the previously produced opposite layer. The previously produced layer could be made by any means including extrusion, solvent casting, etc. Total film thickness of approximately 1.25 mm was typical (maximum 2.0 mm). The ratio of drug reservoir layer to backing layer was approximately 2.75:1.

A second document discussed improvements to this technology [65]. In this case, HME was used to produce mono- and bi-layer films incorporating alkaline labile APIs and utilized PEO as the primary polymeric component. The key to the technology was neutralization or moderate acidification of the PEO prior to blending with the alkaline labile API. Testosterone was the model drug included in the examples. A typical composition of the drug reservoir layer contained approximately 50% PEO and 15 wt% API. The remainder of the composition included acidifying additives, antioxidants, pigments, etc. A combination of PEO grades was commonly used, up to and including three products. The mono-layer film was produced via conventional film extrusion technology. A key requirement of this reservoir layer is that it had good bioadhesive properties and that it provide controlled API release. The bi-layer composition included a hydrophobic low permeability backing layer. The composition of the backing layer in the lead example was approximately 65 wt% PEO. Approximately 40% of the composition was hydrophobic polymers (EC and a polymethacrylate copolymer). The remainder was pigments and plasticizers. The PEO component was again a combination of PEO grades, utilizing the same grades as noted above. The backing layer is specifically formulated to minimize degradation of the alkaline labile drug at the interface of the layers. The bi-layer composition was produced via multi-layer coextrusion or extrusion of the individual layers followed by a lamination step. Alternately, the reservoir layer could be directly extrusion laminated onto the backing layer or the backing layer could be extrusion laminated onto the previously produced reservoir layer. The final bi-layer film was nominally 1.2 to 1.5 mm in thickness with reservoir layer: backing layer ratios ranging from 2.75:1 to 3:1.

Zeng and Eleuterius report the preparation of API-containing films based on PEO which also incorporate silicon dioxide and at least one additional polymer and one additional additive [66]. Silicon dioxide is included to reduce the adhesion of the extruded film to the extrusion and film calendaring equipment to prevent the film from sticking to itself when wound into a roll and to facilitate the flow characteristics of the material being fed to the extruder. Many different polymers are taught as examples of the second polymer including HPC, acrylic acid copolymers, etc. Additives described include pH modifying or buffering agents, antioxidants, cross linking agents, surfactants, etc. The preferred range of API content in the film is 5–10 wt%. The resulting film can be used for transdermal or transmucosal drug delivery. The sole example utilizes HPC (Klucel EF) and PEO (MW = 100,000) in approximately a 1:1 ratio and 90 wt% of the composition. The silicon dioxide content was 1.5 wt%. Fentanyl citgrate was the API utilized, and it was present at approximately 6 wt%. The remainder of the composition included a third polymer (polyacrylic acid), pH buffer and antioxidant.

Hyroxypropyl cellulose (HPC) and PEO combinations to produce films for the treatment of onychomycosis were described by Mididoddi et al. [67]. The antifungal drug ketoconazole was incorporated at 20 wt% in all films produced. Klucel EF and LF were the HPC grades utilized. A 100,000 molecular weight PEO was used. Films of various compositions were produced, although the composition details are not disclosed. Films ranging in

Table 7.3 Calculated interaction parameters.

Combination	Interaction parameter
HPC–PEO	1.68
HPC–ketoconazole	2.98
PEO–ketoconazole	4.63

thickness 228–330 μm (0.23–0.33 mm) were produced using a Killian model KLB-100 single-screw extruder. Analyses of the extruded films via DSC, SEM and XRD all showed that ketoconazole was present in an amorphous form. Bioadhesion testing of the films showed that PEO addition reduced the hydration rate of the film which ultimately resulted in these films having poorer bioadhesion to the human nail. Hansen solubility parameters for HPC, PEO and ketoconazole were calculated via the Hoftyzer/Van Krefelen and Hoy methods. The averaged results of these calculations were used to determine an interaction parameter for combinations of the components, as shown in Table 7.3.

The interaction parameters are all low, indicating that the various combinations are likely to be miscible. This is consistent with the presence of amorphous ketoconazole in the extruded films.

The combination of hydroxypropyl cellulose (HPC) with HPMC to produce lidocaine containing films was reported by Repka et al. [68]. The purpose of this study was to investigate these compositions to produce oral mucoadhesive films. Films approximately 0.6 mm (27.5 mils) in thickness were produced via a Killion single-screw extruder. The polymeric component was either Klucel GF or an 80/20 combination of Klucel GF with METHOCEL K15. Lidocaine was incorporated at 10 wt% in both films. Bioadhesion testing of the films was performed using a TA.XT2i texture analyzer. These results showed that the HPMC-containing film had improved bioadhesion in terms of peak force (adhesive strength) and work of adhesion than the HPC-only film. The authors speculated that this occurs because HPMC is more hydrophilic than HPC and therefore hydrates faster. The net effect is that the HPMC-containing film interacts more quickly with the mucosa because of the increased polymer chain mobility. Analysis of the extruded films via DSC and XRD indicated that lidocaine was present in an amorphous form. Dissolution studies in simulated saliva showed that the HPMC-containing film had slightly retarded lidocaine release. The authors proposed that this was again caused by the relatively more hydrophilic nature of HPMC which resulted in the formation of a thicker gel layer which increased the overall diffusion pathway.

Extrusion of thin PEO-based films was investigated by Yang et al. [69], who included two examples. The first incorporated PEO combined with HPC as the polymeric excipients. They were used at a 2:1 PEO:HPC ratio and were approximately 50% of the total composition. The second formulation contained PEO as the sole polymeric excipient and it was again present at approximately 50% of the total composition. Dimethicone was the active in both formulations, at 1 wt%. The remainder of the formulation included antiblocks, colorants, flavorants and surfactants. Films were produced using a single-screw extruder. The resulting films were reported to exhibit no stickiness to each other and could be wound onto itself without the need for a backing material.

7.6.5 Unique Dosage Forms

EC, HPMC and PEO have been combined for use in a number of unique dosage forms. Mehuys et al. combined an outer EC cylinder prepared via HME with an HPMC-containing core [70]. The EC cylinder was produced using a laboratory-scale twin-screw co-rotating extruder with an annular die. The core was prepared in a secondary operation by heating the components until molten followed by manual mixing. The resulting material was then manually filled into the EC cylinders. This technology was investigated as a means to eliminate the burst effect occasionally observed in HPMC-based matrix tablets. Multiple APIs were evaluated at 5% content of the core composition. These included theophylline monohydrate (aqueous solubility 8.33 g/L), propanolol HCl (aqueous solubility 50 g/L) and hydrochlorothiazide (0.1 N solubility 0.25 g/L). The authors found that API solubility had a negligible impact on release rate. Erosion-controlled zero-order release rates were observed for all three APIs.

A follow-on study found that the bioavailability of propanolol HCl could be increased by an EC cylinder HPMC/Gelucire core [71]. The EC cylinders were prepared as described above. They had a 5 mm internal diameter, 1 mm wall thickness and were 12 mm long. The core contained propanol HCl, HPMC and Gelucire. The core components were heated until molten and then homogenized. The EC cylinders were then manually filled with this mixture. In vivo studies (dog model) were performed and the results compared an alternative sustained-release formulation (Inderal, Wyeth). A significant increase in propanolol HCl bioavailability was observed, $\sim$ 400% greater $AUC_{0\text{-}24}$ compared to Inderal.

McGinity and Schilling developed technology wherein modified-release multi-particulates are embedded in a polymer matrix produced via HME [72]. They proposed that the advantage of this approach is that the final dosage form combines the benefits of a single dosage form that ultimately releases multiple-unit dosage systems after administration. The multi-particulates can be produced by any conventional means including extrusion spheronization, hot-melt extrusion followed by spheronization, wet granulation, drug layering of non-pareils, etc. The preferred size of these multi-particulates is 300–500 microns. The multi-particulates described utilize an enteric polymer or a water-insoluble modified-release polymer that releases the drug via a diffusion mechanism. These multi-particulates are then blended with the matrix-forming polymer and other additives that will make up the final composition. The multi-particulates can be 5–70 wt% of the composition. Extrusion must be performed under conditions such that the multi-particulates suffer minimal damage. Both mechanical and thermal damage can occur during the extrusion process. Mechanical damage was addressed by proper selection of screw design, rpm, etc. Thermal damage was addressed by selecting matrix-forming polymers that can be processed at low temperatures (ideally less than 100°C). Low molecular weight grades of PEO (MW = 100,000 and 200,000) were cited as preferred polymers to use.

7.6.6 Abuse Resistance

HME of PEO has an application in the preparation of abuse-deterrent dosage forms. Such dosage forms have become of interest to prevent abuse of medications typically sold for pain management (e.g. opiods). These include medications such as oxymorphone, oxycodone, fentnayl, etc. There has been much interest in developing technologies which prevent the misuse of these medications. Crushing tablets for snorting or injecting and chewing are common means of abuse.

For example, Arkenau-Maric et al. [73] utilize high molecular weight grades of PEO (MW > 500,000) to produce a dosage form that would not crush when exposed to forces as high as 500 N. The formulations components are pre-blended and then extruded to produce a strand. This strand was then sliced to produce pre-forms which are compressed into tablets using a tablet press. The resulting tablets exhibited controlled-release behavior, due to the presence of the high molecular weight PEO. Alternately, abuse-resistant tablets of similar compositions can be produced by tabletting at elevated temperature [74]. The primary method discussed is to pre-heat the tablet punch components to temperatures of 80°C or above, and then perform the molding within a chamber maintained at this elevated temperature. Tabletting a simple powder blend or the extruded tablet pre-forms as noted above is described.

Arkenau-Maric et al. also describe a process for production of these abuse-resistant dosage forms [75]. In this application, a planetary gear extruder is used to produce an 8 mm diameter extrudate. The extrudate is then cut into disks and these disks are converted to tablets using a tablet press. Alternately, the extrudate can be cut into multi-particulates for filling into capsules. Further, the inventors claim that the extrudate can be shaped directly into the final dosage form using counter-rotating rolls with mutually opposing tablet shapes.

The combination of acrylic copolymers with HPMC via extrusion to produce an alcohol-resistant dosage form is taught by Roth et al. [76]. The technology utilizes ammonioalkyl methacrylate or methacrylate copolymers (Eudragit products) with HPMC 2208 grades as the carrier polymers/controlled-release polymers. Combinations of HPMC grades were typically used to customize the controlled-release behavior of the composition. Hypromellose 2208 grades of 100 and 100,000 cps viscosity were most commonly cited. The dosage form consisted of a core containing the carrier polymer/controlled-release polymers noted above with the abuse-relevant API (e.g. verapamil or hydrocodone). The core typically contains other excipients and may also contain a second API (e.g. acetaminophen). The carrier/controlled-release polymers were approximately 28 wt% of the composition, with a 1:1 ratio of acrylic copolymer:HPMC. The cores were extruded and the extrudate was directly shaped to produce the desired table shape. The cores were coated via spraying. The coating typically contained additional acetaminophen and provided immediate drug release. The publication contains data indicating the dosage form is resistant to mechanical breakage, has significantly reduced extraction of the abuse-relevant API in 40% ethanol/water and exhibits many different controlled release profiles of the abuse-relevant API.

7.6.7 Controlled Release

Coppens et al. presented HME technology to produce amorphous solid dispersions of poorly soluble drugs that incorporate the combination of a water-insoluble polymer and a second excipient (dissolution promoter) [77]. The technology was based on the observation that HME of a poorly water-soluble API with a water-insoluble excipient tended to produce an amorphous solid dispersion. Unfortunately, drug release from compositions such as these was very slow. The solution presented was to include another excipient in the formulation to

Table 7.4 *Ketoprofen formulations, process temperatures, crystalline state and dissolution results.*

Formulation	Process temp. (°C)	Extrudate ketoprofen morphology		Ketoprofen dissolution @ 1440 min (%)
		XRPD	DSC	
EC STD 10/HPMC E5/Ketoprofen (35/45/20)	150	Amorphous	Amorphous	50.6
EC STD 10/stearyl alcohol/Ketoprofen (35/45/20)	120	No data	Amorphous	2.7
EC STD 10/cetyl alcohol/Ketoprofen (35/45/20)	120	No data	Amorphous	12.3
EC STD 10/mannitol/Ketoprofen (35/45/20)	150	No data	Amorphous	13.9
EC STD 10/mannitol/Ketoprofen (35/45/20)	175	No data	Amorphous	7.9
EC STD 10/lactose/Ketoprofen (35/45/20)	150	No data	Amorphous	12.5
EC STD 10/Ac-Di-Sol/Ketoprofen (35/45/20)	175	Amorphous	Amorphous	68.9
EC STD 10/PEO N-10/Ketoprofen (35/45/20)	150	Amorphous	Amorphous	63.8
EC STD 10/PEO 301/Ketoprofen (35/45/20)	150	Amorphous	Amorphous	66.8
EC STD 10/MCA 15/Ketoprofen (45/35/20)	150	No data	Amorphous	11.4
EC STD 10/MCA 15/Ketoprofen (35/45/20)	150	No data	Amorphous	23.6
EC STD 10/MCA 15/Ketoprofen (25/55/20)	150	Amorphous	Amorphous	37.4
EC STD 4/MCA 15/Ketoprofen (35/45/20)	150	Amorphous	Amorphous	26.7
EC STD 10/Ketoprofen (80/20)	150	Amorphous	Amorphous	0.88

accelerate drug release, which they called a dissolution promoter. Many types of excipients were suggested as dissolution promoters including water-soluble polymers, disintegrants, plasticizers, surfactants, etc. EC was used as the example water-insoluble polymer in these studies, with ketoprofen and nifedipine as the model poorly water-soluble drugs. Table 7.4 documents their results for ketoprofen.

Amorphous ketoprofen was observed in all formulations. Significant improvements in ketoprofen dissolution was observed over the control composition (EC STD 10/ketoprofen 80/20). The control had only 0.88% ketoprofen dissolved at 1440 minutes while Ac-Di-Sol, PEO N-10 and PEO 301 all had ketoprofen release greater than 60%. These results were generated on 300 mg tablets formed directly from the extrudate. One composition was milled to produce a powder. Dissolution of this powder was performed using a capsule. These results are shown in Figure 7.3.

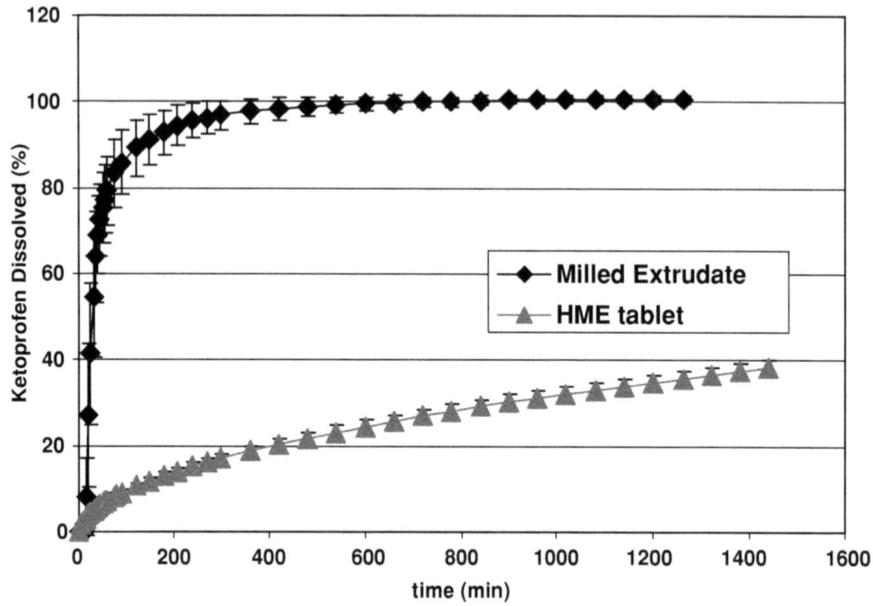

Figure 7.3 *Dissolution profile for EC STD 10/MC A15/ketoprofen (25/55/20): extruded tablet versus milled extrudate.*

The authors noted that the milled extrudate exhibited the expected acceleration in drug release. They concluded that this technology was a viable approach to produce amorphous solid dispersions of poorly water-soluble APIs.

An HME approach to produce matrix tablets with high API loading that exhibit controlled-release behavior was presented by Coppens et al. [78]. The authors noted a number of limitations for conventional approaches to address this need. These included the reduced amount of rate controlling and other excipients that can be included in the formulation. This often has a negative impact on the tabletting process and can lead to tablets with poor mechanical properties. Formulations utilizing EC, HPMC and PEO were all presented. The extrudate was shaped directly into the final tablet upon exiting the extruder. Figure 7.4 shows representative dissolution data for formulations containing 50 wt% acetaminophen.

Table 7.5 contains tablet physical property data for these compositions. Comparative data for direct compression tablets of the same compositions are included.

The HME-produced tablets had excellent hardness. In fact, most the formulation exhibited a hardness that exceeded the capability of the instrument. The HME tablets also exhibited superior friability results, with much less weight loss observed compared to the direct compression tablets.

7.6.8 Solubility Parameters

The use of solubility parameters as a means to predict the crystalline state of an API in an extruded matrix composed of EC, HPMC or PEO was presented by Chan et al. [79]. Hansen solubility parameters were used as the basis for the study. The solubility parameters for the

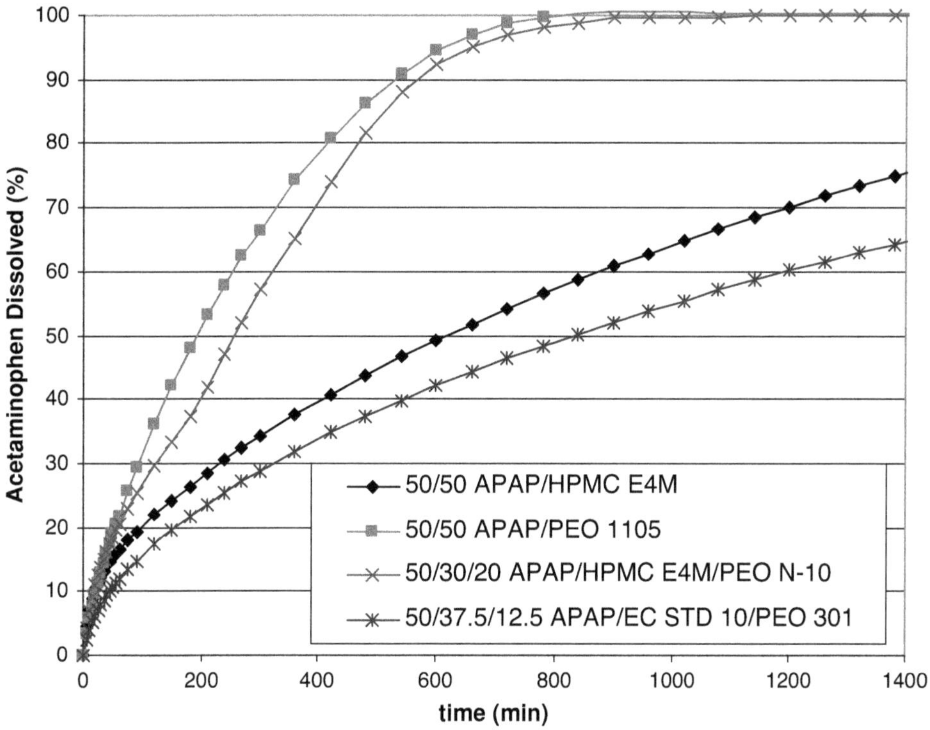

Figure 7.4 *Dissolution results for various formulations containing 50% acetaminophen.*

Table 7.5 *Comparison of physical properties for tablets produced via HME and direct compression.*

Formulation and source	Method	Hardness (SD) (scu)	Friability (% weight loss @ 168 rotations)
50/50 APAP/HPMC E4M	HME	>30 (N/A)[a]	0
	Direct compression	6.7 (1.1)	3.1
50/50 APAP/PEO 1105	HME	N/A (N/A)[b]	N/A[b]
	Direct compression	9.6 (0.3)	0.04
50/30/20 APAP/HPMC E4M/PEO N-10	HME	29.2 (3.9)	0.42
	Direct compression	6.7 (0.4)	0.93
50/37.5/12.5 APAP/EC STD 10/PEO 301	HME	>30 (N/A)[a]	N/A[b]
	Direct compression	11.3 (0.5)	0.49

[a]Tablet hardness exceeded test capabilities
[b]Insufficient number of tablets

Table 7.6 Experimentally derived Hansen solubility parameters for polymeric excipients.

Excipient	δ_d (MPa)$^{1/2}$	δ_p (MPa)$^{1/2}$	δ_h (MPa)$^{1/2}$	δ (g/cc)	Effective solubility parameter (MPa)$^{1/2}$
Ethylcellulose	18.1	13.7	17.6	1	21.0
Hypromellose	16.95	8.55	9.04	1	28.7
Polyethylene oxide	17.3	3	9.4	1.125	22.2

excipients studied were experimentally determined and the solubility parameters for the model drugs (acetaminophen, ketoprofen and nifedipine) were calculated using the group contribution method. The experimentally derived solubility parameters for EC, HPMC and PEO are shown in Table 7.6. This table also includes the effective solubility parameters for these excipients, which is calculated using the formula:

$$\text{Effective Solubility Paramater} = \sqrt{\delta_d^2 + \delta_p^2 + \delta_h^2}$$

where δ denotes the solubility parameter and the subscripts d, p and h represent dispersion, polar and hydrogen bonding, respectively.

A computer model was then constructed which generated predicted solubility curves of the API in the excipient. The predicted results as per the model were compared with those observed in extrudate. The results are summarized in Table 7.7.

In general, a good agreement between predicted and experimental results was observed. The authors concluded that the use of solubility parameters is a reasonable approach to predict API morphology in extruded formulations incorporating EC, HPMC or PEO.

7.7 Milling of EC, HPMC and PEO Extrudate

A cryogenic milling procedure for EC, HPMC and PEO extrudate was described by Coppens et al. [80]. The authors noted that milling of drug containing extrudate by conventional means was challenging, even for excipients with high glass transition temperatures. Many techniques were investigated including ball mill, knife mill, centrifugal mill and hammer mill. These all failed to reliably produce powder of the desired particle size range (150–250 microns). They reported that a hammer mill (Fitzmill) operated under cryogenic conditions was found to be the best option.

The optimum procedure included the following steps:

- mill thoroughly clean and absolutely dry;
- mill configured with impact blades, rotation speed = 9200 rpm (max), 0.5 mm screen on the mill outlet;
- liquid nitrogen fed continuously through the mill;
- milling chamber cooled to –10°C (or colder) prior to beginning the run (surface temperature);

Table 7.7 *Comparison of predicted and actual API morphology for formulations produced via HME.*

API	API loading (%)	Predicted API morphology In EC	Actual API morphology In EC	Predicted API morphology in HPMC	Actual API morphology in HPMC	Predicted API morphology in PEO	Actual API morphology in PEO
Acetaminophen	25	Crystalline	Crystalline	Amorphous	—	Crystalline	Amorphous
	50	Crystalline	—	Amorphous	Amorphous	Crystalline	Crystalline
	75	Crystalline	—	Amorphous	—	Crystalline	Crystalline
Ketoprofen	5	Amorphous	Amorphous	Crystalline	—	Amorphous	Amorphous
	20	Amorphous	Amorphous	Crystalline	—	Amorphous	Amorphous
	50	Crystalline	Amorphous	Crystalline	Crystalline	Crystalline	Amorphous
Nifedipine	5	Amorphous	Amorphous	Crystalline	—	Amorphous	Amorphous
	20	Amorphous	Amorphous	Crystalline	—	Amorphous	Amorphous
	50	Amorphous	Crystalline	Crystalline	Crystalline	Crystalline	Crystalline

- feed material of extrudate pellets (0.5–7 mm in diameter; pre-conditioned in liquid nitrogen for approximately 10 minutes; extrudate fed in a controlled fashion to milling chamber); and
- feed rate controlled such that the milling chamber temperature was maintained at 0°C or colder throughout the run.

By following this procedure, powder with average particle size less than 250 microns for every EC, HPMC or PEO composition evaluated could be produced. Both DSC and XRD analyses showed that milling did not significantly change the morphology of the API in formulations.

References

(1) ETHOCEL Premium Polymers for Pharmaceutical Applications (1998) The Dow Chemical Company, Dow Form No. 198-02002-1096.
(2) Using METHOCEL Cellulose Ethers for Controlled Release of Drugs in Hydrophilic Matrix Systems (2006) The Dow Chemical Company, Dow Form No.198-02075-0702 AMS.
(3) METHOCEL as a Granulation Binding Agent for Immediate Release Tablet and Capsule Products (2003) The Dow Chemical Company, Dow Form No. 198-02017-400XGW.
(4) METHOCEL Cellulose Ethers in Aqueous Systems for Tablet Coating (2002) The Dow Chemical Company, Dow Form No. 198-00755-0702 AMS.
(5) Pharmaceutical Technology Overview (2009) The Dow Chemical Company, Dow Form No. 198-02199 EST.
(6) POLYOX Water-Soluble Resins NF in Pharmaceutical Applications (2004) The Dow Chemical Company, Dow Form No. 326-00013-0802 AMS.
(7) Burrows, L., Lawson, W. and VanWinter, C. (1941) Extrusion Method for Organic Materials. US Patent No. 2,257,104A, Sept. 30 1941.
(8) Chalasani, D. and Johnson, R. (1999) Method for Rapid Stiffening of Extrudates, US Patent No. 5,966,582A, Oct. 12, 1999.
(9) Miller, W., Shaw R. and King, P. (1976) The Facile Extrusion of Ethylene Oxide Resins has been Effected by Employing Water as an Extrusion Aid, US Patent No. 3,941,865, March 2, 1976.
(10) Coppens, K., Hall, M., Larsen, P., Mitchell, S., Nguyen, P., Read, M., Shrestha, U. and Walia, P. (2004) Thermal and rheological evaluation of pharmaceutical excipients for hot melt extrusion. Paper presented at 2004 AAPS Annual Meeting and Exposition, Baltimore, MD.
(11) Alderman, D. and Wolford, T. (1987) Sustained Release Dosage Form Based on Highly Plasticized Cellulose Ether Gels, US Patent No. 4,678,516, July 7, 1987.
(12) Gachter, R. and Muller, H. (eds) (1983) *Plastics Additives Handbook*. Hanser Publishers, Munich, Germany.
(13) Sears, J. and Darby, J. (1982) *The Technology of Plasticizers*. John Wiley & Sons, New York, NY.
(14) Repka, M. and McGinity, J. (2000) Influcence of vitamin E TPGS on the properties of hydrophilic films produced by hot-melt extrusion. *International Journal of Pharmaceutics*, **202**(1–2), 63–70.

(15) Crowley, M., Zhang, F., Koleng, J. and McGinity, J. (2002) Stability of polyethylene oxide in matrix tablets prepared by hot melt extrusion. *Biomaterials*, **23**(21), 4241–4248.
(16) Coppens, K.A., Hall, M.J., Mitchell, S.A. and Read, M.D. (2006) Hypromellose, ethylcellulose, and polyethylene oxide use in hot melt extrusion. *Pharmaceutical Technology*, January 1, 2006.
(17) Zhang, F. and McGinity, J. (1999) Properties of sustained release tablets prepared by hot melt extrusion. *Pharmaceutical Development & Technology*, **42**(2), 241–250.
(18) DeBrabander, C., van den Mooter, G., Vervaet, C. and Remon, J. (2002) Characterization of ibuprofen as a nontraditional plasticizer of ethyl cellulose. *Journal of Pharmaceutical Science*, **91**(7), 1678–1685.
(19) Rambali, B., Verreck, G., Baert, L. and Massart, D. (2003) Itraconazole formulation studies of the melt extrusion process with mixture design. *Drug Development & Industrial Pharmacy*, **29**(6), 641–652.
(20) Coppens, K., Hall, M., He, V., Larsen, P., Koblinski, B., Read, M. and Shrestha, U. (2006) Excipient blends in hot melt extrusion. Poster Presented at the 2006 Annual Meeting and Exposition of the American Association of Pharmaceutical Scientists, San Antonio, TX, USA, Oct 29–Nov 2.
(21) Coppens, K., Hall, M., He, V., Koblinski, B., Larsen, P., Read, M. and Shrestha, U. (2006) Hypromellose and polyethylene oxide blends in hot melt extrusion. Poster Presented at the 33rd Annual Meeting of the Controlled Release Society, Viena, Austria, July 22–26.
(22) Verreck, G., Decorte, A., Li, H., Tomasko, D., Arien, A., Peeters, J., Rombaut, P., Van den Mooter, G. and Brewster, M. (2006) The effect of pressurized carbon dioxide as a plasticizer and foaming agent on the hot melt extrusion process and extrudate properties of pharmaceutical polymers. *Journal of Supercritical Fluids*, **38**(3), 383–391.
(23) Verreck, G., Decorte, A., Heymans, K., Adriensen, J., Liu, D., Tomasko, D., Arien, A., Peeters, J., Van den Mooter, G. and Brewster, M.E. (2006) Hot stage extrusion of *p*-amino salicylic acid with EC using CO_2 as a temporary plasticizer. *International Journal of Pharmaceutics*, **327**(1–2), 45–50.
(24) Lyons, J.G., Hallinan, M., Kennedy, J.E., Devine, D.M., Geeve, L.M., Blackie, P. and Higginbotham, C.L. (2007) Preparation of monolithic matrices for oral drug delivery using a supercritical fluid assisted hot melt extrusion process. *International Journal of Pharmaceutics*, **329**(1–2), 62–71.
(25) Crowley, M., Schroeder, B., Fredersdorf, A., Obara, S., Talarico, M., Kucera, S. and McGinity, J. (2004) Physiochemical properties and mechanism of drug release from ethylcellulose matrix tablets prepared by direct compression and hot-melt extrusion. *International Journal of Pharmaceutics*, **269**(2), 509–522.
(26) DeBrabander, C., Vervaet, C. and Remon, J.P. (2003) Development and evaluation of sustained release mini-matrices prepared via hot melt extrusion. *Journal of Controlled Release*, **89**(2), 235–247.
(27) DeBrabander, C., Vervaet, C., Van Bortel, L. and Remon, J. (2004) Bioavailability of ibuprofen from hot melt extruded mini-matrices. *International Journal of Pharmaceutics*, **271**(1–2), 77–84.
(28) Verhoeven, E., Vervaet, C. and Remon, J.P. (2006) Xanthan gum to tailor drug release of sustained-release ethylcellulose mini-matrices prepared via hot melt extrusion:

in vitro and in vivo evaluation. *European Journal of Pharmaceutics and Biopharmaceutics*, **63**(3), 320–330.

(29) Verhoeven, E., De Beer, T.R.M., Ven Den Mooter, G., Remon, J.P. and Vervaet, C. (2008) Influence of formulation and process parameters on the release characteristics of ethylcellulose sustained release mini matrices produced by hot melt extrusion. *European Journal of Pharmaceutics and Biopharmaceutics*, **69**(1), 312–319.

(30) Verhoeven, E., DeBeer, T.R.M., Schacht, E., Van Den Mooter, G., Remon, J.P. and Vervaet, C. (2009) Influence of polyethylene glycol/polyethylene oxide on the release characteristics of sustained release ethylcellulose mini-matices produced by hot melt extrusion: in vitro and in vivo evaluations. *European Journal of Phamaceutics and Biopharmaceutics*, **72**(2), 463–470.

(31) Hall, M.J., Coppens, K.A., Larsen, P.S., Mitchell, S.A., Read, M.D. and Shrestha, U. (2006) Biologically active composition comprising ethylcellulose. International Patent Application WO 2007/084212 A2, Filed November 17, 2006.

(32) Verhoeven, E., Siepmann, F., De Beer, T.R.M., Van Loo, D., Van den Mooter, G., Remon, J.P., Siepmann, J. and Vervaet, C. (2009) Modeling drug release from hot melt extruded mini-matrices with constant and non-constant diffusivities. *European Journal of Phamaceutics and Biopharmaceutics*, **73**(2), 292–301.

(33) Fukuda, M., Peppas, N.A. and McGinity, J.W. (2006) Properties of sustained release hot melt extruded tablets containing chitosan and xanthan gum. *International Journal of Pharmaceutics*, **310**(1–2), 90–100.

(34) Bruce, C., Fegely, K.A., Rajabi-Siahboomi, A.R. and McGinity, J.W. (2007) Crystal growth formation in melt extrudates. *International Journal of Pharmaceutics*, **341**(1–2), 162–172.

(35) Young, C.R., Dietzsch, C., Cerea, M., Farrell, T., Fegely, K.A., Rajabi-Siahboomi, A. and McGinity, J.W. (2005) Physiochemical characterization and mechanisms of release of theophylline from melt extruded dosage forms based on a methacrylic acid copolymer. *International Journal of Pharmaceutics*, **301**(1–2), 112–120.

(36) Coppens, K., Hall, M., He, H., Koblinski, B., Larsen, P., Read, M. and Shrestha, U. (2006) Hypromellose and polyethylene oxide blends in hot melt extrusion. Poster presented at the 33rd Annual Meeting of the Controlled Release Society, Vienna, Austria July 22–26.

(37) Lyons, J.G., Holehonnur, H., Devine, D.M., Kennedy, J.E., Geever, L.M., Blackie, P. and Higginbotham, C.L. (2007) The incorporation of an organically modified layered silicate in monolithic polymeric matrices produced using hot melt extrusion. *Materials Chemistry and Physics*, **103**(2–3), 419–426.

(38) Hall, M.J., Coppens, K.A., Larsen, P.S., Mitchell, S.A., Read, M.D. and Shrestha, U. (2006) Biologically active composition comprising ethylcellulose. WO 2007/084221 A2, Filed November 17.

(39) Chiou, W.L. and Riegelman, S. (1971) Pharmaceutical applications of solid dispersions. *Journal of Pharmaceutical Science*, **60**, 1281–1302.

(40) Schachter, D., Tirol, G. and Schmitt, R. (2003) Solid solution of a poorly soluble model drug in a phase separated polymer matrix: melt prepared dispersions based on POLYOX WSR. Presented at the 30th Annual Meeting of the Controlled Release Society, Glasgow.

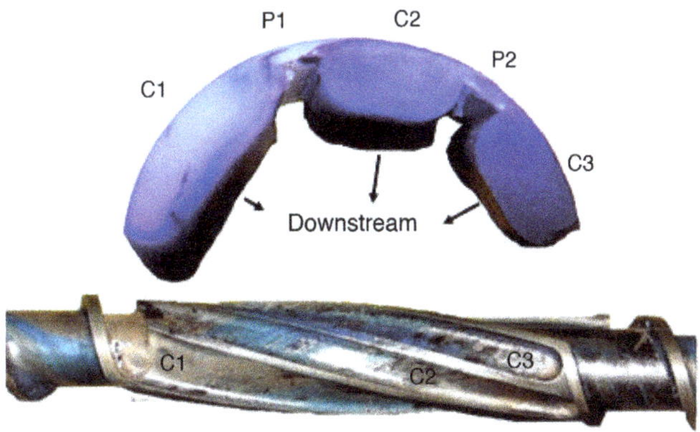

Figure 1.20 A thick frozen section that was cooled and then removed from the screw.

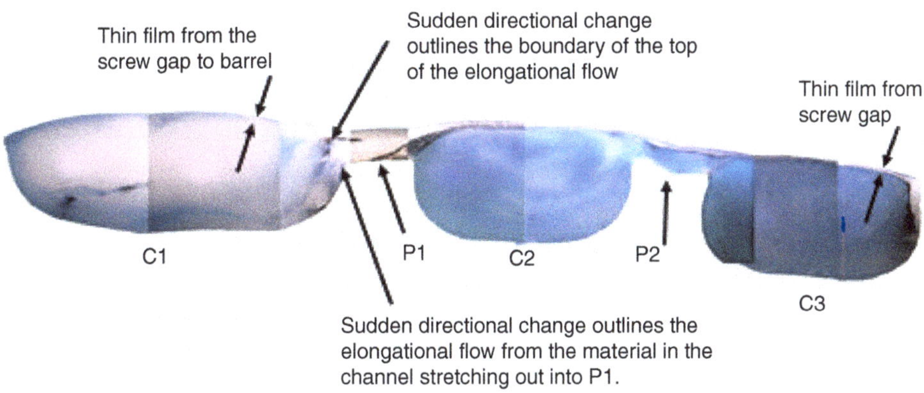

Figure 1.21 A composite of nine pictures shows the elongational mixing flows.

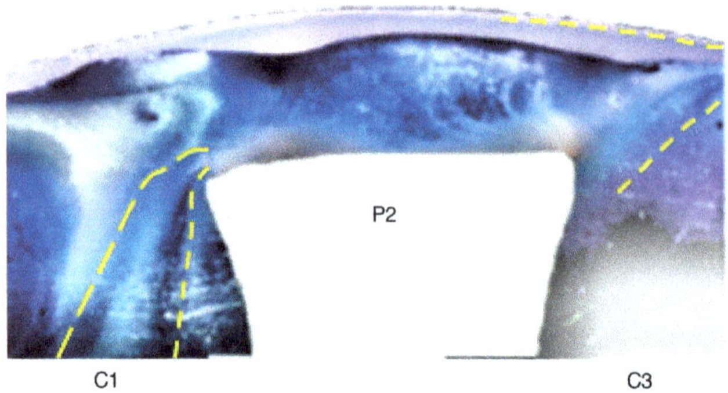

Figure 1.22 *Dotted lines show the converging (extensional) mixing flows at P2.*

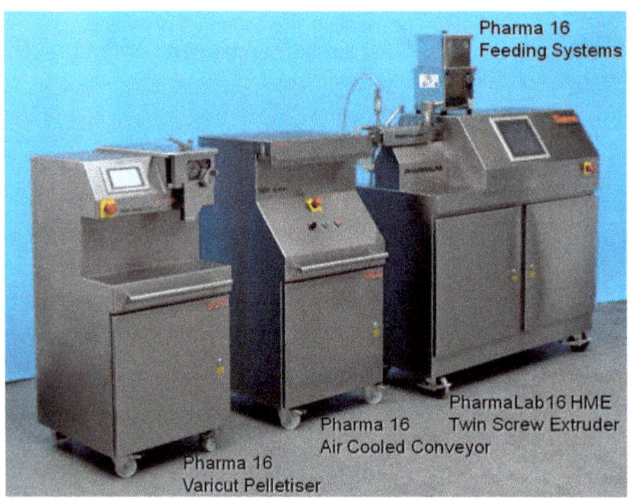

Figure 2.10 *16mm TSE for pharmaceutical usage complete with strand pelletizing unit.*

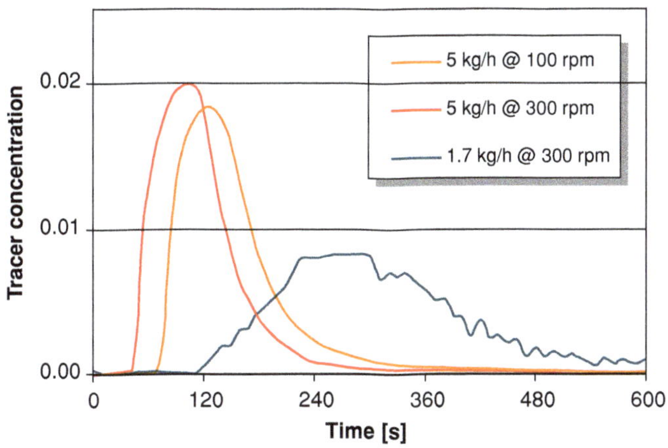

Figure 2.17 *RTD depending on throughput and rpm.*

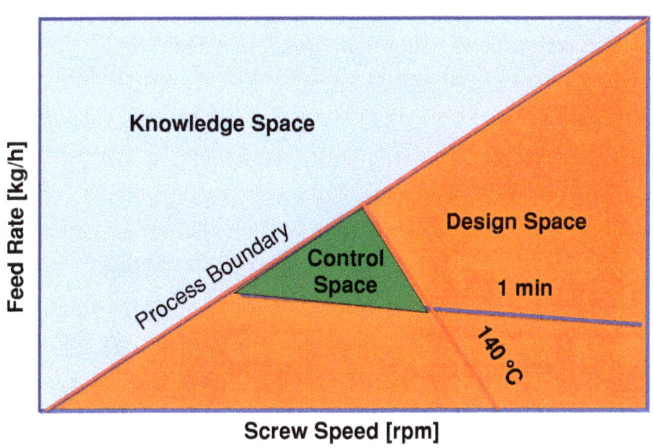

Figure 2.21 *Determining the control space.*

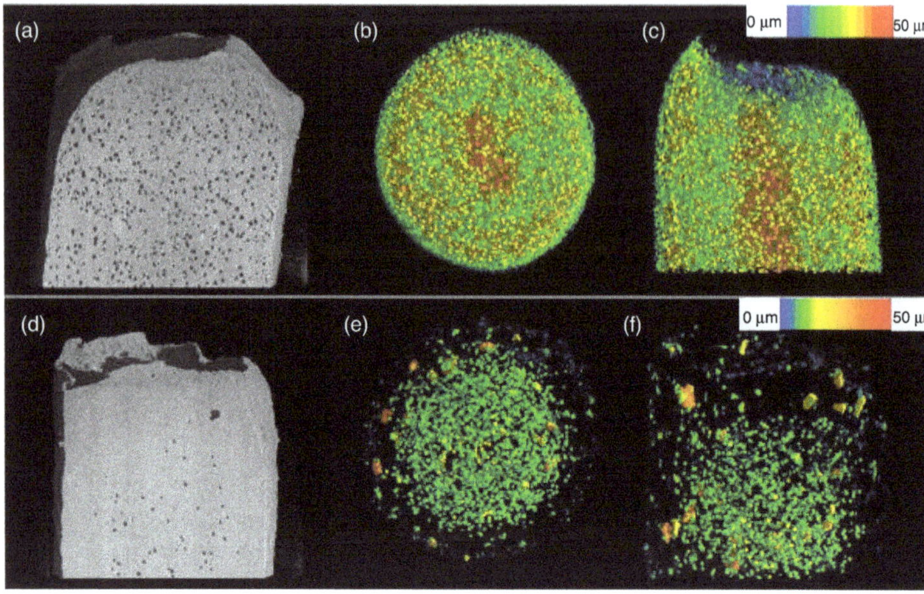

Figure 3.3 *X-ray tomography renderings of EVA40 matrices with 50% of metoprolol tartrate (MPT). (a, c) Axial and (b) radial cross-sections before dissolution. (d) Axial cross-section after 24 h dissolution and (e) radial and (f) axial cross-sections after 72 h dissolution. Black spots (a, d) indicate pores. The color scale used in (b), (c), (e) and (f) represents the pore size (maximum opening) where blue represents small pores and red represents larger pores. S. Almeida et al. 2011, reproduced with the permission of Elsevier.*

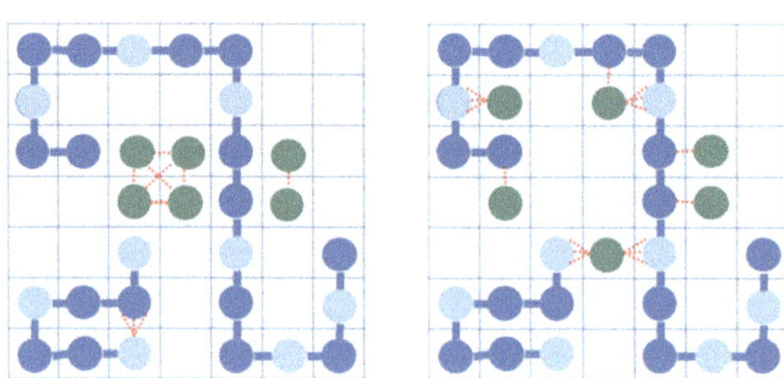

Figure 4.2 *Lattice model of a solid glassy suspension (left) and a solid glassy solution (right).*

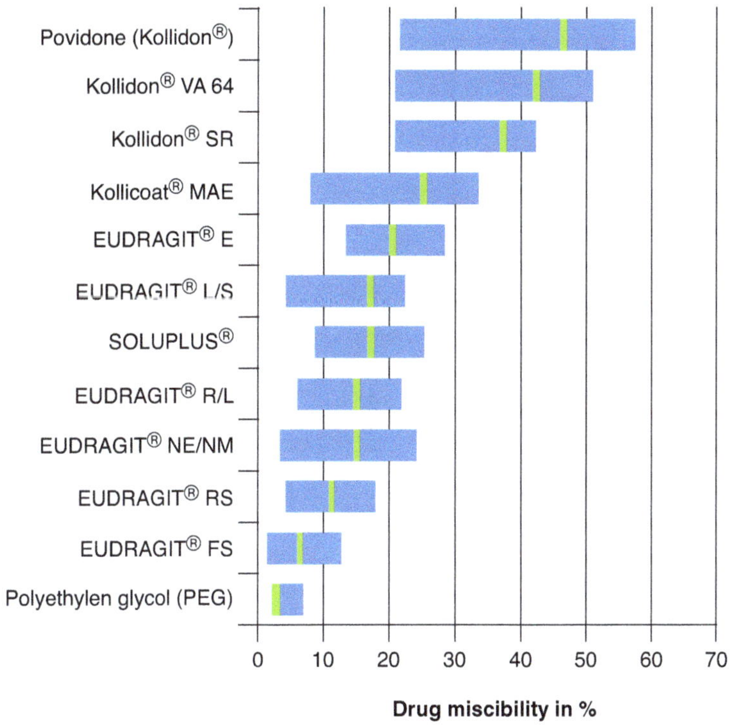

Figure 4.6 *Predicted miscibility of itraconazole in different polymers.*

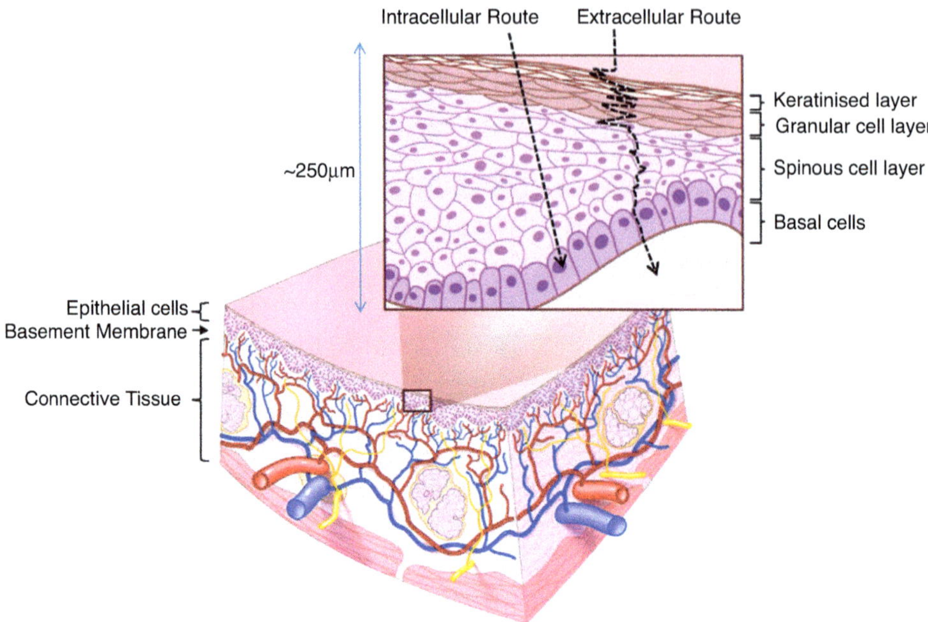

Figure 8.1 *Cartoon of the structure of the oral mucosa. Insert also shows different routes by which drugs can cross the oral mucosa. V. Hearnden et al. 2011, reproduced with the permission of Elsevier.*

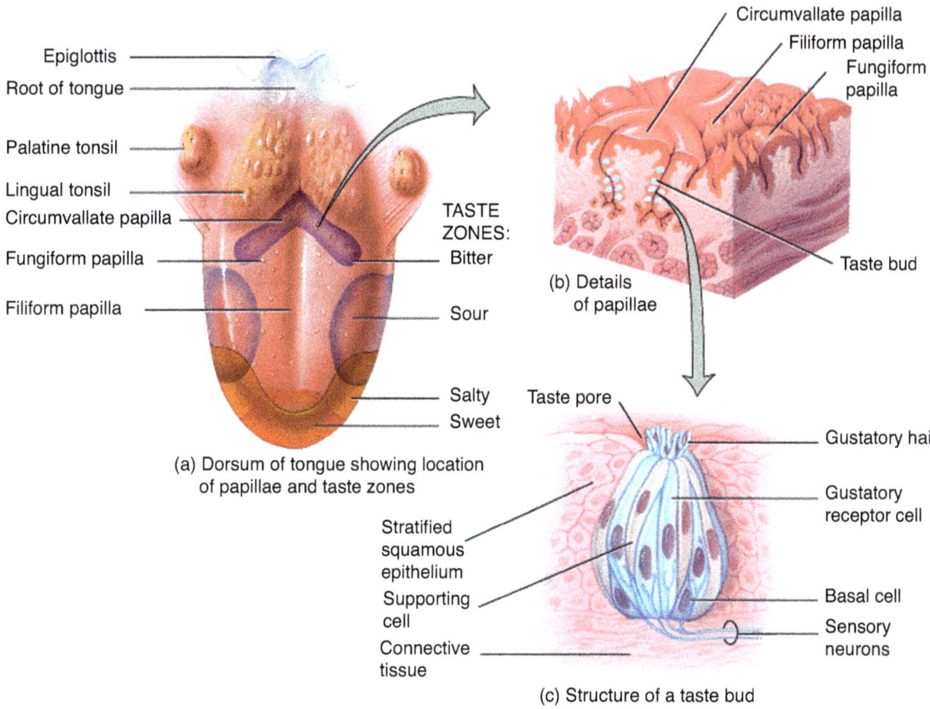

Figure 9.1 *Taste buds and the peripheral innervation of the tongue. (a) Distribution of taste papillae on the dorsal surface of the tongue. Different responses to sweet, salty, sour and bitter tastants recorded in the three cranial nerves that innervate the tongue and epiglottis are indicated at left. The size of the circles representing sucrose, NaCl, HCl, quinine and water corresponds to the relative response of the papillae to these stimuli. (b) Circumvallate papilla showing location of individual taste buds. (c) Diagram of a taste bud, showing various types of taste cells and the associated gustatory nerves. The apical surface of the receptor cells have microvilli that are oriented toward the taste pore.*

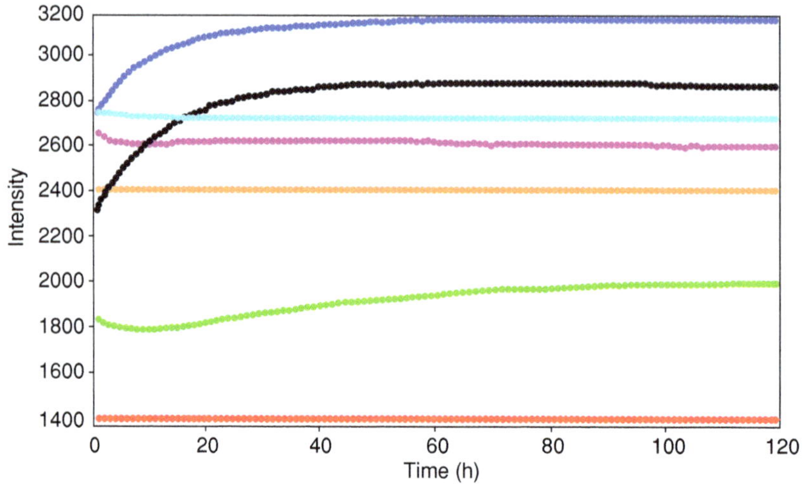

Figure 9.4 *Sensor signals obtained with Astree e-tongue.*

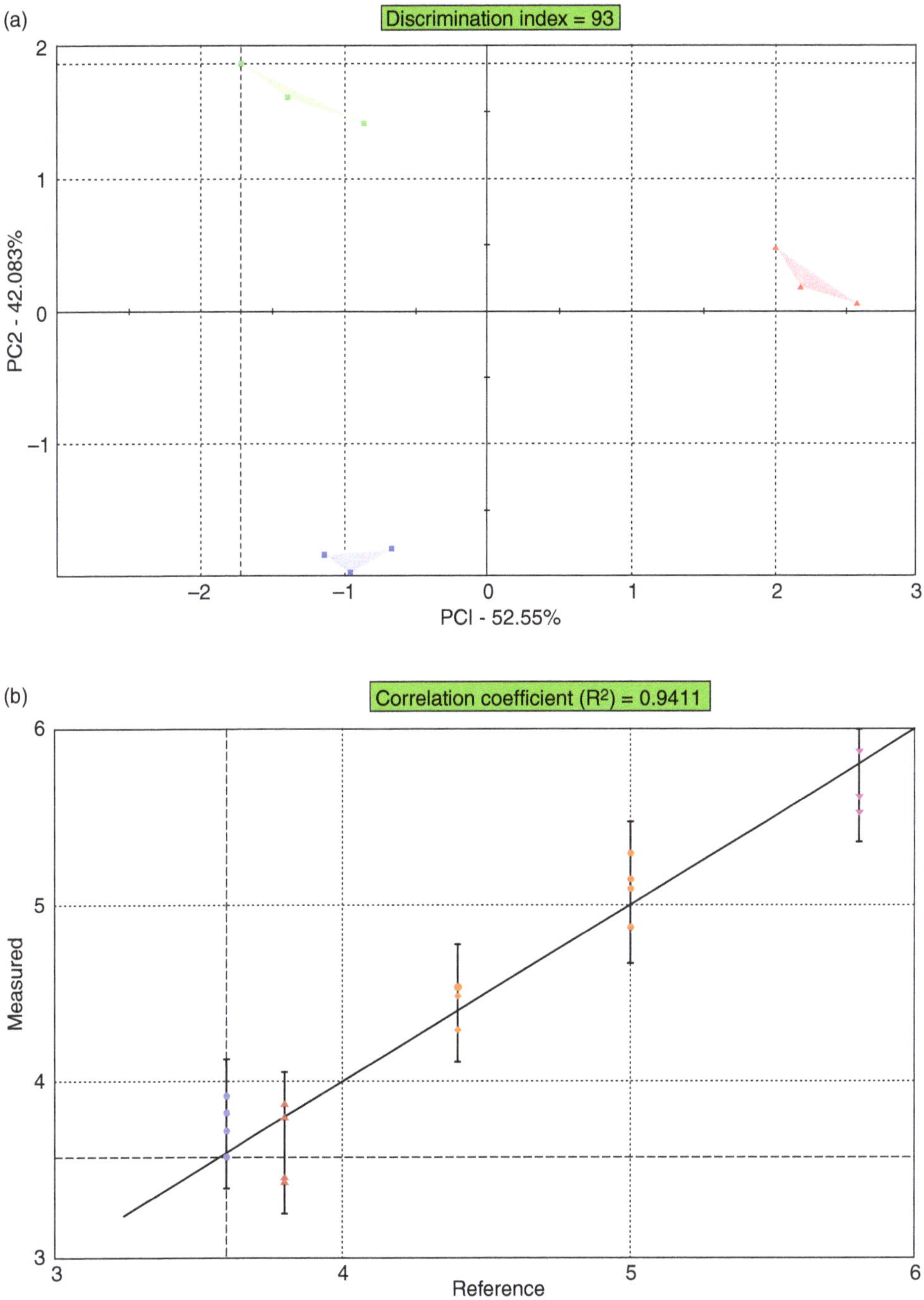

Figure 9.5 *(a) Taste map of three products of different qualities obtained through Principal Components Analysis with ASTREE e-tongue; (b) correlation model between sensory evaluation and e-tongue measurements for the determination of a sensory attribute score (partial least square model); and (c) statistical quality control model showing the area of acceptable quality (green band) and out of specification grade (white area).*

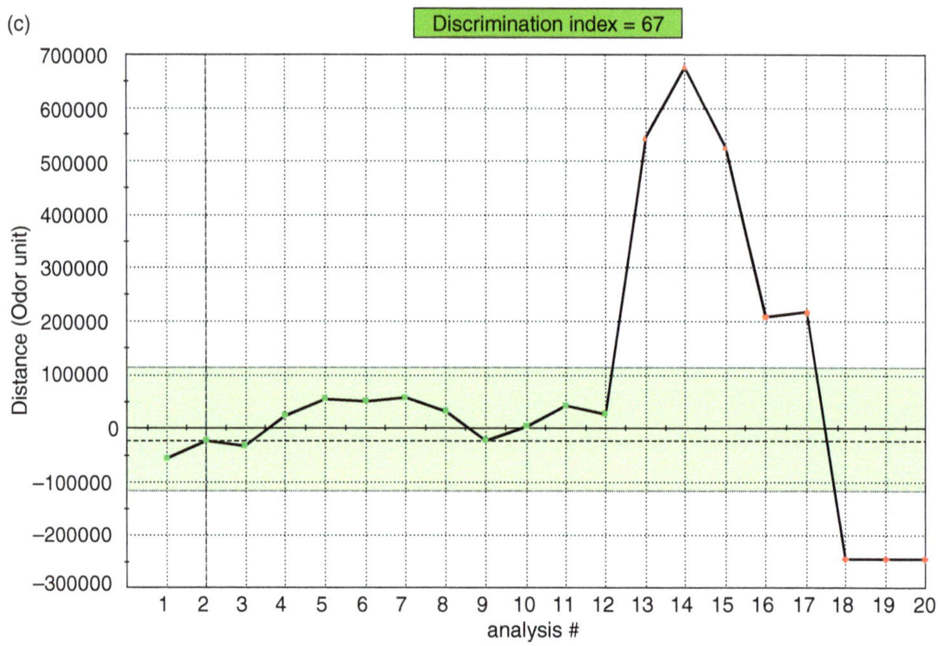

Figure 9.5 *(Continued)*

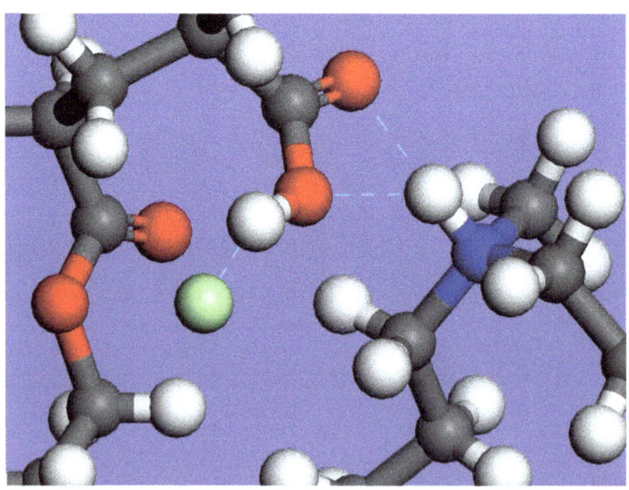

Figure 9.7 *Computerized simulation of the hydrogen bridge bonding between Verapamil HCl and Eudragit® L100–55. Reproduced with permission form Evonik Industries AG.*

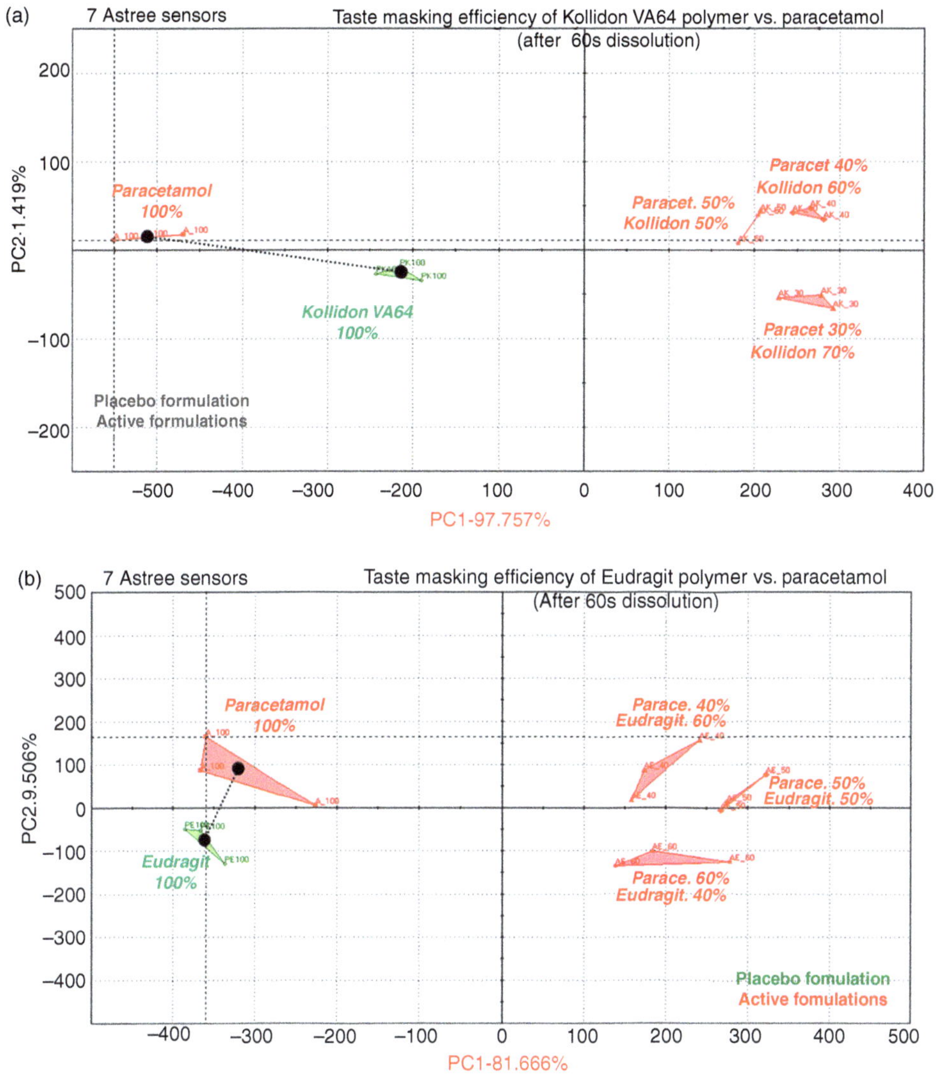

Figure 9.8 *Electronic tongue 'taste map'. Global signal comparison (PCA analysis of the electrode responses) of pure PMOL and extruded formulations to (a) VA64 polymer and (b) EPO polymer after dissolution for 60 s. M. Maniruzzaman et al. 2012, reproduced with permission of Elsevier.*

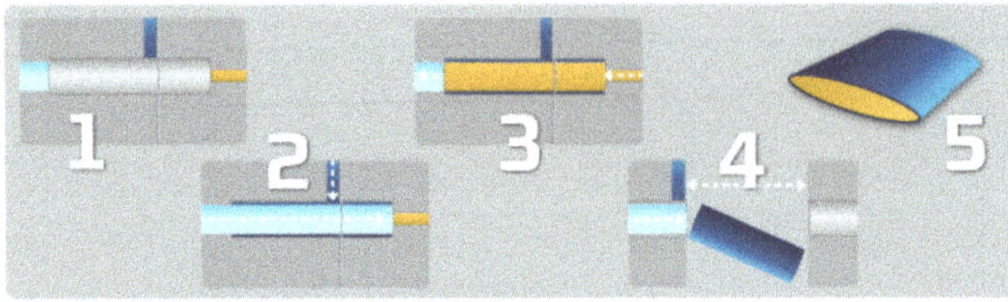

Figure 11.8 *Manufacturing process of Egalet®: (1) cavity is empty; (2) piston moves forward, coat material is injected; (3) coat material hardens, piston recedes while matrix material (containing drug) is injected into the cavity; (4) matrix hardens, piston moves forward ejecting the finished Egalet® tablet; and (5) finished Egalet® tablet.*

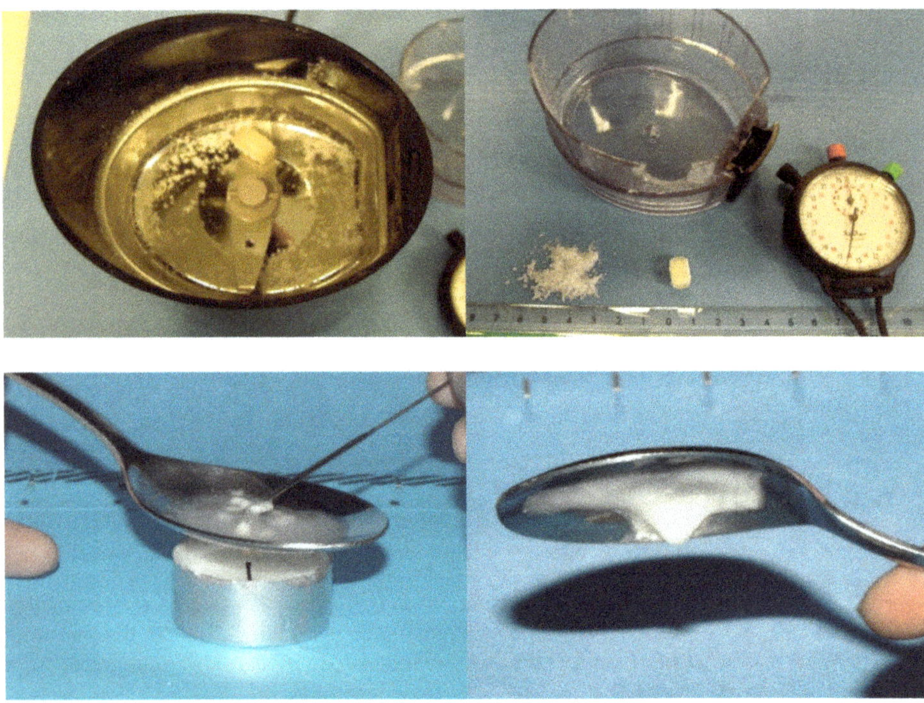

Figure 11.15 *Tests of tamperablity. Top views: test of particle size reduction experiment in a coffee mill until either tablet or lid brakes. In this case the lid broke first. Bottom view: test of injectability. The tablet is dissolved in 2 ml water and forms a gel that is impossible to inject.*

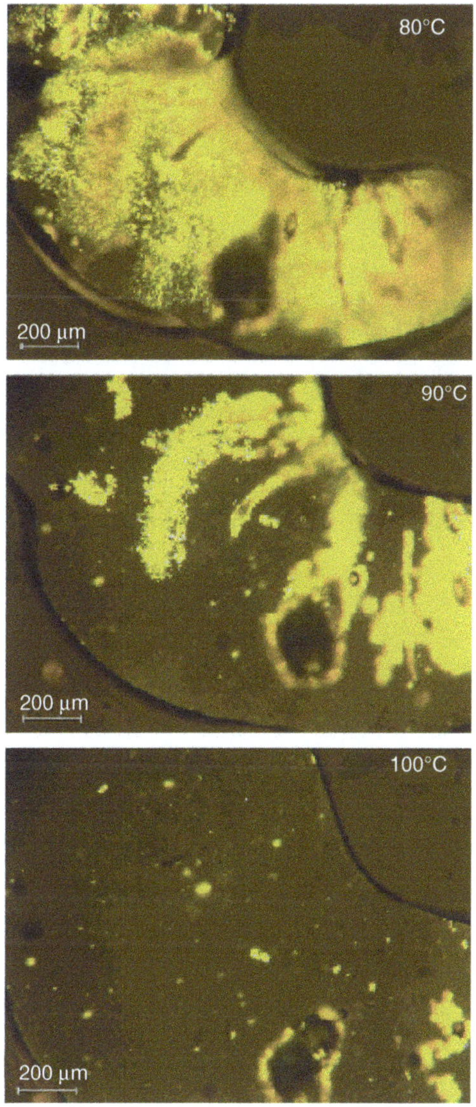

Figure 12.14 *Polarized microscopic images of an APAP-PEO mixture on the hot stage at 80, 90 and 100°C.*

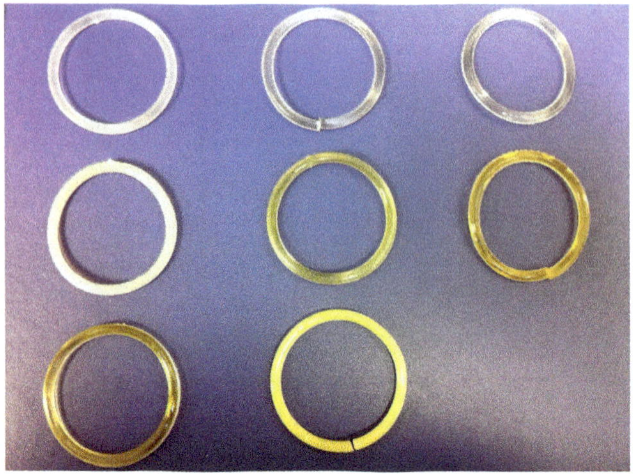

Figure 14.8 *Appearance of IVRs made from various polymers.*

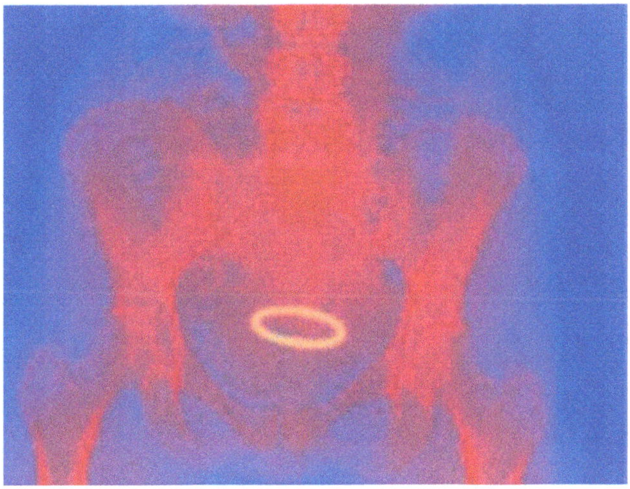

Figure 14.11 *X-ray showing an IVR in position. Reproduced with kind permission of Karl Malcolm, Queens University Belfast.*

Figure 16.8 *1/2-20 UNF Dynisco diffuse reflection probe installed on the last port of a HME.*

(41) Schachter, D., Xiong, J. and Tirol, G. (2003) Solid state nuclear magnetic resonance characterization of melt prepared dispersions based on POLYOX WSR. Presented at the 30th Annual Meeting of the Controlled Release Society, Glasgow.
(42) Coppens, K., Hall, M., He, V., Larsen, P., Koblinski, B., Read, R. and Shresha, U. (2006) Excipient blends in hot melt extrusion. Poster presented at the 2006 Annual Meeting and Exposition for the American Association of Pharmaceutical Scientists, San Antonio, TX, USA, Oct 29–Nov 2.
(43) Li, L., AbuBaker, O. and Shao, Z.J. (2006) Characterization of poly(ethylene oxide) as a drug carrier in hot melt extrusion. *Drug Development and Industrial Pharmacy*, **95**(11), 2473–2485.
(44) Miller, D.A., McConville, J.T., Yang, W., Willams III, R.O. and McGinity, J.W. (2006) Hot melt extrusion for enhanced delivery of drug particles. *Journal of Pharmaceutical Sciences*, **96**(2), 361–376.
(45) Suzuki, H. and Sunada, H. (1998) Influence of water soluble polymers on the dissolution of nifedipine solid dispersions with combined carriers. *Chemical & Pharmaceutical Bulletin*, **46**, 482–487.
(46) Yamashita, K., Nakate, T., Okimoto, K., Ohike, A., Tokunaga, Y., Ibuki, R., Higaki, R. and Kimura, T. (2003) Establishment of new preparation method for solid dispersion formulation of tacrolimus. *International Journal of Pharmaceutics*, **267**, 79–91.
(47) Verreck, G., Six, K., Van den Mooter, G., Baert, L., Peeters, J. and Brewster, M. (2003) Characterization of solid dispersions of itraconazole and hydroxypropylmethylcellulose prepared by melt extrusion – Part I. *International Journal of Pharmaceutics*, **251**(1–2) 165–174.
(48) Six, K., Berghans, H., Leuner, C., Dressman, J., Van Werde, K., Mullens, J., Benoist, L., Thimon, M., Meublat, L., Verreck, G., Peeters, J., Brewster, M. and Van den Mooter, G. (2003) Characterization of solid dispersions of itraconazole and hydroxypropylmethylcellulose prepared by melt extrusion – Part II. *Pharmaceutical Research*, **20**(7), 1047–1054.
(49) Miller, D.A., DiNunzio, J.C., Yang, W., McGinity, J.W. and Williams, R.O. III (2008) Enhanced in vivo absorption of itraconazole via stabilization of supersaturation following acidic-to-neutral ph transition. *Drug Development and Industrial Pharmacy*, **34**(8), 890–902.
(50) Dinunzio, J.C., Brough, C., Williams, D.A., Williams, R.O. III and McGinity, J.W. (2010) Fusion processing of itraconazole solid dispersions by kinetisol dispersing: a comparative study to hot melt extrusion. *Journal of Pharmaceutical Sciences*, **99**(3), 1239–1253.
(51) DiNunzio, J., Brough, C., Hughley, J.R., Miller, D.A., Williams, R.O. III and McGinity, J.W. (2010) Fusion production of solid dispersions containing a heat sensitive active ingredient by hot melt extrusion and kinetisol dispersing. *European Journal of Pharmaceutics and Biopharmaceutics*, **74**(2), 340–351.
(52) Zheng, X., Yang, R., Tang, X. and Zheng, L. (2007) Part I: Characterization of solid dispersions of nimodipine prepared by hot melt extrusion. *Drug Development and Industrial Pharmacy*, **33**(7), 791–802.
(53) Zheng, X., Yang, R., Zhang, Y., Wang, Z., Tang, X. and Zheng, L. (2007) Part II. Bioavailability in beagle dogs on nimodipine solid dispersions prepared by hot melt extrusion. *Drug Development and Industrial Pharmacy*, **33**(7), 783–789.

(54) Ghebremeskel, A.N., Vemavarapu, C. and Lodaya, M. (2007) Use of surfactants as plasticizers in preparing solid dispersions of poorly soluble API: selection of polymer-surfactant combinations using solubility parameters and testing the processability. *International Journal of Pharmaceutics*, **328**(2), 119–129.
(55) Ghebremeskel, A.N., Vemavarapu, C. and Lodaya, M. (2006) Use of surfactants as plasticizers in preparing solid dispersions of poorly soluble API: stability testing of selected solid dispersions. *Pharmaceutical Research*, **23**(8), 1928–1936.
(56) Yokoi, Y. and Yonemochi, E. and Terada, K. (2005) Effects of sugar ester and hydroxypropylmethylcellulose on the physicochemical stability of amorphous cefditoren pivoxil in aqueous suspension. *International Journal of Pharmaceutics*, **290**, 91–99.
(57) Okimoto, K., Miyake, M., Ibuki, R., Yasumura, M., Ohnishi, N. and Nakai, T. (1997) Dissolution mechanism and rate of solid dispersion particles of nilvadipine with hydroxypropylmethylcellulose. *International Journal of Pharmaceutics*, **159**, 85–93.
(58) Gao, P., Guyton, M.E., Huand, T., Bauer, J.M., Stefanski, K.J. and Lu, Q. (2004) Enhanced oral bioavailability of a poorly water soluble drug PNU-91325 by supersaturatable formulations. *Drug Development & Industrial Pharmaceutics*, **30**, 221–229.
(59) Repka, M., Prodduturi, S. and Stodghill, S. (2003) Production and characterization of hot melt extruded films containing clotrimazole. *Drug Development & Industrial Pharmaceutics*, **29**(7), 757–765.
(60) Prodduturi, S., Manek, R.V., Kolling, W.M., Stodghill, S.P. and Repka, M.A. (2005) Solid state stability and characterization of hot melt extruded polyethylene oxide films. *Journal of Pharmaceutical Sciences*, **94**(10), 2232–2245.
(61) Prodduturi, S., Urman, K.L., Otaigbe, J.U. and Repka, M.A. (2007) Stabilization of hot melt extrusion formulations containing solid solutions using polymer blends. *AAPS PharmSciTech*, **8**(2), Article 50.
(62) Munjal, M., Stodghill, S.P., Elsohly, M.A. and Repka, M.A. (2006) Polymeric systems for amorphous Δ^9-tetrahydrocannabinol produced by a hot melt method. Part I: chemical and thermal stability during processing. *Journal of Pharmaceutical Sciences*, **95**(8), 1841–1853.
(63) Munjal, M., Elsohly, M.A. and Repka, M.A. (2006) Polymeric systems for amorphous Δ^9-tetrahydrocannabinol produced by a hot melt method. Part II: Effect of oxidation mechanisms and chemical interactions on stability. *Journal of Pharmaceutical Sciences*, **95**(11), 2473–2485.
(64) Crowley, M.M., Keen, J.M., Koleng, J.J., Mack, R., Sebree, T., Smith, T.M. and Zhang, F. (2007) HME transdermal compositions containing testosterone. WO 2007/112287 A2, Filed March 22.
(65) Crowley, M.M., Keen, J.M., Koleng, J.J. and Zhang, F. (2007) Stabilized compositions containing alkaline labile drugs. WO2007/112286 A2, filed March 22.
(66) Zeng, F. and Eleuterius, L. (2009) Hot melt extruded film containing silicon dioxide. WO 2009/151574 A1, Filed June 8.
(67) Mididoddi, P.K. and Repka, M.A. (2007) Characterization of hot melt extruded drug delivery systems for onychomycosis. *European Journal of Pharmaceutics and Biopharmaceutics*, **66**(1), 95–105.
(68) Repka, M.A., Gutta, K., Prodduturi, S., Junjal, M. and Stodghill, S.P. (2005) Characterization of cellulosic hot melt extruded films containing lidocaine. *European Journal of Pharmaceutics and Biopharmaceutics*, **59**(1), 189–196.

(69) Yang, R.K., Fuisz, R.C., Myers, G.L. and Fuisz, J.M. (2004) Polyethylene oxide based films and drug delivery systems made therefrom. US 2005/0037055A1, Filed May 28.
(70) Mehuys, E., Vervaet, C. and Remon, J. (2004) Hot melt extruded ethylcellulose cylinders containing a HPMC-gelucire core for sustained drug delivery. *Journal of Controlled Release*, **94**(2–3), 273–280.
(71) Mehuys, E., Vervaet, C., Gielen, I., Van Bree, H. and Remon, J. (2004) In vitro and in vivo evaluation of a matrix-in-cylinder system for sustained drug delivery. *Journal of Controlled Release*, **96**(2), 261–271.
(72) McGinity, J.W. and Schilling, S. (2009) Hot melt extrusion of modified release multiparticulates. WO 2010/022193 A2, Filed August 20.
(73) Arkenau-Maric, E., Bartholomaus, J. and Kugelmann, H. (2008) Abuse proof dosage form. US Patent Application US2008/0311049 A1, Filed June 17.
(74) Ashworth, J., Arkenau-Maric, E. and Bartholomaus, J. (2008) Crush resistant delayed release dosage form. US Patent Application US 2008/0311187 A1, Filed June 17.
(75) Arkenau-Maric, E. and Bartholomaus, J. (2008) Process for the production of an abuse proofed solid dosage form. US Patent Application US 2008/0312264 A1, Filed June 17.
(76) Roth, W., Burst, A. and Zietsch, M. (2009) Abuse resistant melt extruded formulation having reduced alcohol interaction. US2009/0317355 A1, Filed January 26.
(77) Coppens, K.A., Hall, M.J., Koblinski, B.D., Larsen, P.S., Read, M.D. and Shrestha, U. (2009) Controlled release of amorphous solid dispersions utilizing hot melt extrusion. Poster Presented at the 2009 Annual Meeting and Exposition of the American Association of Pharmaceutical Scientists, Los Angeles, CA, November 8–12.
(78) Coppens, K.A., Hall, M.J., He, V., Koblinski, B.D., Larsen, P.S., Mandare, P., Read, M.D. and Shrestha, U. (2007) Evaluation of formulations produced via hot melt extrusion that contain high API loading and exhibit controlled release. Poster Presented at the 2007 Annual Meeting and Exposition of the American Association of Pharmaceutical Scientists, San Diego, CA, November 11–15.
(79) Chan, A., Coppens, K.A., Hall, M.J., He, V., Jog, P., Larsen, P.A., Koblinski, B.D., Read, M.D., Rothe, D., Somasi, S. and Shrestha, U. (2006) Solubility parameters as a tool to predict API morphology in hot melt extruded (HME) formulations containing ethylcellulose, hypromellose, and polyethylene oxide. Poster Presented at the 2006 Annual Meeting and Exposition of the American Association of Pharmaceutical Scientists, San Antonio, TX, Oct 29–Nov 2.
(80) Coppens, K.A., Hall, M.J., He, V., Koblinski, B.D., Larsen, P.A., Read, M.D. and Shrestha, U. (2010) Cryogenic milling of extrudate containing ethylcellulose, hypromellose, and polyethylene oxide. Poster Presented at the 2010 Annual Meeting and Exposition of the American Association of Pharmaceutical Scientists, New Orleans, LA, Nov 14–18.

8

Bioadhesion Properties of Polymeric Films Produced by Hot-melt Extrusion

Joshua Boateng and Dennis Douroumis
University of Greenwich, School of Science, Medway Campus, Chatham Maritime, Kent

8.1 Introduction

Films are thin sheets which are generally developed from polymers and can be transparent or opaque, depending on their use. Films are used in mucosal formulations for applications such as wound dressings and, more recently, as oral strips where they deliver drugs to exert mainly a local effect [1]. Films for pharmaceutical applications should be soft, flexible, elastic and strong enough to bear stress during handling, storage and after application, and these are determined largely by the composition and the method of preparation of the films [2]. Film development is a multistep process involving the selection of appropriate formulations or formula, method of preparation and evaluation of films [3, 4] and these parameters should be selected with critical care. Selection of method depends on the polymer, active ingredient, formulation additives and the application. The physical properties of films such as strength, flexibility and stability are affected by process-related factors including equipment, solvent evaporation, drying time, film integrity and uniformity of thickness [4]. Process optimization is therefore important to control the process-related factors. A wide range of natural and synthetic polymers are used for film preparation (Table 8.1) and can be generally divided into cellulosic, vinyl, acrylic derivatives and natural (biomaterial) polymers. Both natural and synthetic polymers may be biodegradable or non-biodegradable.

Hot-melt Extrusion: Pharmaceutical Applications, First Edition. Edited by Dennis Douroumis.

Table 8.1 *Examples of commercially available film-forming polymers [5–8].*

Type	Name of derivative/polymer
Acrylic derivatives	Acrylate/octylacrylamide copolymer Polyurethane-14 and AMP-acrylates copolymer Poly(butyl methacrylate, (2-dimethylaminoethyl) Methacrylate, methyl methacrylate) 1:2:1 Poly(ethyl acrylate, methyl methacrylate)2:1 Poly(ethyl acrylate, (2-trimethylaminoethyl)methacrylate, methyl methacrylate) 1:0:2:2 chloride Poly(methacrylic acid, methyl methacrylate) 1:2
Vinyl derivatives	Polyvinylpyrrolidone Polyvinylpyrrolidone-vinyl acetate copolymer Polyvinyl alcohol Carbopol
Cellulose derivatives	Hydroxypropylymethylcellulose Hydroxypropylcellulose Carboxymethylcellulose Methylcellulose Ethylcellulose
Natural polymers	Chitosan Sodium alginate Xanthan gum Carrageenan

Increasingly, a combination of the above polymers are used in the preparation of films [4] to take advantage of the unique properties of each type of polymer.

The use of oral thin films or strips for the delivery of active agents in the oral cavity commenced in the 1970s when lidocaine was formulated in polymer films for dental applications [9, 10]. This administration route is especially convenient for pediatric and geriatric patients with difficulties in swallowing or chewing solid dosage forms. These films or strips are designed to either release the active substances rapidly or to modify the release patterns, depending on the choice of the polymer matrix and the film thickness. Typical film/strip dosage forms incorporate water-soluble and bioadhesive polymers which hydrate rapidly upon contact with saliva, adhere on the oral mucosa (e.g. buccal, gingival, sublingual) and subsequently dissolve to release the active agents. Thin films or strips are also characterized by their non-invasive administration, localization and patient compliance. The film utilization as an alternative oral dosage form presents several advantages compared to conventional tablets. The mucosal mucosa has a relatively small drug adsorptive surface area ($100\,cm^2$) but exhibits higher drug permeability than other routes such as the gastrointestinal tract (GIT) or the transdermal. Furthermore, the mucosal tissue presents low enzymatic activity, drug stability (e.g. proteins, peptides) and high vascular supply, rendering it an attractive site for drug administration.

It also presents the possibility of administering drugs to unconscious or incapacitated patients. In addition, the risk of choking via accidental entry into the respiratory tract during swallowing is diminished [11]. The variability in mucosal lining is less between patients and therefore exhibits lower inter-subject variation. However, current technologies present many challenges. The rapid elimination of drugs due to the flushing action of saliva or involuntary swallowing alone or with food may require frequent dosing in case of local delivery.

Furthermore, distribution of drugs into saliva is non-uniform due to uneven release from solid or semisolid delivery systems; this may result in lower amounts reaching the mucosal tissues and subsequently the systemic circulation. There is also the possibility of chemical modifications of certain drugs by saliva. Another disadvantage is patient non-compliance due to possible irritation and bitter taste of drugs, requiring the need for taste masking and flavoring agents. The latter disadvantage can however be overcome with the use of hot-melt extrusion which is known to help mask the bitter taste of drugs. Absorption is limited to non-keratinized areas in the cheek and underneath the tongue. Furthermore, there is limited permeation in the absence of external stimuli with only few milligrams of drug able to cross the oral mucosa.

In 2003 an oral strip was introduced by Zengen, Inc. [12] for the delivery of active substances such as benzocaine (3 mg) and menthol (3 mg) in Chloraseptic® Relief Strips™. This new bilayer film consisted of a substrate active layer and a second dry layer manufactured to provide rapid or sustained release of the active substance over time. The dry layer is usually applied by spray-coating methods. Both layers were composed of water-soluble bioadhesive polymers such a cellulose or pullulan, a flavoring agent (cherry flavor), plasticizer (glycerin) and sweetener (sucralose). Similarly, thin-film patented technologies have been developed by McNeil-PPC to deliver active agents under the trade names of Triaminic Thin Strips® and Benadryl® quick-dissolve strips [13, 14].

The current film manufacturing technologies are mainly based on solvent-casting approaches where water-soluble inactive ingredients are dissolved until they form clear viscous solution (gel). The active substances are dissolved in smaller solution quantities and finally mixed in a tank to form the casting solution. The solution is then metered onto a moving web and dried in temperature-controlled multizone ovens to produce dried films that are die-cut and packaged. Solvent-casting film technologies provide film uniformity, clarity, flexibility and adjustable thickness to accommodate drug loadings. It is also a relatively easy scalable process with low production costs, which makes it widely used. However, casting techniques are related to decreased elongation or elasticity and increased film tensile strength when physical aging is applied [15]. Other studies showed that the grade and level of plasticizers, curing times or temperatures have a substantial effect on drug dissolution rates when films are prepared from aqueous dispersions.

Another major aspect in the manufacture of films is the use of organic solvents for water-insoluble polymers. The hazardous nature of the organic solvents and the residual solvents after drying affect the selection of the appropriate solvent [16–19]. Furthermore, the drug uniformity and heterogeneity is also a challenging matter for medicated solvent cast films. Initially, the drug non-uniformity was associated with their monolayer structure [20], but further studies revealed the importance of the drying times in prohibiting film aggregation [21]. For the above reasons, the employment of novel film manufacture technologies that will overcome the aforementioned difficulties is of paramount interest to the pharmaceutical industry.

8.2 Anatomy of the Oral Cavity and Modes of Drug Transport

8.2.1 Structure

The oral cavity forms part of the digestive system and is bordered by the base of the mouth, cheeks, palate, lips, the uvula and the palatine arches on each side of the uvula. It is divided into the *cavum oris* (main oral cavity) – which comprises the front and lateral areas surrounded by the rows of teeth and contains the tongue – and the region between the teeth and lips known as the oral vestibule (*vestibulum oris*). For purposes of oral mucosal drug delivery, the target is the cavum oris on both sides of the cheek and under the tongue in the sublingual region.

The entire oral cavity is lined by the mucosa and is kept moist by saliva produced by the salivary glands. The oral mucosa consists of a non-keratinized area (sublingual and buccal mucosa) and the keratinized area (the gum or gingiva, the palatal mucosa and the inner side of the lips). The non-keratinized regions are generally more permeable compared to the keratinized areas [22]. The oral cavity offers a large surface area for absorption (100–200 cm^2) and is richly vascularized with little proteolytic activity. Blood reaches the buccal mucosa via the maxillary artery at a faster and richer blood flow rate (2.4 $ml/min/cm^2$) than that in the sublingual, gingival and palatal regions, thus facilitating passive diffusion of drug molecules across the mucosa. The estimated turnover time (cell regeneration) for the buccal epithelium is between 5 and 6 days [23].

The buccal mucosa is composed of several layers of different cells as shown in Figure 8.1 [24]. The epithelium is about 40–50 cell layers thick and is similar to stratified squamous epithelia found in the rest of the body. Lining the epithelium of the buccal mucosa is the non-keratinized stratified squamous epithelium that has a thickness of about 500–800 μm and surface area of 50.2 cm^2. The rough textured buccal mucosa is thus suitable for retentive delivery systems [25]. Basement membrane and lamina propria followed by the submucosa are found below the epithelial layer [26]. The lamina propria is rich with blood vessels and capillaries which open to the internal jugular vein. Lipids present in the buccal tissues include phospholipid (76.3%), glucosphingolipid (23.0%) and ceramide NS (0.72%). The buccal epithelium is primarily designed to provide protection of the underlying tissue. In non-keratinized regions, lipid-based permeability barriers in the outer epithelial layers protect the underlying tissues against fluid loss and entry of potentially harmful environmental agents such as antigens, carcinogens, microbial toxins and enzymes from foods and beverages [27].

8.2.2 Modes of Drug Transport and Kinetics

Various mechanisms including passive diffusion, facilitated passive diffusion, active transport and pinocytosis have been proposed for the transport of drugs across the oral mucosal membrane, as summarized in Table 8.2. Compared to other routes such as transdermal, the oral mucosa surface does not have a stratum corneum; the main barrier to drug transport is therefore removed [28].

There are two major routes reported to be involved in drug permeation across epithelial membranes: transcellular route and paracellular route. Transcellular transport is the transfer of molecules across cells and paracellular transport is the transport of molecules around

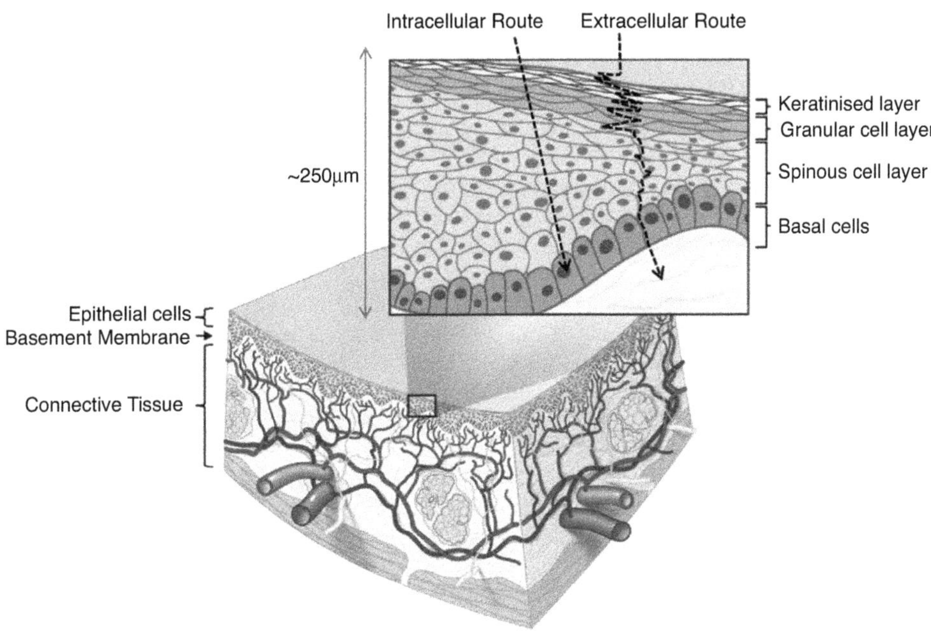

Figure 8.1 *Cartoon of the structure of the oral mucosa. Insert also shows different routes by which drugs can cross the oral mucosa. V. Hearnden et al. 2011, reproduced with the permission of Elsevier. For a better understanding of the figure, please refer to the color section.*

or between cells via tight junctions that exist between the cells. Such intercellular tight junctions form a key barrier to paracellular transport of macromolecules and polar compounds [28]. In most cases, absorption enhancers which facilitate temporary opening of the tight junctions are required to cross this barrier to absorption. Permeation enhancers include well-known polymers such as methylated cyclodextrins, polyacrylic acid derivatives and chitosans [29–32]. In the case of transcellular routes, the enhancers normally facilitate absorption by modifying membrane, lipid–protein interactions and lipid bilayer; for the paracellular route, the absorption enhancer disrupts intracellular occluding junctional complexes [33].

8.2.3 Factors Affecting Drug Absorption

There are various factors that affect drug absorption including the oral mucosa, physiological conditions within the mouth, properties of the drug and the type of drug delivery system. These parameters affect the stability, solubility and bioavailability of the drug of interest. In some cases, the basic application of this dosage form is releasing drug gradually and over a long time period to maintain the level of the drug and prevent the frequency of administration. To achieve this goal various forms of oral mucosa dosage forms are developed such as films and patches [34], gels and ointments [35] and tablets [36].

Another notable factor is the pH conditions within the mouth, as this may affect the administration of certain lipophilic and hydrophilic drugs via the mucosal route. Previous studies showed that the optimum uptake of drugs through the mucosal tissues normally

Table 8.2 Mechanisms of drug absorption across mucosal membranes.

Absorption	Mechanism of absorption
Passive diffusion (via paracellular and transcellular routes)	The driving force in the process of passive diffusion is the concentration gradient which determines and controls the transfer of molecules across the cell membrane. Diffusion occurs following the accumulation of a higher concentration in a region which induces molecules to pass through membranes towards the lower concentration region.
Facilitated passive diffusion	If the drug molecule is transferred through the membrane with the help of a carrier protein, the process is referred to as facilitated passive diffusion. However, it is limited to particular types of molecules as the carrier proteins only bind to specific molecules. The process also depends upon the accessibility of the carriers which can become saturated at some stage during the transport process.
Active transport	In active transport, the transport of the molecules and ions occur against the concentration gradient i.e. from lower to higher concentrations which require excess energy from the relevant cells in the form of ATP (adenosine triphosphate).
Pinocytosis	Pinocytosis is a form of endocytosis in which cells surround large molecules in the extracellular area. The cell membrane folds inside, surrounds the particle and then fuses to produce a vesicle which separates from the membrane and into the interior of the cell. This mechanism has a key role in protein drug transport.

occurs when the drug is in the unionized form and pH variation alters the percentage of unionized drug at a particular point in time. The mouth pH conditions can therefore change the efficiency of certain drugs administered buccally or sublingually [37]. The optimum pH of saliva is usually expected to be between 6.5 and 6.9 and suitable for drugs with a high pK_a [38].

Another critical parameter is molecular size and lipophilic properties of the drug molecule. Non-lipophilic drugs are not readily transported across the mucosal tissues. Protein and peptide drugs with very large molecular weights and electrically charged functional groups do not easily partition into mucosal tissues.

8.3 Mucoadhesive Mechanisms

The term 'bioahesion' is defined as the 'attachment of a synthetic or natural macromolecule to mucus and/or an epithelial surface' [39]; the mechanisms of polymer attachment to mucosal surfaces are well studied. However, the actual mechanisms are not yet fully understood and various theories have been proposed [40]. The 'adsorption theory' involves primary and secondary chemical bonds of the covalent (electrostatic and Van der Waals forces) and non-covalent (hydrogen and hydrophobic bonds) that take place with the contact of the mucus and the mucoadhesive polymer. The fundamental principle of this mechanism is that the substrates must be in intimate contact to maximize the adhesive strength. The 'diffusion theory' describes the chain entanglement between glycoproteins of the mucus and the mucoadhesive polymer. The polymer chains diffuse into the mucus structure, creating an entangled network between the two surfaces. In this case, the driving force is the concentration gradient across the interface. The inter-diffusion is then affected by the polymer chain flexibility, polymer exposure on the mucus membrane, similarity of the chemical structures and the diffusion coefficients of the polymers [41].

A third 'electronic theory' is related to the different electronic properties of the mucoadhesive polymers and the mucus glycoproteins, suggesting that electron transfer occurs between the two surfaces [42]. As a result, a charged double layer is formed at the interface of the mucus and the polymer leading to attraction forces and inter-diffusion of the two surfaces.

The 'wetting theory' is applicable to liquid bioadhesive systems and assumes that, if the contact angle of liquids on the substrate surface is lower, then there is a greater affinity for the liquid to the substrate surface. The wetting capacity of the polymer plays a key role for the occurrence of this mechanism. Other mucoadhesive mechanisms have also been proposed such as the 'interlocking' or 'fracture' theories with limited contribution in the adhesion process.

8.4 Factors Affecting Mucoadhesion in the Oral Cavity

There are a variety of factors that affect the mucoadhesive characteristics of the polymers used to manufacture films or strips. The molecular weight influences the bioadhesive strength which increases with the polymer molecular weight. Bioadhesion becomes stronger with molecular weight greater than 10 kDa, as shown for polyoxyethylene polymers [43]. Polymer flexibility is also important to achieve chain entanglement with the mucus. During bioadhesion, the polymer chains fuse in the interfacial region and it is therefore important that the polymer possesses a considerable degree of flexibility. The chain flexibility is related to polymer viscosity and diffusion coefficient, and higher flexibility induces greater diffusion.

The hydrogen bonding capacity is another important factor. Mucoadhesion increases when polymers have functional groups able to form hydrogen bonds (e.g. polymethacrylic acids, polyvinyl alcohols and hydroxylated methacrylates). It has been shown that increased cross-linking polymer densities reduce the diffusion rates, resulting in insufficient polymer swelling and subsequently reduced interpenetration with the mucus [44]. The polymer charge affects bioadhesion where anionic polymers appear to endure greater adhesion compared to non-ionic polymers [41]. In neutral or alkaline medium, cationic polymers were found to display excellent mucoadhesion. Finally, the polymer hydration (swelling) rate at which the polymer takes up water and swells is critical to bioadhesive properties. Theoretically, if a polymer swells very quickly, it will also quickly interact with the mucin and thereby ensure good adhesion.

8.5 Determination of Mucoadhesion and Mechanical Properties of Films

A wide variety of methods have been reported in the literature for the measurement of bioadhesion of polymers [19,45,46]. In most of these studies, indirect measurements were used to measure time of adhesion or retention time of the dosage form rather than the actual force of adhesion. Such approaches are characterized by a lack of reproducibility, and a range of tests need to be reported in order to collect reproducible results. For the first time, Guo et al. [47, 48] reported the recording of the detachment force which was then plotted against time or distance. A range of polymers were used to extract the maximum peeling strength through a texture analyzer. Thereafter, texture analyzers have been frequently used for ex vivo bioadhesion studies. For instance, TA.TX2 texture analyzers (Stable Micro Systems) have been used to measure bioadhesion of buccal films [49] or tablets [50]. In a typical measurement the tested material is attached to the mobile arm of the texture analyzer and is lowered at a steady rate until contact with the membrane is made for a short period of time (~120 sec). The probe is then withdrawn from the membrane at a steady rate and the detachment force is recorded. Mucoadhesion can be evaluated from the peak detachment force and the work of adhesion (area under curve), which is related to increased physical entanglement of the polymers.

The robustness of manufactured films or strips is evaluated through the determination of the mechanical properties. The film tensile strength is one of the critical properties, and is defined as the maximum stress applied to a point at which the film/strip specimen breaks [51]. The tensile strength is described by the following equation where the applied load at rapture is divided by the cross-sectional area:

$$\text{Tensile strength} = \frac{\text{Load at failure} \times 100}{\text{Strip thickness} \times \text{Strip width}}$$

The elongation at break is a second mechanical property and is the maximum deformation the film can undergo when stress is applied until finally tearing apart. The elongation (strain) increases with increased plasticizer amounts [52].

$$\%\,\text{Elongation at break} = \frac{\text{Increase in length of strip} \times 100}{\text{Initial length of strip}}$$

Young's modulus (tensile modulus) determines the film stiffness or deformation in the elastic region. It is related to the initial elastic deformation and is derived from the ratio of the tensile stress to the tensile strain in the elastic (linear) portion of a stress–strain curve:

$$\text{Young's modulus} = \frac{\text{Tensile stress}}{\text{Tensile strain}} = \frac{\text{Slope} \times 100}{\text{Strip thickness} \times \text{Cross} - \text{head speed}}$$

The film tear resistance and folding endurance are two more tests used to provide a full profile of the mechanical properties of a film.

8.6 Bioadhesive Films Prepared by HME

The utilization of HME as a feasible technology for the manufacture of thin films was first investigated by Aitken-Nichol et al. [53]. In this approach the authors dried Eudragit® E100 pellets to remove moisture and then melted them at 120°C. Afterwards, E100 was mixed with plasticizers in the nip between two rollers, ground with a ball mill and finally extruded through a single-screw Brabender extruder. The extruded films were compared to cast films prepared by dissolving the drugs and the polymer in ethanol and drying for one week. The authors observed that the nature of the plasticizer has an effect on the film quality. Triethylcitrate increases film ductility compared to triacetin, while both plasticizers significantly reduced the glass transition temperature of the E100. Lidocaine HCl was found to act as a plasticizer in the film and thus the film elongation was affected. Higher lidocaine concentrations produced softer and ductile films. In addition, the glass transition temperature of Basic Butylated Methacrylate Copolymer (EPO) was reduced by 10°C due to a better intermolecular mixing. The processing temperatures were above the melting point of lidocaine, resulting in drug solubilization and increased plasticizing effect. However, complete drug–polymer miscibility was not obtained. The plasticizing effect was higher for cast films which presented higher elongation at break values and lower peak stress compared to the extruded films. Furthermore, the lidocaine release from the extruded films at low drug loadings (5%) was slower than the cast films, but similar patterns were obtained for both film types at higher lidocaine loadings (10%). Interestingly, lidocaine release of the hot-melt extruded films was not complete after 24 hours, suggesting drug entrapment in the polymer. Nevertheless, this was the first attempt and demonstrated the viability of HME for preparing thin films.

HME was also utilized to extrude thin films of lidocaine embedded in hydroxypropyl cellulose (HPC) and hydroxapropyl methyl cellulose (HPMC) [54]. Two loaded formulations with only HPC and HPC:HPMC (80:20) produced homogenous films with an average thickness of 0.66 mm (±0.01), with HPC being the matrix-forming polymer and HPMC as the drug-release modifier. The formulations also included PEG 3350 to act as a plasticizer and polyethylene oxide (PEO) as the bioadhesive polymer. All formulations were stable after 6 months storage and lidocaine content was approximately 99%. Bioadhesion studies were carried out thoroughly and the HPC:HPMC films presented a higher area under curve (work of adhesion) and peak adhesion force compared to HPC films. In general, polymer hydration resulted in chain mobilization followed by interpenetration and physical entanglement upon contact with the mucin membrane due to the hydroxyl group interactions. However, HPMC has a more hydrophilic character which allows faster hydration and interaction with the mucin membrane. The peak adhesion forces and the work of adhesion for both formulations showed a considerable increase when the time of application of the force increased. The DSC characterization of the films showed lidocaine solubilization in the presence of EPO and PEG 3350 during the heating cycle. It was therefore not clear if lidocaine was in amorphous state in the film matrices.

This assumption was supported with X-ray studies where the extruded films showed one broad peak which is typical of amorphous materials. In the X-ray diffractograms of both film formulations, no crystalline peaks of lidocaine were observed compared to the physical mixtures. The dissolution studies of the extruded films demonstrated sustained-release patterns without significant differences, as estimated by the $f1$ (difference factor) value (9.2) and the $f2$ (similarity factor) value (57). Nevertheless, the presence of HPMC in the film formulations showed slight lidocaine retardation due to the higher HPMC swelling capacity. As a result, a thicker swollen gel is produced in the HPMC-containing films that increases the lidocaine diffusion distance to the film surface. The predominant drug-release mechanism was diffusion and both formulations fitted the Higuchi model.

The first study of Aitken-Nichol [53] demonstrated the importance of the plasticizer in HME processing. In a later study, Repka et al. investigated the effect of various plasticizers on hot-melt-extruded films of hydroxypropylcellulose (HPC) by using hydrocortisone (HC) and chlorpheniramine maleate (CPM) as model drugs [55]. It was not possible to extrude HPC films in the absence of plasticizers due to the excessive stress on the extruder. Film percentage elongation decreased with time for some of the plasticizers, but a significant reduction was observed after six months only for PEG400. The authors attributed the film ductility to the molecular weight (MW) of the plasticizer; by increasing the MW and thus the size, the mole fraction of the available hydroxyl groups to interact with HPC is decreased leading to less elastic films. The PEG400 films showed significant increase in T_g after six months, which could explain the reduced ductility. When HC was loaded in HME films, the ductility remained the same due to particle aggregation of the hydrophobic HC. All films showed high tensile strength values with a significant drop after three and six month storage.

The acetyltributyl citrate (ATBC) films showed a large drop in tensile strength and Young's modulus (YM) during storage. Initially, the manufactured films containing PEG400, PEG8000, Triethyl citrate (TEC) and ATBC presented similar YM values but the HC incorporation showed high YM and brittle films, probably due to the hydrophobic character of HC. The authors observed that the percentage elongation at break of the drug-loaded films was statistically increased when testing performed perpendicular to flow as opposed to in the direction of flow.

These results were inconsistent with the previous study with hot-melt-extruded EPO/Lidocaine HCl films. The authors attributed this behavior to the cellulose structure itself and consequently cross-linking by the various plasticizers and drugs embedded within the films. The most stable films in terms of tensile strength and thickness were the films containing PEG8000 and TEC. Mechanically and chemically stable films were obtained from 1% CPM extruded films, with CPM showing strong plasticizing effect. In contrast, HC showed acceptable plasticizing effect but not chemical stability at high processing temperatures (200°C). Only 63% of HC remained intact at 200°C; at 170°C over 91% of HC was found stable after 12 months.

HPC was further explored for the manufacture of bioadhesive films in terms of moisture absorption, physical-mechanical properties and bioadhesiveness [56]. The extruded HPC films contained polymer additives including polyethylene glycol (PEG) 5%, polycarbophil 5%, carbomer 5%, Eudragit E-100 5% and sodium starch glycolate (SSG) 5%. After two weeks storage, the HPC/PEG 3350 showed a threefold increase in water content as relative humidity increased (7.2% at 0% RH and 21.3% at 100% RH, 25°C). The rest of the polymers exhibited a 6–16-fold increase in water (%) at 0 and 100% RH (25°C), respectively. The SSG-containing films presented the highest water content with considerable increase from the 25°C/100% RH to 40°C/100% RH. This is probably due to the moisture absorption properties of the SSG, which is a well-known super-disintegrant.

All the other film formulations showed insignificant moisture uptake when tested at 25°C/100% RH and 40°C/100% RH. The tensile strength of all films determined at 80% RH decreased in comparison to those at 50%RH. This behavior was expected as water acts as plasticizer in hydrophilic polymers by weakening the inter-molecular attractions between polymer chains, thus altering film mechanical properties (tensile strength). The authors observed a 4–5-fold increase in the percentage elongation for the HPC/Polycarbophil films compared to the other films, attributed to the intermolecular interactions between the two polymers and, more precisely, to the hydrogen bonding of the hydroxyl groups.

It is worth mentioning that the HPC/PEG films were strongly affected by the heat processing conditions, showing sensitivity to thermal history. The bioadhesion of extruded films was evaluated in vivo on human epidermis by recording the force-deflection profiles. These profiles are very similar to stress–strain charts and the adhesive force represents the force applied to remove the films from the skin. The optimum bioadhesion performance was observed for films containing 5% polycarbophil which presented high detachment forces (2.5 N/cm^2 $\pm$ 0.08) followed by the HPC/polycarbophil/PEG (1.66 N/cm^2 $\pm$ 0.09). The HPC/PEG films demonstrated poor bioadhesion (0.64 N/cm^2 $\pm$ 0.05), indicating low interaction with skin. The excellent HPMC/polycarbophil profiles were attributed to the increased hydrogen bonding between the epidermal layer and the polymers due to the high content of carboxylic groups in the acrylic acid copolymer.

Based on the previous studies, the same group developed mucoadhesive matrix films containing 10% w/w clotrimazole (CT) for the prophylaxis and treatment of oral candidiasis [57]. A 6-inch flexlip die was utilized to extrude films with a thickness range of 0.34–0.36 mm. The extruded films consisted of HPC and PEO as polymeric carriers, polycarpbophil as adhesive polymer, PEG 3350 as plasticizer and antioxidants (butylated hydroxyl-toluene, propyl gallate). Because of the thermal treatment, the authors investigated the chemical stability, which was found to be 93.3%. The extruded films showed small CT degradation of 6.8% as a result of the imidazole group hydrolysis, that led to an o-(chlorophenyl)diphenyl methanol degradation product.

Nevertheless, the HME films were further assessed for bioadhesion on rabbit intestinal mucosa using a TA.XT2i texture analyzer. The recorded force-deflection profiles confirmed the significant film adhesion. Further, film characterization suggested the presence of molecularly dispersed CT within the extruded films. The absence of a CT melting endothermic peak was related to the creation of solid solution due to the solubilization of CT in the PEO/PEG polymers. X-ray studies confirmed the absence of crystalline CT compared to the physical mixtures of identical formulations.

The authors conducted a second, more comprehensive, study [58] to investigate the solid-state characteristics, moisture-sorption, bioadhesivity, mechanical properties and physicochemical stability of HME films which contained PEOs (N–80, N–750) of different molecular weights. Bioadhesion studies carried out with a texture analyzer showed that PEO N-750 had a higher peak adhesion force and work of adhesion than PEO N–80. This observation was related to the polymer molecular weight; high molecular weight resins showed deeper and more rapid chain penetration into the membrane network. The stress–strain curves showed flexible films with low Young's modulus and ductile films with high% strain. The low molecular weight PEO N–750 presented higher tensile strength and percent elongation values to those of the lower molecular weight PEO N–80. The different mechanical properties of the low molecular weight resins can be attributed to the low number of disordered chain units and the small thickness of the lamellar region which cannot sustain large deformations when undergoing mechanical stress, inducing a break in the extruded films [59].

CT release studies demonstrated faster dissolution rates with decreased polymer molecular weight, while the release mechanism was mainly erosion controlled irrespective of the polymer molecular weight. As observed in the previous study, CT was susceptible to thermal treatment but this time PEO polymers reduced the extent of degradation at 1.8% and 2.1% for N–80 and N–750, respectively. The degradation was significantly increased under accelerated storage conditions at 25°C/60% RH and 40°C/75% RH.

In a continuation of the previous studies, the authors attempted to further develop CT thin films by enhancing the physical stability of the extrudates. For this purpose, they used polymer blends of HPC and PEO [60]. By adjusting the PEO amounts (0–55%) it was feasible to modify the drug release, mechanical and bioadhesive properties and polymer stability. The physical stability of the CT and PEO in the extruded films increased with HPC concentration at the expense of other properties. Bioadhesion and flexibility of the PEO films decreased with increasing HPC concentration. The optimum physical-mechanical, bioadhesive and release properties were achieved for films consisting of HPC:PEO:CT at 55:35:10 weight ratios. This study revealed that the glass transition temperature of PEO plays an important role in the physical stability of CT and also in the mechanical properties of the film. As mentioned in previous studies, increasing PEO amounts enhanced film flexibility (Young's modulus). The release profiles suggested sustained release of CT for all film formulations, with release rates being slower when the EPO concentration increased.

Similar studies were carried out with films containing HPC of different molecular weights (140kDa, 370kDa and 850kDa) [61]. The film moisture sorption isotherms were recorded and equilibrium moisture content (EMC) was independent of the polymer molecular weight. Indeed, EMC values at 90% and 50% RH were 17% and 3.4%, respectively, compared to 20% and 4%, respectively, for the bulk HPC. The lower EMC film values are attributed to the HME process which increases the molecular packing of glassy polymer leading to less polymer hydration [62]. The shear forces developed during the HME high pressure reduces the free volume occupied by the polymer molecules and increased packing is achieved. Stress–strain curves showed that the tensile strength and percent elongation at break depend on the molecular weight.

In-depth analysis of the film tensile strength showed that films tested perpendicular to the direction of the melt flow have lower values than those tested parallel to the flow direction. At increased moisture conditions the tensile strength and Young's modulus were significantly decreased, while percent elongation was increased. At 75% RH the stress–strain curves suggest film transformation from brittle to ductile. This phenomenon was again attributed to the plasticizing effect of the absorbed water that depresses the polymer intermolecular attractions and facilitates chain flexibility.

All film formulations demonstrated sustained release of CT with different release rates. The CT release rates decreased when HPC molecular weight increased, showing 45% (MF grade), 80% (GF grade) and 100% (JF grade) release patterns after 24 hours. The authors suggested that the HME process is carrier controlled and the effect of a drug's properties are negligible. The drug-release profiles can therefore be tuned without affecting the release mechanism. However, the CT release profiles showed strong dependence on the moisture content when films were stored at 25°C/75% RH. The CT release patterns were analyzed by using different kinetic models, and it was found that erosion is the predominant release mechanism irrespective of the polymer's molecular weight. As was observed in the previous study, CT was susceptible under heat treatment and low degradation products were detected at 25°C/60% RH after three months. When films were stored at accelerated stability conditions, CT degradation was accelerated.

Crowley et al. (2004) investigated the influence of guaifenesin (GFN) or ketoprofen (KTP) on polyethylene oxide (PEO) hot-melt-extruded films [63]. The physicochemical and mechanical properties of empty EPO films, GNF/PEO films loaded from 5 to 30% and KTP/PEO loaded from 5 to 15% were assessed. All films were clear, homogeneous and flexible, except the 30% KTP which were soft and tacky and did not solidify upon cooling.

The Hansen solubility parameters (δt) were used to predict drug–polymer miscibility using the Hoftyzer and van Krevelen approach [64]. The difference of the solubility parameters indicated KTP/EPO miscibility but not for GFN/EPO blends. The Hansen solubility estimations were later confirmed by the film characterization. By increasing GFN load in PEO films from 10 to 30%, the X-ray diffractograms showed increased peak intensity of GFN peaks. The authors assumed that preferential crystal growth takes place on the film surface.

Differential scanning calorimetry (DSC) studies of GFN/EPO films exhibited depression and broadening of the PEO melting endothermic peak, and only a single peak was observed in the DSC thermograms. Interestingly, the DSC thermograms of the GFN/PEO physical mixtures also exhibited a single broad peak (67.6°C), which implies melted PEO solubilized GFN. These results were contradictory to the X-ray observations, but the absence of the GFN thermal transition was associated with the drug–polymer miscibility at high temperatures. However, it appears that GFN crystallized into a separate phase upon cooling. In the case of KTP/PEO films no crystalline peaks related to KTP could be identified in the X-ray diffractograms. The reduced KTP crystallinity implied partial drug–polymer miscibility. These findings were confirmed by the DSC thermograms, where KTP melting endotherm disappeared while EPO melting peak became broader and suppressed at lower temperatures.

The morphology of film surfaces were examined by scanning electron microscopy (SEM). Unloaded EPO films showed clear and homogenous surfaces with sporadic visible crystals. The GFN/PEO films exhibited crystal formation, which was magnified with increasing GFN concentrations. In contrast, KTP crystallization was not observed on the film surfaces until KTF concentrations reached 15%.

The authors conducted gel permeation chromatography studies where PEO molecular weight was decreased by approximately 6.8% after extrusion. In contrast, GFN and KTF diminished the PEO molecular weight loss and the load on the extruder motor, especially at high drug loadings. The film mechanical studies revealed significant reduction in percentage elongation and tensile strength of GFN loaded films. Surprisingly, the tensile strength reduction was inversely proportional to GFN concentration. The mechanical properties of the KTP/PEO deviated from those of GFN films and the film elongation was greatly increased as the KTF concentration increased due to the plasticizing effect of KTP on the PEO polymer. Another reason for the different mechanical properties of GFN and KTP films was the presence of crystals on the film surface. Crystals were observed at every GFN concentration, and it is believed that they cause discontinuities in the polymer network by breaching the hydrogen bonds of the polymer segments.

Thumma et al. [65] investigated the manufacturing of a prodrug of Δ^9- tetrahydrocannabinol (THC-HG) films in PEO polymeric matrices produced by hot-melt extrusion for systemic delivery of THC through the oral transmucosal route [65]. A hemiglutarate ester prodrug, THC is currently available in soft gelatin capsules under the trade name of Marinol® but presents limited stability and low bioavailability due to its first-pass metabolism and poor solubility. The hemiglutarate ester prodrug THC-HG was developed to improve pharmacokinetic performance and to improve the physicochemical properties of the parent drug.

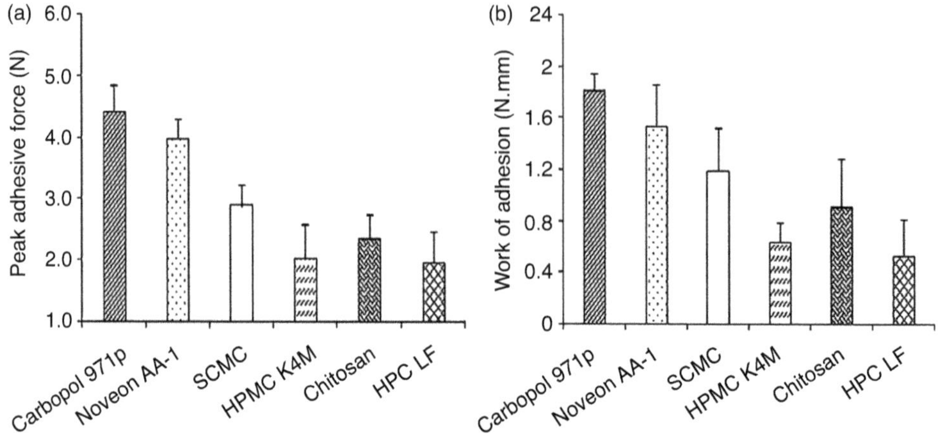

Figure 8.2 *Killion extruder Model KLB–100. S. Prodduturi et al. 2004, reproduced with the permission of John Wiley & Sons.*

In this comprehensive study [65], the influence of pH modifiers and antioxidants employed as stabilizing agents was examined. In addition, the bioadhesion of PEO matrices was studied as a function of bioadhesive polymer type and concentration, contact time, drug loading and wetting time. The extruded matrices consisted of 5% THC-HG and were processed at 110°C for 5–10 min. A wide variety of bioadhesive polymers were blended with PEO and evaluated at contact times of 60 sec.

The polymers demonstrated different bioadhesion performance due to their different chemical nature, molecular structure and hydration ability. Carbopol® 971P outperformed all polymers, probably due to the increased content of carboxylic groups and their ability to interact with mucin membrane through hydrogen bonding. The bioadhesion performance was arranged in a descending order of Carbopol® 971P>Noveon® polycarbophil (AA-1)>sodium carboxymethyl cellulose (SCMC)>chitosan>HPMC>HPC, as illustrated in Figure 8.2.

The difference between Carbopol and Noveon can be rationalized by Noveon's high cross-linking which prevents certain groups interacting with the mucin membrane. Both Carbopol® 971P and Noveon® AA-1 were extruded at concentrations of 2%, 4%, 7% and 10% to investigate their effect on bioadhesion. Interestingly, increased Carbopol® content from 0 to 4% resulted in a higher peak of adhesion forces and work of adhesion values. When Carbopol® exceeds 4% concentrations, the bioadhesion strength was reduced. It is believed that, at high polymer concentrations, the number of –COOH available for interaction were shielded inside the coils due to the intra-molecular hydrogen bonding, thus reducing the available groups for interaction.

The effect of contact time and drug loading was further investigated for THC-HG-PEO films. Both peak adhesion force and area work of adhesion values were significantly increased with an increase in contact time. Films examined at 120 sec showed a 1.6-fold higher peak of adhesion values compared to those examined at 15 sec. Similarly, the work of adhesion values was doubled for the same contact times. The films were also examined at higher drug loads (7.5 and 10.0 wt%) and 120 sec contact times. Under these conditions, both bioadhesion properties showed a statistical increase with increased drug load. For matrices tested at 30 and 60 sec contact times, bioadhesion showed the same profiles at low drug loadings (2.5 and 5.0 wt%) and increased profiles at higher drug loadings (7.5 and 10.0 wt%). This behavior was attributed to the highly lipophilic and sticky resinous nature of the prodrug at ambient temperature. This explains the fact that, at low drug concentration, there was an effect on the film bioadhesiveness.

The authors observed that THC-HG degraded significantly at 40°C/75%RH (90.3%) due to the hydrolysis of the ester bond and thus they incorporated pH modifiers to protect THC-TG. For this purpose, they examined the performance of acidic and basic pH modifiers. Further analysis demonstrated that acidic pH modifiers reduced drug degradation in comparison to the basic modifiers by 20–70%. Citric acid was the most effective protecting agent and THC-HG was stable in a pH range of 3.0–4.0, while higher pH values enhanced drug degradation.

In order to reduce THC-HG susceptibility to oxidation, a number of antioxidants was tested in PEO films. All of the investigated antioxidants (e.g. EDTA, ascorbic acid, propyl gallate) reduce the drug oxidation at levels lower than 1%. Ascorbic acid (1%) was found to reduce degradation at 8.2% after three months at 25°C/60%RH.

These results helped the authors to develop two formulations of THC-HG extrude films, tested at 40°C/75RH. Formulations I and II consisted of THC-HG (5.0%, w/w), PEO N-80 (78.3%, w/w), vitamin E succinate (10.0%, w/w, VES), citric acid (2.5%, w/w), Noveon® AA-1 (4.0%, w/w), butylated hydroxytoluene (BHT) (0.2%, w/w), THC-HG (5.0%, w/w), PEO N-80 (78.3%, w/w), VES (10.0%, w/w), citric acid (2.5%, w/w), Carbopol® 971p (4.0%, w/w) and BHT (0.2%, w/w). The degradation of THC-HG was further reduced for both formulations at levels below 5%, proving that the selection of the appropriate excipients facilitates drug stability against oxidation and hydrolysis.

The same group investigated the effect of processing conditions (temperature and heating duration), plasticizer type and concentration and storage conditions on the stability of HPC-HG [67]. The processed film formulation included plasticizers such as vitamin E succinate (VES), acetyltributyl citrate (ATBC), triethyl citrate (TEC), triacetin and polyethylene glycol 8000 (PEG8000). All processing parameters were found to affect the stability of THC-HG in PEO matrices. The process optimization provided acceptable films when processed at 110°C for 7 min with post-processing drug content of 95%, while significant degradation of THC-HG (42%) was observed when films were processed at 200°C for 15 min. Nevertheless, the addition of plasticizers reduced the THC-HG degradation considerably during the HME processing and also under storage conditions. VES was proved the most effective plasticizer compared to the other at concentrations of 10, 20 and 30%. The plasticizer grade and concentration affected the release profiles of THC-HG where faster release patterns were obtained from water-soluble plasticizers, PEG 8000 and triacetin, and with increasing concentration. In contrast, slower release rates were observed with an increase in concentration of water-insoluble plasticizers (VES, ATBC).

In a recent study, HME was employed to prepare oral fast-disintegrating films and was compared with films prepared by casting and solvent evaporation technology [68]. The film formulations processed by HME composed of maltodextrin (MDX), microcrystalline cellulose (MCC), sorbitan monoleate and a model drug piroxicam (PXC) with loading varying from 9 to 15%. The batches were granulated prior to the extrusion process, while the addition of MCC was crucial in order to obtain no-sticking films. Preliminary studies for the plasticizer selection indicated a better performance of glycerin (GLY) compared to poly(ethyleneglycol) (PEG400). GLY films exhibited a 2-fold higher elongation at break values to those of PEG400.

In addition, an in vivo assessment of films by human volunteers for taste requirements gave excellent taste scores for GLY, but unpleasant taste for PEG400. A comparison of the placebo showed that HME films were more fragile to those prepared by casting and the percentage elongation was about 100-fold lower for certain HME formulations. This difference in the mechanical properties was attributed to the presence of MCC in the HME films which caused the formation of a non-continuous sheet of film. Furthermore, for casting films, the in vivo disintegration times (10 sec) were faster when compared to the disintegration times of the extruded films (45 sec). According to the authors, the swelling of MCC retarded the film hydration and consequently increased the disintegration times. In addition, MCC left an unpleasant sensation in the buccal cavity after disintegration. Similar results were observed for the PXC-loaded films, with casting films showing faster disintegration/dissolution times from the HME films. The authors concluded that film casting was a more suitable process for film preparation compared to HME. Nevertheless, film manufacture by HME requires a more complicated approach, especially when taste masking of the active ingredient is required. Formulations designed for casting technologies cannot always be employed for HME due to the different operational principles of the two technologies.

Hot-melt-extruded films have also been developed for transdermal drug delivery to treat fungal infections. Onychomycosis is a fungal disease that affects fingernails or toenails leading to thickening, discoloration, disuration and splitting. The treatment of antifungal diseases is challenging because oral administration is associated with limited success, toxicity, drug interactions, microbe resistance and increased medication costs. Repka et al. [66, 69] investigated the physicochemical properties of hot-melt-extruded different HPC and/or PEO films containing ketoconazole (20%). The influence of 'nail etching' on film bioadhesion and drug permeability was also studied. Different batches were extruded using a Killion KLB–100 extruder (Figure 8.3) with processing temperatures varying from 115 to 160°C depending on the batch composition. The post-extrusion content of ketoconazole (KTZ) was estimated by sampling from different areas of the films, and showed very good content uniformity.

Despite the fact that samples were extruded at temperatures above the melting point of KTZ, no degradation was observed. The negligible degradation was attributed to KTZ thermal stability and the short residence time of the processed material in the extrusion chamber. DSC studies of extruded films showed the presence of molecularly dispersed KTZ in formulations containing PEO. The melting endothermic peaks of KTZ disappeared in both HME formulations and physical mixtures, suggesting that KTZ was either amorphous or solubilized by EPO. The absence of distinct crystalline peaks corresponding to KTZ in the HME films confirmed the existence of the drug in amorphous state.

Bioadhesion studies were conducted by using the peel test at 10–30 sec contact time to determine the peak area force and the work of adhesion. For the purposes of the study, non-treated (control) or treated with phosphoric acid (etched) nail samples were used. The etched samples showed 2.5-fold higher peak adhesion forces compared to the control at 10 and 30 sec contact times. The high surface area of the etched nails led to an increase of the polymer interpenetration/entanglement, and subsequently to the

Figure 8.3 Influence of various bioadhesive polymer on the (a) peak adhesive force and (b) work of adhesion of THC-HG–PEO matrices (n = 5). M.A. Repka et al. 2004, reproduced with permission of Elsevier.

formation of secondary covalent bonds between the chains. Further wetting of the film samples facilitated better intimate contact between the nail and the film and developed stronger adhesion.

It is believed that film hydration advances the chain segment mobility and hence polymer inter-diffusion. The successful treatment of onychomycosis requires drug release from the films and sufficient drug permeation through the nail barrier. A Franz cells system composed of 9 diffusion cells was therefore employed to investigate KTZ permeability through a set of nail plates. The KTZ permeability was assessed by using a KTZ (0.125%) gel prepared from Carbopol® 974P NF as a control and the extruded film formulations. The control studies of the gel formulation showed a 60% higher KTZ permeation for the 'etched' nail plates compared to the untreated nails. However, the KTZ dose level in the Carbopol® gel was very low and KTZ permeability levels were approximately 140 mg and 240 μg for the untreated and 'etched' nails, respectively. In contrast, the HME films exhibited a 6-fold increase of KTZ permeability for the 'etched' nail plates in comparison to the control nails. The KTZ amounts detected in the acceptor compartment of the Franz cells were above 1400 mg.

The authors explained the increase in permeability of the HME films through the controlled disruption of the dorsal surface which decreased the 'effective membrane thickness' for drug permeation. Furthermore, the nail 'etching' increased the surface area by creating microporosities. This assumption was supported by atomic force microscopy (AFM) studies with 'etched' nail plates showing a rough surface [70].

In a later study, the authors introduced tartaric acid (TTA) as a surface modifier incorporated into HPC polymers of different molecular weights and KTZ [71, 72]. The addition of TTA demonstrated a significant effect on the bioadhesion and mechanical properties of the extruded films. Force-deflection profiles of TTA-containing films increased the peak adhesion force (2–3-fold) and work of adhesion (12-fold), thus providing a better surface modification of human nail. The TTA/HPC films exhibited lower tensile strength and higher percent elongation compared to films without TTA. Moisture content in TTA/HPC films was higher than in films without TTA, due to its moisture absorption properties.

8.7 Summary

We have discussed the advantages of HME as a processing technology for the manufacture of films or strips by reporting several case studies. HME has attracted considerable attention for the development of oral mucosal or transdermal films. Indeed, HME can be efficiently employed to produce robust film formulations compared to conventional film techniques such as solvent casting. As mentioned above, HME processing offers several advantages in terms of increased drug loading, incorporation of both hydrophilic/hydrophobic active substances, enhanced drug–polymer interactions to produce solid dispersions and reduced drug degradation due to hydrolysis or oxidation. Further advantages of HME include the absence of solvents, limited processing steps, easy scale-up and continuous production. In the case of oral thin films, the potential to mask the unpleasant taste of bitter active pharmaceutical ingredients (APIs), avoiding the need to add flavoring or sweetening agents, is a very attractive option.

References

(1) Boateng, J.S., Stevens, H.N.E., Eccleston G.M., Auffret, A., Humphrey, M. and Matthews K.H. (2009) Development and mechanical characterization of solvent-cast polymeric films as potential drug delivery systems to mucosal surfaces. *Drug Development & Industrial Pharmacy*, **35**, 986–996.
(2) Rai, D, Manirruzzaman M. and Boateng J.S. (2010) Development and characterisation of sodium alginate and HPMC films for mucosal drug delivery. *International Journal of Biotechnology*, **11**, 169–181.
(3) Janos J., Feher, M. and Pintye-Hodi, K. (2007) Effects of plasticizer on surface of free films prepared from aqueous solutions of salts of cationic polymers with different plasticizers. *Applied Surface Science*, **253**, 7303–7308.
(4) Shahrzad, M. and Reza, F. (2004) A novel approach in the assessment of polymeric film formation and film adhesion on different pharmaceutical solid substrates. *AAPS PharmSciTech*, **5**, 29.
(5) Wurster, D.E., Bhattacharjya, S. and Flanagan, R. (2007) Effect of curing on water diffusivities in acrylate free films as measured via a sorption technique. *AAPS PharmSciTech*, **8**(3), Article 71.
(6) Aulton, M.E. (2002) *Pharmaceutics: Coating of Tablet and Multiparticulate, The Science of Dosage Form Design*. Churchill Livingstone, Second Edition, pp. 441–446.

(7) Hoffman, A.S. (2006). Selecting the right polymer for biomaterial application. In Uchegbu, I.F. and Schatzlein, A.G. (eds) *Polymers in Drug Delivery*, Taylor & Francis, Florida, 7–19.

(8) Ghaari, A., Navaee, K., Oskoui, M., Bayati K. and Ra.ee-Tehrani, M. (2007) Preparation and characterization of free mixed-film of pectin/chitosan/Eudragit_RS intended for sigmoidal drug delivery. *European Journal of Pharmaceutics & Biopharmaceutics*, **67**, 175–186.

(9) Roller, N.W. (1975) Lidocaine topical film strip for oral mucosal biopsies. *Journal of Oral Medicine*, **30**, 55–58.

(10) Danjo, K., Higuchi, F. and Otsuka, A. (1995) Release of lidocaine from polymer film dosage forms. *Chemical & Pharmaceutical Bulletin* (Tokyo), **43**, 1759–1763.

(11) Cilurzo, F., Cupone, I.E., Minghetti, P., Selmin, F. and Montanari, L. (2008) Fast dissolving films made of maltodextrins. *European Journal of Pharmaceutics & Biopharmaceutics*, **70**, 895–900.

(12) Zengen, Inc. (2004) Edible film for relief of cough or symptoms associated with pharyngitis. WO2004045537A2.

(13) McNeil-PPC, Inc. (2009) Fast dissolving orally consumable films. US7491406.

(14) McNeil-PPC, *Inc.* (2010) *Fast dissolving orally consumable films* **con**taining a taste masking agent. US7648712.

(15) Gutierrez-Rocca J.C. and McGinity J.W. (1993) Influence of aging on the physical-mechanical properties of acrylic resin films cast from aqueous dispersions and organic solutions. *Drug Development & Industrial Pharmacy*, **19**, 315–332.

(16) Steuernagel C.R. (1997) Latex emulsions for controlled drug delivery. In McGinity, J.W. (ed.), *Aqueous Polymeric Coatings for Pharmaceutical Dosage Forms*, **79**. Marcel Dekker Inc., New York.

(17) Barnhart S., (2008) Thin film oral dosage forms. In Rathbone, M.J., Hadgraft, J., Roberts, M.S. and Lane, M.E. (eds), *Modified-release Drug Delivery Technology*, Informa Healthcare, 209–216.

(18) International Conference on Harmonization (2009) ICH topic Q3C(R3) Impurities: Residual Solvents. Available at http://www.emea.europa.eu/pdfs/human/ich/028395en.pdf.

(19) Morales, J.O. and McConville, J.T. (2011) Manufacture and characterization of mucoadhesive buccal films. *European Journal of Pharmaceutics & Biopharmaceutics*, **77**, 187–199.

(20) Schmidt, W. (1989) Process for producing an administration or dosage form for drugs, reagents or other active ingredients. US Patent 4849246.

(21) Yang, R., Fuisz, R., Myers, G. and Fuisz, J. (2008) Thin film with non-self-aggregating uniform heterogeneity and drug delivery systems made therefrom. US Patent 7425292.

(22) Veuillez, F., Kaliaa, Y.N., Jacquesa, Y., Deshussesb, J. and Buriac, P. (2001) Factors and strategies for improving buccal absorption of peptides. *European Journal of Pharmaceutics and Biopharmaceutics*, **51**, 93–109.

(23) Harris, D. and Robinson, J.R. (1992) Drug delivery via the mucous membranes of the oral cavity. *Journal of Pharmaceutical Science*, **81**, 1–10.

(24) Hearnden, V. Sankar, V., Hull, K., Juras, D.V., Greenberg, M., Kerr, A.R., Lockhart, P.B., Patton, L.L., Porter S. and Thornhill, M.H. (2011) New developments and

opportunities in oral mucosal drug delivery for local and systemic disease. *Advanced Drug Delivery Review*, doi:10.1016/j.addr.2011.02.008.
(25) Rathbone, M.J., Ponchel, G. and Ghazali, F.A. (1996) Systemic and oral mucosal drug delivery and delivery systems. In Rathbone, M.J. (ed.), *Oral Mucosal Drug Delivery*, **74**. Marcel Dekker Inc., New York, 241–284.
(26) Gandhi, R.B. and Robinson J.R. (1988) Bioadhesion in drug delivery. *Indian Journal of Pharmaceutical Science*, **50**(3), 145–152.
(27) Squier, C.A. and Finkelstein, M.W. (1998) Oral mucosa. In Ten Cate, A.R. (ed.), *Oral Histology, Development, Structure and Function*. C.V. Mosby, St Louis, 345–385.
(28) Madhav, N.V.S., Shakya, A.K., Shakya, P. and Singh K. (2009) Orotransmucosal drug delivery systems: A review. *Journal of Controlled Release*, **140**, 2–11.
(29) Nusrat, A.M. and Bi-Botti, C.Y. (2006) The quest for non-invasive delivery of bioactive macromolecules: a focus on heparins. *Journal of Controlled Release*, **113**, 91–101.
(30) Lu, Y. and Low, P.S. (2002) Folate-mediated delivery of macromolecular anticancer therapeutic agents. *Advanced Drug Delivery Review*, **54**, 675–693.
(31) Indu, K. and Meenakhi, K. (2002) Ocular preparation: the formulation approach. *Drug Development & Industrial Pharmacy*, **28**, 473–493.
(32) Giacomo, D., Ylenia, Z. and Chiara, Z. (2005) Review: polymeric enhancers of mucosal epithelia permeability: synthesis, transepithelial penetration-enhancing properties, mechanism of action,safety issues. *Journal of Pharmaceutical Sciences*, **97**, 680.
(33) Lorenza, G.M., Sandra, H. and Jeesus, V. (2008) Inventions designed to enhance drug delivery across pithelial and endothelial cells through the paracellular pathways. *Recent Patents on Drug Delivery Formulation*, **2**, 145–147.
(34) He, H., Cao, X. and Lee, I.J. (2004) Design of a novel hydrogel-based intelligent system for controlled drug release. *Journal of Controlled Release*, **95**, 391–402.
(35) Codd, J.E. and Deasy, P.B. (1998) Formulation development and in vivo evaluation of a novel bioadhesive lozenge containing a synergistic combination of antifungal agents. *International Journal of Pharmaceutics*, **173**, 13–24.
(36) Junginger, H., Hoogstraate, J. and Verhoef, J.C. (1999) Recent advances in buccal drug delivery and absorption – in vitro and in vivo studies. *Journal of Controlled Release*, **62**, 149–159.
(37) Degim, T., Eglen, B. and Ocak, O. (2006) A sustained release dosage form of acyclovir for buccal application: An experimental study in dogs. *Journal of Drug Targeting*, **14**, 35–44.
(38) Sasaki, S., Koumi, S., Sato, R., Murata, M., Nagasawa, K., Sakurai, E., Hikichi, N. and Hayakawa, H. (1998) Kinetics of buccal absorption of propafenone single oral loading dose in healthy humans. *General Pharmacology*, **31**, 589–591.
(39) Longer, M.A. and Robinson, J.R. (1986) Fundamental aspects of bioadhesion. *Pharmaceutical International*, **7**, 114–117.
(40) Salamat-Miller, N., Chittchang, M. and Johnston, T.P. (2005) The use of mucoadhesive polymers in buccal drug delivery. *Advanced Drug Delivery Review*, **57**, 1666–1691,
(41) Peppas, N.A. and Buri, P.A. (1985) Surface, interfacial and molecular aspects of polymer bioadhesion on soft tissues. *Journal of Controlled Release*, **2**, 257–275.

(42) Derjaguin, B.V., Toporov, Y.P., Muller, V.M. and Aleinikova, I.N. (1977) On the relationship between the electrostatic and molecular component of the adhesion of elastic particles to a solid surface. *Journal of Colloid & Interface Science*, **58**, 526–533.
(43) Tiwar, D., Goldman, D., Sause, R. and Madan, P.L. (1999) Evaluation of polyoxyethylene homopolymers for buccal bioadhesive drug delivery device formulations. *AAPS PharmSci*, **1**, E13.
(44) Gu, J.M., Robinson, J.R. and Leung, S.H.S. (1998) Binding of acrylic polymers to mucin/epithelial surfaces: structure–property relationships. *Critical Reviews in Therapeutic Drug Carrier Systems*, **5**, 21– 67.
(45) Nafee, N.A., Ismail, F.A., Boraie, N.A. and Mortada, L.M. Mucoadhesive buccal patches of miconazole nitrate: in vitro/in vivo performance and effect of ageing. *International Journal of Pharmaceutics*, **264**, 1–14.
(46) Anders, R. and Merkle, H. (1989) Evaluation of laminated muco-adhesive patches for buccal drug delivery. *International Journal of Pharmaceutics*, **49**, 231–240.
(47) Guo, J. (1994) Bioadhesive polymer buccal patches for buprenorphine controlled delivery: formulation, in-vitro adhesion and release properties. *Drug Development & Industrial Pharmacy*, **20**, 2809–2821.
(48) Guo, J.H. (1994) Investigating the surface properties and bioadhesion of buccal patches. *Journal of Pharmacy & Pharmacology*, **6**, 647–650.
(49) Wong, C., Yuen, K. and Peh, K. (1999) Formulation and evaluation of controlled release Eudragit buccal patches. *International Journal of Pharmaceutics*, **178**, 11–22.
(50) Boyapally, H., Nukala, R.K., Bhujbal, P. and Douroumis, D. (2010) Controlled release from directly compressible theophylline buccal tablets. *Colloids Surf B Biointerfaces*, **7**, 227–233.
(51) Felton, L., Donnell, P.O. and McGinity, J. (2008) Mechanical properties of polymeric films prepared from aqueous dispersions. In McGinity, J. and Felton, L. (eds) *Aqueous Polymeric Coatings for Pharmaceutical Dosage Forms*, 3rd edition. Informa Healthcare, New York, pp. 108.
(52) Felton, L., Donnell, P.O. and McGinity, J. (2008) Mechanical properties of polymeric films prepared from aqueous dispersions. In McGinity, J. and Felton, L. (eds) *Aqueous Polymeric Coatings for Pharmaceutical Dosage Forms*, 3rd edition. Informa Healthcare, New York, pp. 105–128.
(53) Aitken-Nichol, C., Zhang, F. and McGinity, J.W. (1996) Hot melt extrusion of acrylic films. *Pharmaceutical Research*, **13**, 804–808.
(54) Repka, M.A., Gutta, K., Prodduturi, S., Munjal, M. and Stodghill, S.P. (2005) Characterization of cellulosic hot-melt extruded films containing lidocaine. *European Journal of Pharmaceutics & Biopharmaceutics*, **59**, 189–196.
(55) Repka, M.A., Gerding, T.G., Repka, S.L. and McGinity, J.W. (1999). Influence of plasticizers and drugs on the physical-mechanical properties of hydroxypropylcellulose films prepared by hot melt extrusion. *Drug Development & Industrial Pharmacy*, **25**, 625–633.
(56) Repka, M.A. and McGinity, J.W. (2000) Physical-mechanical, moisture absorption and bioadhesive properties of hydroxypropylcellulose hot-melt extruded films. *Biomaterials*, **21**, 1509–1517.

(57) Repka, M.A., Prodduturi, S. and Stodghill, S.P. (2003) Production and characterization of hot-melt extruded films containing clotrimazole. *Drug Development & Industrial Pharmacy*, **29**, 757–765.
(58) Prodduturi, S., Manek, R.V., Kolling, W.M., Stodghill, S.P. and Repka, M.A. (2005) Solid-state stability and characterization of hot-melt extruded poly(ethylene oxide) films. *Journal of Pharmaceutical Sciences*, **94**, 2232–2245.
(59) Ranade, V.V. (1996) Role of polymers in drug delivery. In Ranade, V.V. and Hollinger, M.A. (eds) *Drug Delivery Systems*. CRC Press, Boca Raton, Florida.
(60) Prodduturi, S., Urman, K.L., Otaigbe, J.U. and Repka, M.A. (2007) Stabilization of hot-melt extrusion formulations containing solid solutions using polymer blends. *AAPS PharmSciTech*, **8**(2), Article 50.
(61) Prodduturi, S., Manek, R.V., Kolling, W.M., Stodghill, S.P. and Repka, M.A. (2004) Water vapor sorption of hot-melt extruded hydroxypropyl cellulose films: effect on physico-mechanical properties, release characteristics, and stability. *Journal of Pharmaceutical Sciences*, **93**, 3047–3056.
(62) Rehage, G. and Borehard, W. (1973) Molecular dynamics modeling of amorphous polymers. In Haward, R.N. (ed.) *The Physics of Glassy Polymers*, Applied Science Publishers, London, pp. 54–107.
(63) Crowley, M.M., Fredersdorf, A., Schroeder, B., Kucera, S., Prodduturi, S., Repka, M.A. and McGinity, J.W. (2004) The influence of guaifenesin and ketoprofen on the properties of hot-melt extruded polyethylene oxide films. *European Journal of Pharmaceutical Sciences*, **22**, 409–418.
(64) van Krevelen, D.W. and Hoftyzer, P.H. (1976) *Properties of Polymers, their Estimation and Correlation with Chemical Structure*. Elsevier Scientific Publication Co., Inc., Amsterdam.
(65) Thumma, S., Majumdar, S., ElSohly, M.A., Gul, W. and Repka, M.A. (2008) Chemical stability and bioadhesive properties of an ester prodrug of Delta 9-tetrahydrocannabinol in poly(ethylene oxide) matrices: effect of formulation additives. *International Journal of Pharmaceutics*, **362**, 126–132.
(66) Repka, M.A., Mididoddi, P.K. and Stodghill, S.P. (2004) Influence of human nail etching for the assessment of topical onychomycosis therapies. *International Journal of Pharmaceutics*, **282**, 95–106.
(67) Thumma, S., ElSohly, M.A., Zhang, S.Q., Gul, W. and Repka, M.A. (2008) Influence of plasticizers on the stability and release of a prodrug of Delta(9)-tetrahydrocannabinol incorporated in poly (ethylene oxide) matrices. *European Journal of Pharmaceutics & Biopharmaceutics*, **70**, 605–614.
(68) Cilurzo, F., Cupone, I.E., Minghetti, P., Selmin, F. and Montanari, L. (2008) Fast dissolving films made of maltodextrins. *European Journal of Pharmaceutics & Biopharmaceutics*, **70**, 895–900.
(69) Prodduturi, S., Manek, R.V., Kolling, W.M., Stodghill, S.P. and Repka, M.A. (2005) Solid-state stability and characterization of hot-melt extruded poly(ethylene oxide) films. *Journal of Pharmaceutical Sciences*, **94**, 2232–2245.
(70) Repka, M.A., O'Haver, J., See, C.H., Gutta, K. and Munjal, M. (2002) Nail morphology studies as assessments for onychomycosis treatment modalities. *International Journal of Pharmaceutics*, **245**, 25–36.

(71) Mididoddi, P.K. and Repka, M.A. (2004) Influence of etching and instrument variables on the bioadhesion of hot-melt extruded drug delivery systems for the human nail. *American Association of Pharmaceutical Scientists*, **6**(4), 95–106.
(72) Mididoddi, P.K., Prodduturi, S. and Repka, M.A. (2006) Influence of tartaric acid on the bioadhesion and mechanical properties of hot-melt extruded hydroxypropyl cellulose films for the human nail. *Drug Development & Industrial Pharmacy*, **32**, 1059–1066.

9

Taste Masking Using Hot-melt Extrusion

Dennis Douroumis
University of Greenwich, School of Science, Medway Campus, Chatham Maritime, Kent
Marion Bonnefille and Attila Aranyos
Alpha MOS, Toulouse, France

9.1 The Need and Challenges for Masking Bitter APIs

Taste masking of poorly palatable drugs is a major facet for the development of oral dosage forms. An increased number of active substances are already in the pharmaceutical pipeline, with many of those stimulating an unpleasant taste such as bitter, sour or metallic. Taste masking is therefore a prerequisite for the improvement of palatability and compliance of pediatric or geriatric orally administrated products. Interestingly, palatability is not always considered the key defining characteristic and in some cases is omitted. For example, the US Food and Drug Administration in the Guidance for Industry: Orally Disintegrating Tablets [1] has not included palatability criteria. In contrast, the new Pediatric Regulation (EMEA 2007) describes planned measures for the pediatric development where taste masking or palatability assessment is included in the proposed studies.

The increased industrial interest in new taste-masking technologies indicates that palatability plays an important role in the commercial success of finished dosage forms. Therefore, the combination of taste masking and formulation technologies [2, 3] such as orally disintegrating tablets (ODTs), quick-dissolving films/strips or pellets aims to address patient palatability. The major objective is to minimize the bitter taste intensity and duration of the active substance by leaving a pleasant taste and mouth feel.

Hot-melt Extrusion: Pharmaceutical Applications, First Edition. Edited by Dennis Douroumis.

Taste masking of oral solid dosage forms and especially pediatric medicines [4] can be challenging and formulators need to take into account a variety of formulation attributes. From a technical point of view, a taste-masking approach that does not impinge the other characteristics of the dosage form is a huge challenge. The taste masking of active substances with high solubility is often a difficult task as drugs dissolve rapidly in the oral cavity. High effective doses can compromise taste-masking efficiency due to the limitation in the drug loading of the used masking technology. Specific consideration needs to be given to potential excipient-related adverse effects in pediatric formulations, particularly for chronic conditions where repeated administration is required. The excipient types and levels should therefore be in accordance to the regulatory guidance without exceeding the established limits. In particular, carcinogenic sweeteners and artificial coloring/flavoring agents should be avoided. The selection of the appropriate excipient is paramount as the taste masking of an active substance must be balanced with the dissolution and the organoleptic properties of the finished product. Taste masking should be stable during manufacturing and disintegration in the mouth, but provide the appropriate profile in the gastrointestinal tract and consequently retaining the desired pharmacokinetic profile. The in vivo performance of a dosage form can be altered depending on the taste-masking approach (e.g. use of ion exchange resins, prodrugs or salt forms) [5].

Furthermore, there are several challenges related to the taste assessment methods. Currently the existing in vitro and in vivo techniques include the electronic tongue and animal and human models that are used to evaluate or predict the taste-masking efficiency at the early stage. However, information about the taste of drugs in the current literature is very limited. Recently Cram et al. [5] underlined the need to generate taste data at an early stage during the formulation development to direct pediatric products. Another challenge is the different taste perception in both adults and children or between healthy and sick individuals [6, 7]. Even though in vivo taste-masking evaluation is not an issue for adults, there are several restrictions and concerns. The EU (EMEA 2008) regulatory guidelines [8] strictly forbid the enrolment of children in clinical studies and encourage adult participation. The validity and the interpretation of pediatric palatability tests are quite often questioned. Hence, quantitative taste assessments are conducted by trained evaluators in various steps [9]. Recent advances in taste assessment use multichannel taste sensor instrumentation commonly referred to as 'the electronic tongue' (e-tongue) which has become an established alternative to human sensory analysis panels (SAPs). An increased number of pharmaceutical laboratories have employed the e-tongue to assess the bitterness and masking efficiency of pharmaceutical products. Furthermore, the e-tongue is used in placebo formulation in taste-matching studies and in unknown-to-reference comparisons [10, 11]. The operational principles and features are described later in this chapter.

A wide variety of techniques have been employed to mask the taste of poor-tasting active ingredients such as fluid bed coating, microencapsulation, complexation, freeze-drying or spray congealing [2]. The type of taste-masking platform used is dictated by the physicochemical properties of the active ingredient and its physical form, including water solubility, permeability, polymorphism, hydroscopicity, physical/chemical stability or mechanical properties such as compressibility. An effective taste-masking approach should prevent direct contact of the active substance with the taste buds present on the tongue surface. Several masking technologies (e.g. Zydis®, Durasolv®, Flashtab®) have been successfully commercialized including a wide range of approved products.

Hot-melt extrusion is an emerging technology and, despite the fact that there are no commercial products, it can be efficiently used to mask the taste of very bitter active substances. HME can be proved advantageous compared to the aforementioned techniques as it is a continuous process, easy to scale up, applicable for moisture sensitive actives, provides enhance API stability within the carrier matrix and it is not time consuming. In this chapter we will describe how HME can be employed for taste-masking purposes by producing pellets and granules.

9.2 Organization of the Taste System

The taste system is the second chemosensory system which represents the chemical and physical characteristics of ingested substances. The taste system detects what is present in the mouth and enables discrimination and recognition of the chemical components including the levels. In simple words, it indicates whether or not a food or an active substance is safe to be ingested. When a substance is placed in the mouth, its chemical constitutes interact with receptor proteins on taste cells located on epithelial specializations called taste buds. The taste buds are distributed not only on the tongue but also on the soft palate, pharynx, larynx, cheeks and the upper part of the esophagous [12]. The afferent taste information is transmitted through the taste bud via neural fibers. Primary sensory axons in the chorda tympani branch of cranial nerve VII (facial), the lingual branch of cranial nerve IX (glossopharyngeal) and the superior laryngeal branch of cranial nerve X (vagus) innervate with the neural fibers and carry taste information. The information is transferred into the central nervous system at the level of the brain stem and synapse, initially onto the nucleus of the solitary tract in medulla known as gustatory nucleus [13].

9.2.1 Taste Perception in Humans and Organization of Peripheral System

The taste system recognizes five perceptually distinct categories of tastants such as salt, sour, sweet, bitter and umami [14]. Most taste stimuli include non-volatile, hydrophilic molecules that are soluble in saliva. Some examples include salts such as NaCl which is required for electrolyte balance, amino acids such as glutamate that are needed for pro-energy, sugars and various carbohydrates such as glucose that are required for energy, acids such as citric acid that are related to the acididity (H^+) and thus palatability of foods. Generally, the perceived intensity of taste is directly related to the concentration of the taste stimuli with the high threshold concentrations for most of the ingested tastants.

The tastants are located over the full surface on receptive molecules named taste papillae. Papillae are multi-cellular protuberances surrounded by local invaginations in the tongue epithelium forming a trench to concentrate surfactants. The taste buds are located on the lateral surfaces of the papillar protuberance and the trench walls. Three types of papillae are known including fungiform (contains 25% of the taste buds), circumvalate (50%) and foliate (25%), which are discontinuously distributed on the tongue surface. Fungiform papillae are placed on the anterior two-thirds of the tongue with the highest density (30 per cm^2) being at the tip. They have a mushroom-shaped structure with three taste buds on their apical surface [13]. In contrast, there is an increased number of circumvalate papillae in a chevron at the rear of the tongue and 250 taste buds on the trench walls. Finally, foliate

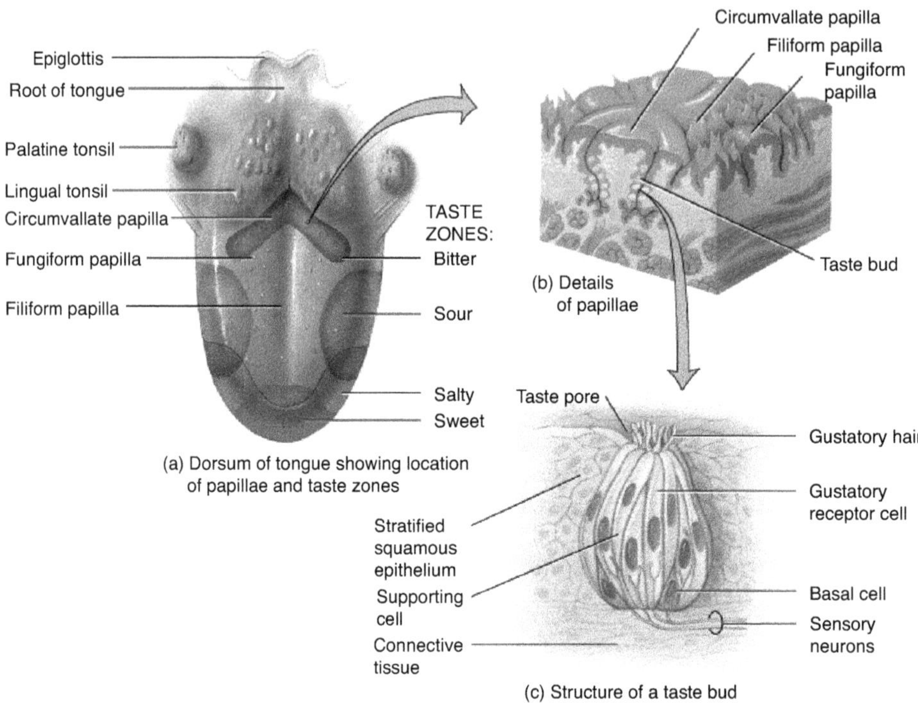

Figure 9.1 *Taste buds and the peripheral innervation of the tongue. (a) Distribution of taste papillae on the dorsal surface of the tongue. Different responses to sweet, salty, sour and bitter tastants recorded in the three cranial nerves that innervate the tongue and epiglottis are indicated at left. The size of the circles representing sucrose, NaCl, HCl, quinine and water corresponds to the relative response of the papillae to these stimuli. (b) Circumvallate papilla showing location of individual taste buds. (c) Diagram of a taste bud, showing various types of taste cells and the associated gustatory nerves. The apical surface of the receptor cells have microvilli that are oriented toward the taste pore. For a better understanding of the figure, please refer to the color section.*

papillae are located on the posterolateral tongue with about 20 parralel ridges and 600 taste buds in their walls.

There are approximately 4000 taste buds in humans distributed within the total oral cavity. They are 50 μm wide and 80 μm long with each of them containing 100 taste cells and a few basal cells (Figure 9.1). The taste bud receptors are composed of at least four cell classes: dark cells (Type I), light cells (Type II), basal cells and stem cells [15]. They are called dark cells because of their affinity for accumulating basophilic dyes and are slender cells extending from the base of a taste bud to its apical surface, where they end in a number of small receptor villi. The light cells are averse to basophilic dyes and extend the length of a taste bud, but their apical surface ends as a single, large, club-shaped villus. Light cells are the main sensory cells and are extensively innervated by afferent fibers while numerous synapses occur on their basal surfaces. Quite often synapses can be found on dark cells, and these cells may represent a second class of sensory cells. In humans, the taste buds include a small number of disk-like basal cells that lie directly on the basement membrane of the epithelium. These cells often come into contact with the entering afferent fibers and basal processes of both dark and light cells. The basal cells also contain several types of small vesicles and are rich in serotonin.

In general, each of the five tastes represented over the tongue surface correspond to distinct classes of receptor molecules expressed in subsets of the taste cells. The categories of taste perception and representation are therefore linked to the molecular biology of taste transduction. The mapping of response to sweet, bitter, salty, sour and umami in healthy humans demonstrated that each taste elicits focal activity in the taste cortex, indicating that information about each remains segregated throughout the taste system.

9.2.2 Transduction of Taste Signals

The molecular machinery for chemosensory transduction has been elucidated and it is well known how gustatory information is encoded by the taste buds. Salty and sour tastes are elicited by ionic stimuli such as the positive ions in salts (Na^+ from NaCl) or the H^+ in acids (e.g. acetic acid). A sensory transduction is initiated via specific ion channels that engage an amiloride-sensitive Na+ channel (ENaC) for salty tastes. Similarly, a H^+-permeant non-selective cation channel, which is a member of the transient receptor (TPR) family, is related to the sour tastes. Parenthetically, the sour receptor channel is related to a similar channel protein which is mutated in polycystic kidney disease (PKD). A distinct subset of taste cells is used to express the sour receptor similar to the receptor proteins for sweet, umami and bitter. In this case, a receptor potential is generated by the positive inward current carried either by Na+ for salt or H+ for sour, that depolarizes the relevant taste cell. Subsequently, this initial depolarization induces the activation of voltage-gated Na+ channels in the basolateral membrane of the taste cells. Furthermore, the additional depolarization activates voltage-gated Ca^+ channels, initiating the release of neurotransmitters and the activation of potentials in ganglion cell axons.

The sweet and umami receptors are heterodiomeric G-protein-coupled receptors that share a common seven-trans-membrane receptor subunit called T1R3 paired with a T1R2 receptor for sweet perception or with a T1R1 for amino acids. Both T1R2 and T1R1 receptors are expressed in different subsets of taste cells. When sugars or other sweet stimuli bind, the T1R2/T1R3 heteroderimer initiates a G-protein cascade that activates the phospholipase C iso-form $PLC_{\beta 2}$, leading to increased concentrations of inostitol triphosphate (IP3) and to the opening of TRP channels. The TRP depolarizes the taste cells through increased intracellular Ca^{2+}. Similarly, the transduction of amino acid stimuli can lead to depolarization of the TRPM5 channel via the $PLC_{\beta 2}$-mediated activation.

Bitter tastes are transuded through another G-protein-coupled receptor (GPCRs) known as the T2R receptor. Various studies confirmed that approximately 30 T2R subtypes encoded by 30 genes in humans and multiple T2R subtypes are expressed in single taste cells. Two of the most commonly used bitter compounds in human bitter taste studies, PTC (a mutation of the human T2R) and its structurally similar proxy PROP, can be recognized via the human receptor T2R38. However, PROP may be able to activate other T2Rs such as the T2R4 receptor. The distribution of T2R receptors among taste cells supports the hypothesis that bitter taste is distinct and encoded specifically in taste receptor cells. T2Rs are not expressed in the same taste cells such as T1R1, 2 and 3 receptors. The transduction of bitter stimuli has a similar mechanism to sweet and amino acid tastes; however, the cell-specific G-protein gustducin (a heterotrimeric guanine-binding protein) is found primarily in T2R expressing cells. Alpha-gustducin shares 80% amino acid identity with alpha-heteromeric and is selectively expressed in about 30% of taste bud cells; all T2Rs have been localized to alpha-gustducin-expressing taste bud cells. The activated alpha-gustducin stimulates phosphodiesterases to hydrolyze cAMP, and the decrease in cAMP levels modulates cyclic nucleo-tide-regulated ion channels and/or kinases. Beta and gamma subunits of gustducin activate phospholipase C of the b2 subtype to generate IP3, which leads to release of Ca^{2+} from internal stores via activation of inositol 1,4,5-trisphosphate receptor type III (IP3R3) [12].

9.3 Taste Sensing Systems (Electronic Tongues) for Pharmaceutical Dosage Forms

Various approaches have been used to evaluate the masking efficiency of developed pharmaceutical dosage forms (e.g. granules, tablets) such as human taste panels, animal models or analytical techniques [16, 17]. Most of these technologies encounter limitations and can lead to significant variations. In a recent publication, Woertz et al. [18] discussed the unknown toxicity of new chemical entities, the differences in taste senses between individual tasters and ethical concerns related to children that have an impact on the taste-masking evaluation. In addition, animal models are considered unreliable as they have totally different taste perception to humans.

Alternatively, electronic taste-sensing systems can be employed to evaluate the taste of pharmaceutical formulations [18–21]. Currently, there are two commercially available electronic tongues – the Astree e-tongue (Alpha MOS, Toulouse, France) and Insent taste sensing system (Atsugi-shi, Japan) – that have been employed and evaluated for taste-masking purposes. The Astree e-tongue can be used to evaluate the bitterness of pure active substances in comparison to formulated products. The e-tongue studies demonstrated excellent good correlation with human taste panels, reproducibility, low detection limits and high sensitivity [21].

9.3.1 Alpha MOS Electronic Tongue: Instrumentation and Operational Principles

The e-tongue is an instrument designed for taste analysis that mimics the working principle of the human sense of taste (Figure 9.2). Instead of measuring and identifying the various compounds responsible for the taste, as many analytical techniques can do, an e-tongue captures the global profile of a taste, as the human tongue does.

The biological mechanism of taste sensing involves three major steps as follows.

- *Detection*: Chemoreceptors for gustation (taste) respond to chemicals in an aqueous environment. Chemicals dissolved in saliva excite the taste receptors of the mouth. The receptors are able to transform this chemical signal into a change of the cell state involving an electrical signal.
- *Transmission*: When this electrical signal reaches a threshold, it generates another electrical signal which is then propagated through nerves to specific areas of the brain.

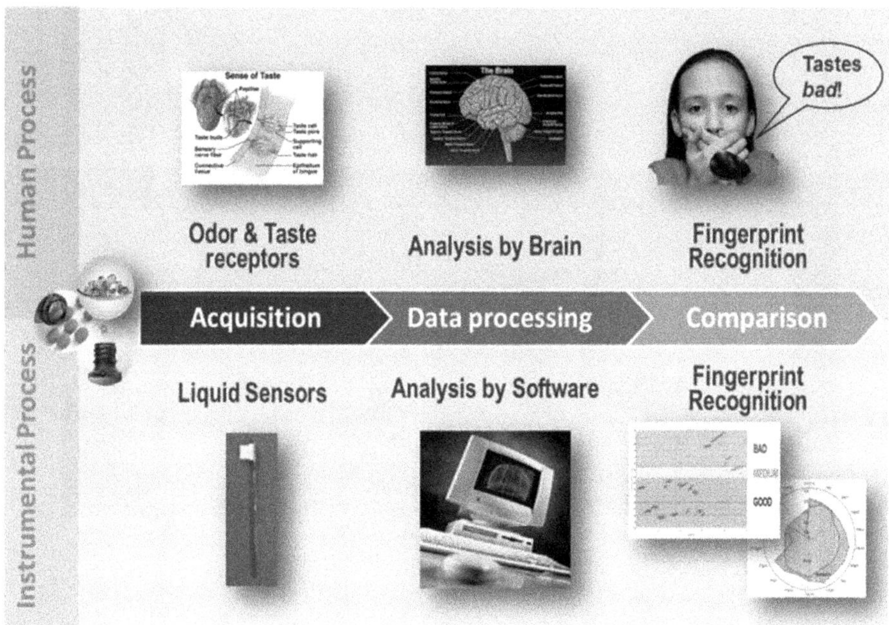

Figure 9.2 *Comparison of the biological mechanism of taste and the working principle of the electronic tongue.*

- *Integration*: Taste quality perception and recognition is based on building or recognition of activated sensory nerve patterns by the brain and on the taste fingerprint of the product.

Similarly, the e-tongue performs those three steps as follows.

- *Detection*: the instrument is equipped with several sensors that detect the dissolved organic and inorganic compounds. Like human receptors, each sensor is cross-selective which means it reacts to many substances responsible for various tastes and its spectrum of reactions is different from the others. The information given by each sensor is complementary and the combination of all sensor results generates a unique fingerprint.

  The Astree e-tongue (Alpha MOS, France) uses ChemFET (Chemical modified Field Effect Transistor) technology which is based on non-covalent and reversible chemical bonds (such as ionic or van der Waals bonds) with dissolved molecules. A potentiometric difference is measured between each of the 7 sensors and the Ag/AgCl reference electrode. Each sensor has a specific organic membrane which interacts with ionic, neutral and chemical compounds present in the liquid sample. Any interaction at the membrane interface is detected by the sensor and converted into an electronic signal. Most of the detection thresholds of ChemFET sensors are similar to or better than those of human receptors.
- *Transmission*: E-tongue sensors generate electric signals as potentiometric variations. Transducers are incorporated in the e-tongue to transform a chemical interaction (between compounds in the tested sample and the sensitive material of the sensors) into an electrical signal.

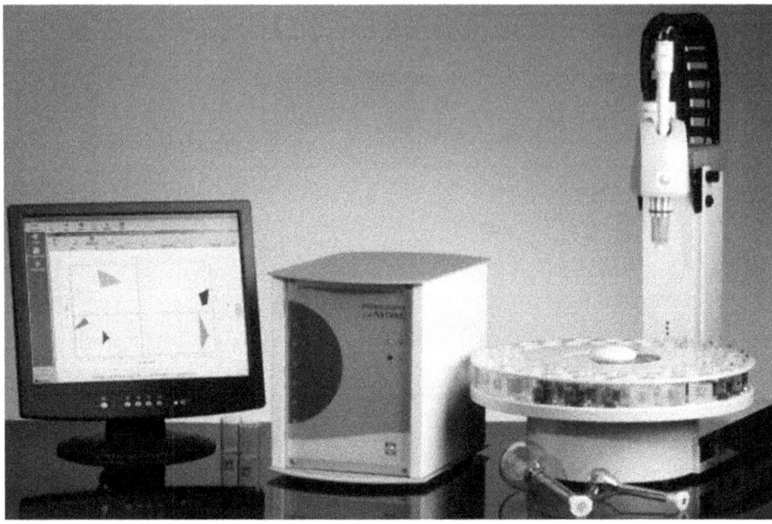

Figure 9.3 Astree e-tongue.

- *Integration*: The interpretation of the e-tongue sensor data into taste patterns is achieved through the instrument software. This software uses multivariate statistics to process the input from the different sensors. This allows easy-to-understand visual representations of the products' taste profiles to be delivered according to qualitative or quantitative criteria.

In practice, the Astree e-tongue system (Figure 9.3) comprises several parts:

1. a sampling system to ensure reproducibility of the analytical conditions;
2. an array of sensors directly mounted on the autosampler arm for automated analysis;
3. an electronic data acquisition system; and
4. a software for system monitoring, data acquisition and data processing.

9.3.2 Taste Analysis

The most widely used measurement method for taste assessment has long been the human sensory test, often conducted by a large trained panel. These human tests are resource-consuming and may require the same safeguards as a clinical trial to avoid health hazards. As an alternative, the e-tongue, which is designed for taste analysis of liquid matrices, can help meet the needs for human safety and measurement rapidity. In the pharmaceutical area, this instrument is particularly suitable for the analysis of oral forms such as gels, syrups, solutions, tablets, lozenges, caps, granulates, etc.

Liquids are directly analyzed without any preparation. Solids require a preliminary dissolution before measurement. Before analysis, the solutions should be filtered to remove particles that could stick to the membrane of the sensor and interfere with the measurement. The reference electrode and the sensors are dipped in a beaker containing the sample for 120 sec. Once sensors have reached a stable equilibrium (Figure 9.4), raw data are recorded in the form of the potentiometric value.

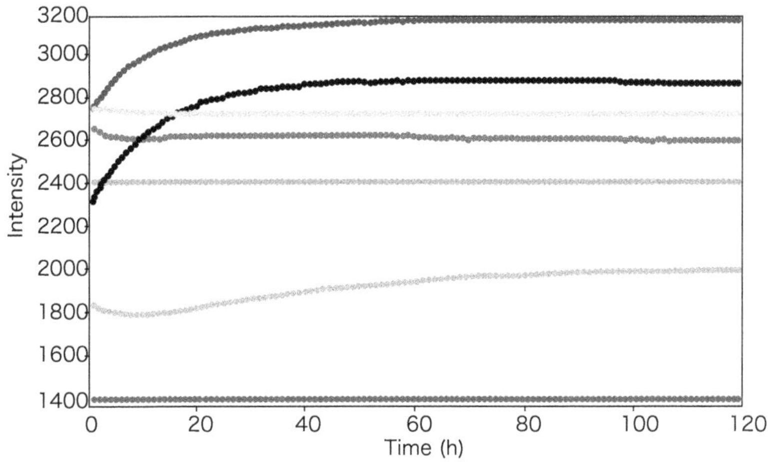

Figure 9.4 *Sensor signals obtained with Astree e-tongue. For a better understanding of the figure, please refer to the color section.*

The set of sensors and reference electrode are rinsed with distilled water between two analyses. Sensor data can be processed using different multivariate statistical models (Figure 9.5), based on the objective of the application:

- qualitative models such as principal component analysis (PCA) or discriminant factorial analysis (DFA) are used to differentiate products based on desired parameters (quality, origin, supplier, batch, etc);
- quality-control charts (statistical quality control model) allow product conformity to be checked;
- quantitative models (partial least square regression) for determining a taste attribute score or a chemical compound concentration; or
- shelf-life model is designed for following and comparing aging profiles and stability of raw materials or end products under various storage conditions or over time.

In the pharmaceutical industry, the e-tongue is principally used by the departments or companies involved in the formulation of oral forms. The main applications concern:

- the evaluation of the bitterness-masking efficiency of a formulation (encapsulation, extruded polymer, coating, flavors, excipients, etc.) towards active principles;
- the assessment of the placebo taste so that it matches the taste of the active formulation as much as possible for double-blind clinical trials;
- the quantification of API (active pharmaceutical ingredient) or NCE (new chemical entities) bitterness for which almost no information is available;
- the follow-up of taste stability of pharmaceutical ingredients over time; or
- the benchmarking of competitive or generic products in terms of taste.

9.3.3 Taste Masking Efficiency Testing

The vast majority of pharmaceutical active ingredients have an unpleasant (often bitter) taste. This unpleasant taste ends up in the drug and, in the worst cases, could prevent the

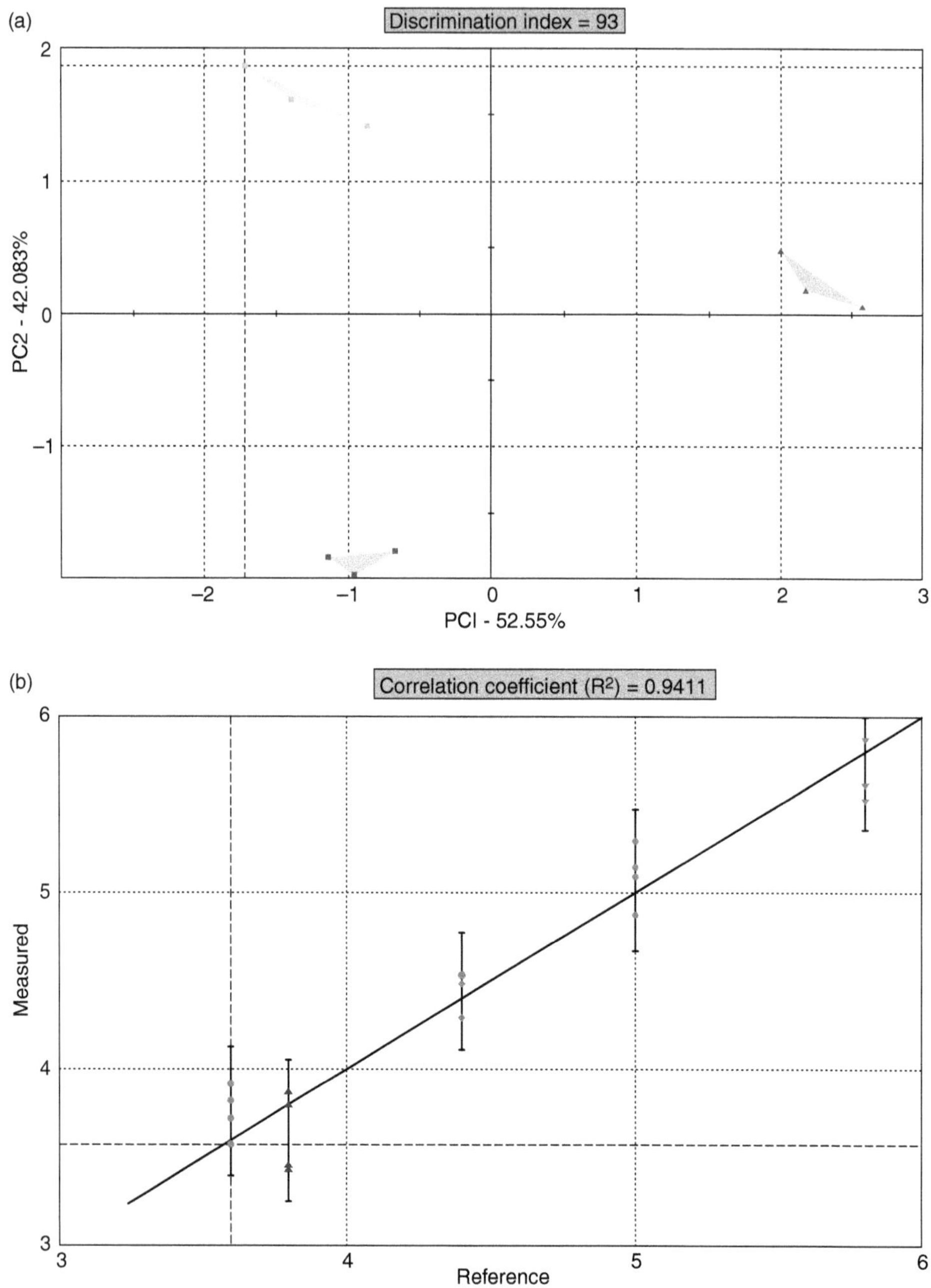

Figure 9.5 (a) Taste map of three products of different qualities obtained through Principal Components Analysis with ASTREE e-tongue; (b) correlation model between sensory evaluation and e-tongue measurements for the determination of a sensory attribute score (partial least square model); and (c) statistical quality control model showing the area of acceptable quality (green band) and out of specification grade (white area). For a better understanding of the figure, please refer to the color section.

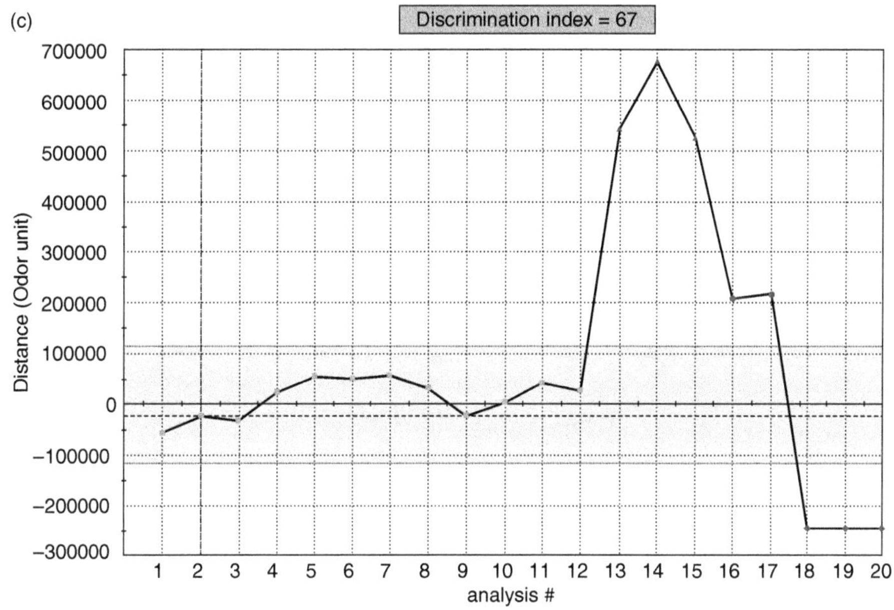

Figure 9.5 *(Continued)*

patient from following his treatment. The taste-masking challenge that formulators need to address consist of finding which, among several formulations, will better mask the bad taste. For that, each pharmaceutical formulation containing the active ingredient (so-called active formulation) is compared to the corresponding placebo formulation (containing exactly the same ingredients at the same concentrations without active ingredient) having a good and acceptable taste. The smaller the difference between the active formulation and the corresponding placebo formulation, the more the active formulation will be detected as similar to the corresponding placebo formulation and better will be the active ingredient taste masking.

The taste fingerprint of all products can first be visualized on a PCA. PCA allows the taste measured to be mapped in seven dimensions (seven sensors) in two principal dimensions using combined data from all the raw data (Figure 9.6). The axes of such a taste map are composed axes based on the original axes where the two most 'informative' (capturing most variance) are selected as the two dimensions. Such axes have no direct physical meaning, other than that they represent variability and difference in taste. In the illustrative example above, Formulation 4' is clearly the best choice to be as close as possible as to the placebo taste. To quantify the masking power, the Euclidean distance inactive-active formulation is then calculated; the smaller the distance between the two products, the more similar their taste.

9.3.4 Advantages of E-tongue Taste Analysis

Some years ago, it was commonly accepted that drugs tasted bad. Some even thought that the worse the remedy tasted, the more effective it was. Nowadays, patients are no longer willing to swallow bad-tasting medicines. Product developers are therefore under pressure to create pleasant-tasting medicines. The development of oral pharmaceuticals, orally disintegrating

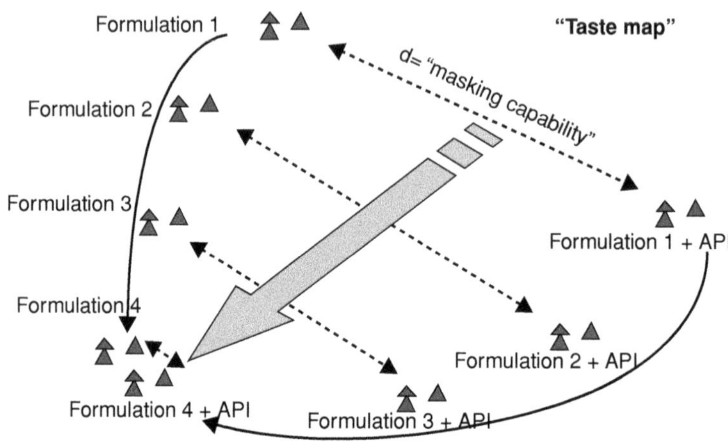

Figure 9.6 PCA map calculated from e-tongue data comparing different formulations.

tablets and dissolvable films has led to an increasing interest in taste-masking strategies with sweeteners and flavorings, vehicles and coatings.

To improve a product's taste palatability and patients' acceptance, sensory-directed formulation development is mandatory. Such formulation developments require a solid base of excipients, fully screened taste-masking agents and a controlled system to fine-tune the process.

In parallel, taste evaluation should not add an excessive additional expense and extra development time. The formulation scientists therefore need reliable, fast and cost-effective methods for taste evaluation. If the sensory panel has long been the reference method, the development of e-tongue analyzers has opened new possibilities and perspectives.

Substituting the e-tongue for human testing brings several advantages. First, the use of a reproducible and reliable instrumental measurement avoids the subjectivity or the bias entailed by fatigue in the evaluation when a panel is employed. Instrumental methods also prevent the health of panelists, who would have to ingest medical substances, from being jeopardized.

Moreover, constituting and maintaining a panel group is resource demanding: recruitment, training, conducting tasting sessions and setting safety measures and controls all require both time and money. By requiring almost no sample preparation and by performing analyses 24 hours a day, the e-tongue significantly reduces development time and costs.

Thanks to the e-tongue, research and development industrials are able to perform fast qualitative and quantitative analyses of medicine sensory properties.

9.4 Hot-melt Extrusion: An Effective Means of Taste Masking

9.4.1 Taste Masking via Polymer Extrusion

Hot-melt extrusion has been demonstrated to be an efficient taste-masking processing technology for bitter active pharmaceutical ingredients. Although there is a limited number of taste-masking studies, various researchers have demonstrated that HME facilitates taste masking. Typically, this can be achieved by selecting the appropriate lipophilic or hydrophilic inactive ingredients and optimizing the processing parameters.

It was firstly Petereit et al. [22] who extruded anionic methacrylate copolymers (e.g. Eudragit L100, Eudragit L100-55) with basic salts of active ingredients to produced taste-masked granules. They observed that HME processing of methacrylate polymers with a high content of anionic radicals can result in drug–polymer complexation when processed with cationic drugs after the addition of the suitable plasticizer. According to the authors, HME processing enables strong interactions between the functional groups of the drug–polymer mixture. For the purposes of the study, Verapamil HCl was used as a model drug with loadings varying form 50–70% (0.38–1.47 molar ratios). In a typical process, each batch was processed though a Leistritz Micro 18 GL-40D extruder with a 1.5 mm nozzle diameter, residence time approximately 4 min, screw speed 150–300 rpm/min and zone temperatures 120–150°C. The produced granules were evaluated in terms of masking efficiency for 30 sec by an experienced panel group. Taste masking was proportional to the Verapamil HCl amounts, showing neutral taste for drug loadings of 30–50% and only a slightly bitter taste for 70% loadings. A comparative extrusion study at 50% Verapamil HCl showed improved taste masking of the produced granules when the processing temperatures were elevated to 140–150°C from 120–130°C while the screw rotation speed was reduced to 150 rpm from 300 rpm. In the latter case, the panelist scores gave neutral taste compared to the slightly bitter taste for batches processed at highest temperatures and lower rotation speeds. Interestingly, these findings demonstrate the HME mixing power where dissolution of drugs in polymeric melts can take place through a convective diffusion process [23]. The masking efficiency can therefore be enhanced by adjusting the temperature and the screw speed. In addition, extrudates presented drug dissolution rates of 55–60% within 60 min.

The Verapamil HCl/Eudragit L100-55 extrudates were further characterized by X-ray photoelectron spectroscopy (XPS) to determine possible drug–polymer interactions. The XPS analysis provided strong evidence of Verapamil HCl interactions with L100-55 via hydrogen bonding. Indeed, the estimated N-coefficient values from XPS data showed low values for the extrudates compared to pure drug, suggesting strong interactions between the tertiary nitrogen atom of Verapamil HCl and the carboxylic groups of the Eudragits (Figure 9.7). Low N-coefficient values indicate higher amounts of protonized nitrogen atoms and thus stronger inter-molecular interactions.

The same principle applied for anionic active substances processed with polymers which have functional tertiary amino groups [24]. In this example, Petereit et al. processed Ibuprofen (IBU, 17–40%), a water-insoluble substance, with Eudragit EPO® copolymer at various ratios. For each sample stearic acid (~15–35%) and talc (8.5–16.5%) were used as processing aids. The samples were processed at 100°C at 60 rpm and a residence time of 20 min. The panelist scores showed effective masking of IBU granules for at least 2 min and drug concentrations up to 25%. At concentrations of 33%, the bitterness of IBU appeared after 1 min. It was also found that stearic acid contributed to the reduced bitterness after comparison of extrudates with and without this ingredient. The authors proposed that the reduced IBU bitterness is due to the strong drug–polymer ionic interactions and, as a result, the drug is molecularly dispersed in the polymer matrix. This assumption was confirmed by differential scanning calorimetry and X-ray diffraction studies. Further, Fourier transform infrared (FTIR) spectroscopy investigations elucidated a possible taste-masking

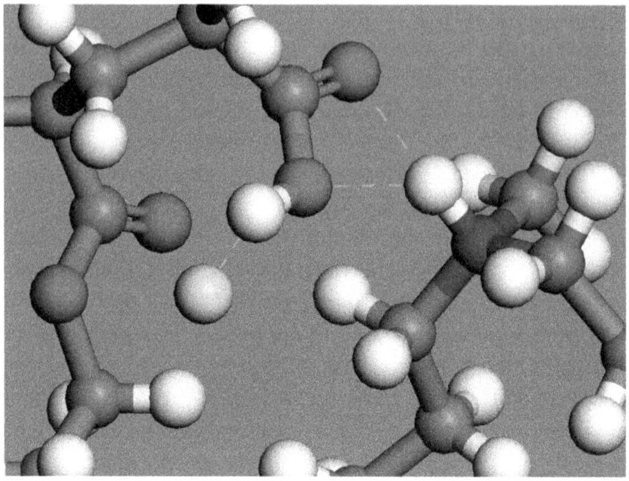

Figure 9.7 Computerized simulation of the hydrogen bridge bonding between Verapamil HCl and Eudragit® L100–55. Reproduced with permission from Evonik Industries AG. For a better understanding of the figure, please refer to the color section.

mechanism attributed to the intermolecular ionic interactions between the IBU functional carboxylic and the EPO dimethylamino groups. IBU can act as a hydrogen donor with the hydrogen bonding acceptor dimethylamino group. The deprotonation of the –COOH facilitates the formation of a carboxylate salt and consequently builds a taste-masking effect on the molecularly dispersed IBU.

In a recent study Gryczke et al. [25] processed 40% IBU with EPO (50%) and talc (10%) but without stearic acid. The authors demonstrated sufficient taste masking for at least 2 min when processing temperatures were elevated to 140°C. It was found that the increased IBU concentrations enhanced drug–polymer interactions, as IBU has been found to present plasticizing effects compared to traditional plasticizers [26]. The presence of a single T_g and the absence of the IBU melting endotherm confirmed the complete miscibility of IBU/EPO and the creation of a glassy solution where IBU was molecularly dispersed within EPO [27, 28]. The obtained granules were incorporated in orally disintegrating tablets (ODTs) using a variety of superdisintegrants while the compressed ODTs were further evaluated and compared to Nurofen® Meltlet ODTs. The ODTs developed by extruded granules showed disintegration times and crushing resistance similar to commercial Nurofen® tablets but improved tablet friability with increased IBU release rates.

More recently, our group conducted a comparative taste-masking study of extruded paracetamol (PMOL) with Eudragit EPO and cross-linked polyvinylpyrrolidone (Kollidon VA64®) at different drug loadings [29]. Extrudates were prepared through a Randcastle (US) 0625 single-screw extruder with temperatures varying over 100–115°C across the five zones and residence time of 5 min. The drug loadings varied over 40–60% for both polymers without the addition of a plasticizer. The taste evaluation of the developed formulations carried out by using an Astree-tongue equipped with seven sensor tests and the generated data were analyzed using multidimensional statistics. The data analysis showed significant suppression of the bitter taste for PMOL for both polymers. However, PMOL masking was strongly dependent on the nature of the polymeric carriers and the drug loading in the final formulation. Both polymers showed excellent taste masking for active concentrations up to 50%.

The control solutions (100% PMOL) and placebo polymers (VA64 & EPO) were well separated on the taste maps (Figure 9.8). Additionally, the taste maps showed significant discrimination between placebo and extruded formulations. The three drug–polymer solutions were close and far from PMOL, indicating a significant taste evolution and a masking improvement for the Kollidon extrudates towards pure PMOL. Despite the lowest distance

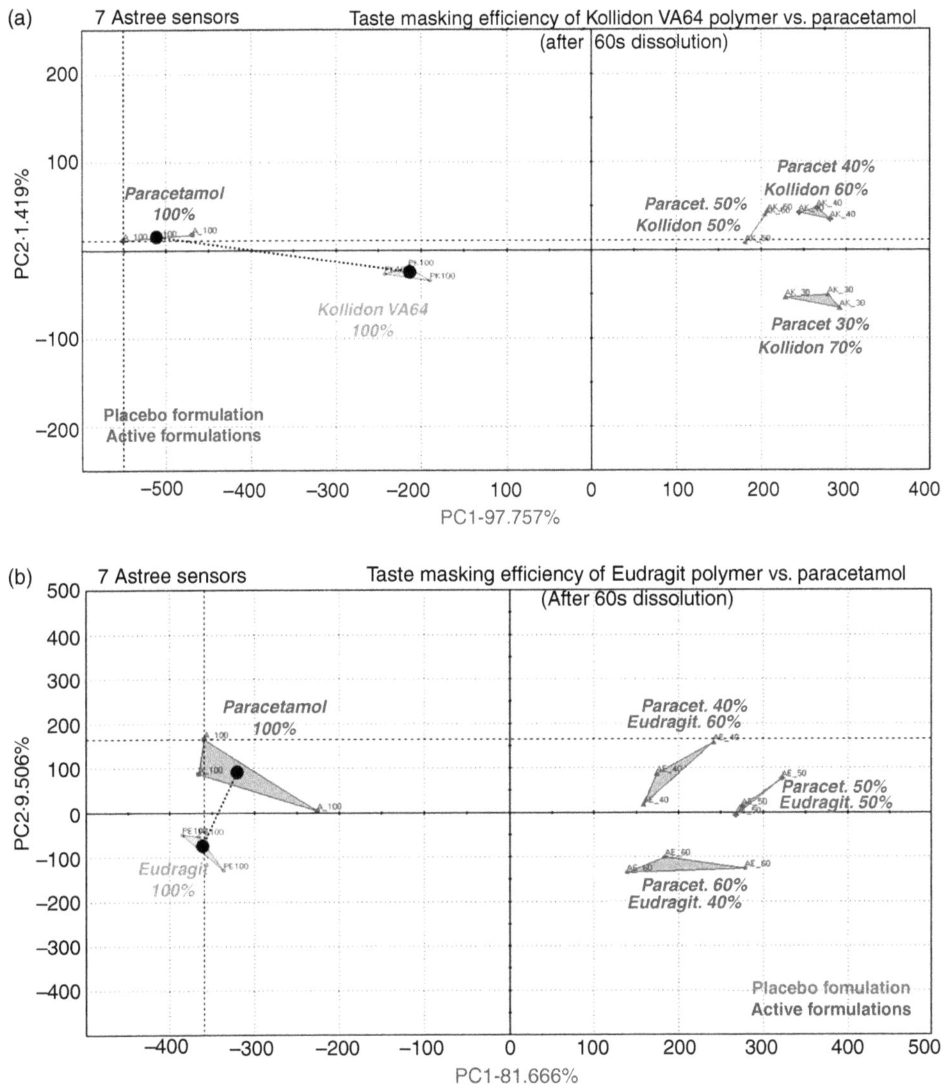

Figure 9.8 *Electronic tongue 'taste map'. Global signal comparison (PCA analysis of the electrode responses) of pure PMOL and extruded formulations to (a) VA64 polymer and (b) EPO polymer after dissolution for 60 s. M. Maniruzzaman et al. 2012, reproduced with permission of Elsevier. For a better understanding of the figure, please refer to the color section.*

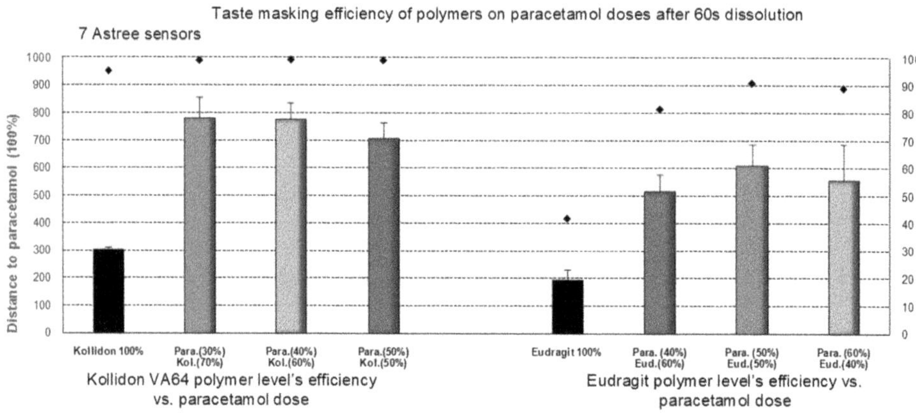

Figure 9.9 *Distance and discrimination comparison between signal of 100% PMOL formulation and each polymer's formulation on Astree e-tongue (after 60 s). M. Maniruzzaman et al. 2012, reproduced with permission of Elsevier.*

from pure active to placebo formulation as VA64 solutions, the same conclusions were observed for EPO polymer. The distance between the pure active and the polymer formulations are indicative of the taste-masking power of each polymer's dose. When it increases, it suggests that taste is farther than pure PCM and thus significant taste improvement was attribute to both drug–polymer extrudates compared to polymer alone (discrimination index or DI > 80%). An improved taste was observed with VA64 polymer in which the highest average distance was obtained for 30% drug loading. Although EPO showed the closest distance and lowest DI to pure API, the masking is not significantly different from that of VA64 but the optimum result was achieved with 50% PCML/EPO loadings (Figure 9.9). The extruded PMOL formulations were investigated in parallel by in vivo taste-masking studies carried out by a group of six healthy human volunteers. The in vivo evaluation was in good agreement with the e-tongue results. As can be seen in Figure 9.10, a sensory correlated model based on partial least square (PLS) was built to evaluate the correlation with sensory scores. The correlation model is valid ($R^2 < 0.8$) despite dispersion and low discrimination between formulations ($p > 0$).

9.4.2 Taste Masking via Solid Lipid Extrusion

An interesting approach that has been employed to mask the taste of bitter active substances via HME is the use of lipidic matrices. Lipid excipients have been used mainly in hot-melt coating processes or solid dispersions to achieve taste masking of active substances [30–32]. It was first Breitkreutz et al. [33, 34] who introduced the utilization of different lipids in a process named 'cold solvent-free extrusion'. In this study glycerol distearate (Precirol® ATO5), hard fat (Witocan® 42/44) and stearic acid were cold extruded to mask the taste of sodium benzoate. The dry powder blends were fed at room temperature and required the addition of polyethylene glycol (because sodium benzoate did not melt or soften under the processing conditions). In this initial attempt, only hard fat extrudates appeared to be superior to glycerol distearate and stearic acid, but a coating process with Eudragit E

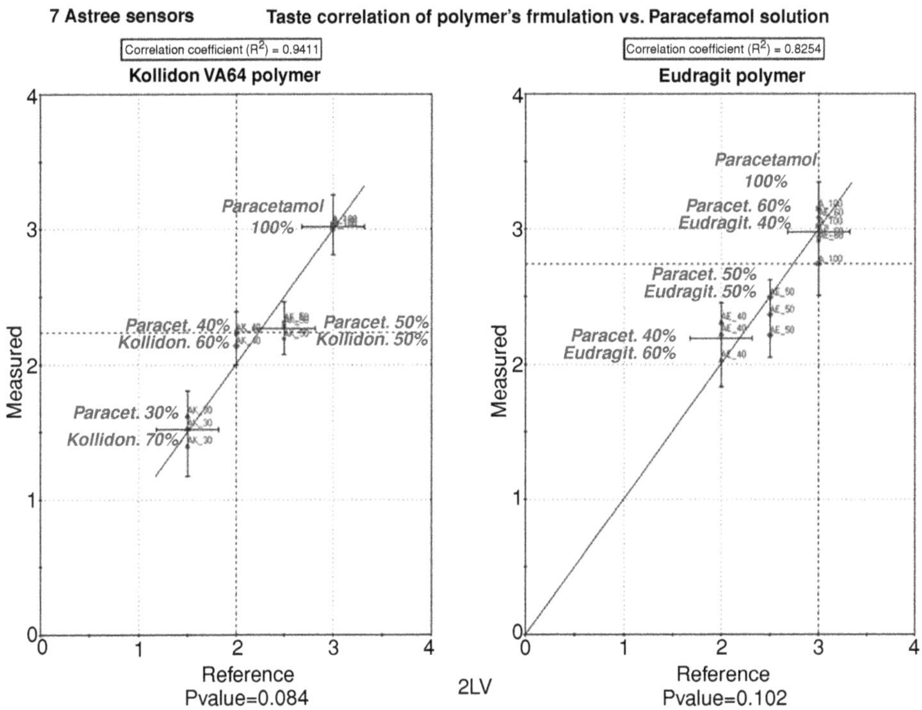

Figure 9.10 *Correlation of human sensory data 'reference' with Astree e-tongue measurements ('measured'). M. Maniruzzaman et al. 2012, reproduced with permission of Elsevier.*

was still necessary to formulate an acceptable pediatric dosage form. Nevertheless, taste masking was effective for at least 5 min in the buccal cavity, mainly due to the Eudragit E coating layer that prevented the release of sodium benzoate. Another disadvantage of this approach was the shape of the obtained lipid granules which was still cylindrical and not spherical, even though a spheronization process had been applied.

In a later study, the same group [35] processed various binary, ternary and quaternary mixtures of powdered lipids with sodium benzoate through solvent-free cold extrusion. The aim of the study was to prepare immediate release pellets with solid lipid binders and compare them to well-known wet extrusion binders such as microcrystalline cellulose and κ-carrageenan. In order to improve the extrudates, the authors combined one or more lipids (Witocan® 42/44, Dynasan® 114, Precirol® ATO5, Compritol® 888ATO) resulting in binary, ternary and quaternary powder mixtures that were fed in a twin-screw extruder at room temperature. The produced extrudates presented rapid-release profiles with 90% sodium benzoate released within 40 min; when stored under accelerating conditions, slower release rates were however observed. This behavior was only observed with samples containing Precirol ATO5 for the first hour of dissolution followed by complete drug release after two hours. However, the storage conditions did not affect the release profiles of the batches with Witosan 42/44 or Witosan 42/44 and Compritol 888ATO. In addition, some of the developed pellets demonstrated spherical shape, narrow particle size distribution and high drug loading of 80% sodium benzoate. The authors did not examine the masking efficiency of the manufactured lipid formulations, which makes the solvent-free cold extrusion process questionable for the taste-masking purposes. Nevertheless, further evaluation of this process is required until it is considered a viable taste-masking approach.

In a similar approach Michalk et al. [36] investigated thoroughly the dependency of the release profile on the original diameters of milled extrudates manufactured by solid lipid extrusion. The authors used different die diameters in the range 0.3–5.0 mm with different lengths and the length to diameter ratio was kept constant for diameters smaller than 0.5 mm [37]. In addition, the effect of the screw speed and temperature range on the extrusion process was assessed. The samples were processed at temperatures below and above the melting point of the lipid. The main lipid carrier was Compritol® 888 ATO (Gattefosse, France), a mixture of mono-, di- and triglycerides of glyceryl behenate combined with enrofloxacin, a derivate of 3-quinolinecarboxylic acid bitter active substance.

The authors [36] observed that, at low pH 1.2, the dissolution profiles of milled extrudates showed insignificant differences. As a result, the authors observed that the released active per time is independent of the surface area of the milled products if the active ingredient is slightly soluble or even more soluble in the dissolution medium. In contrast, the release patterns at pH 7.4 showed increased drug release with increasing surface area. The differences between the extrudates of small original diameters were therefore more distinct than those of the bigger diameter. When the same original die diameter was used at different processing temperatures, the milled extrudates did not show any influence in the release profiles. In this study, the authors highlighted the influence of the enrofloxacin release on the taste perception by assuming that the increase of the drug released increased the probability of taste perception. They proved that the use of a smaller die diameter (0.3 mm versus 0.4 mm) reduces the specific surface area of the milled extrudates and subsequently reduces the released amount of active during short time intervals (15 sec and 1min). As a result, they concluded that low drug-release rates reduce the probability of a bitter taste perception in the mouth.

Recently, Witzleb et al. [38] developed a continuous solid lipid extrusion process that included post-process milling of the extrudates and masked the taste of praziquantel. Various lipids (glyceryl tripalmitate, glyceryl dibehenate, glyceryl monostearate, cetyl palmitate and solid paraffin) were blended with the active ingradient, silicon dioxide and polyethylene glycol (PEG). PEG was selected as a suitable antistatic agent [39, 40] to avoid electrostatic charges created by the dies during processing. Small die diameters (< 0.5 mm) were used to mask the taste of the bitter praziquantel with processing temperatures 6°C lower than the melting point of the lipid. Drug loading was found to be dependent on the die diameter and high drug loadings were achieved at larger diameters. The extrudates of the same size distribution and different drug loads showed faster dissolution rates for the low drug-loading compositions. This was attributed to the low PEG amounts at high drug-load compositions, as the drug:lipid ratio was kept constant.

The authors investigated the taste-masking effect of lipid extrudate formulations though small diameters in a randomized palatability study with cats. The selection was based on the fact that cats are sensitive to bitter tastes and tend to reject food that is given along with bitter-tasting medicine. The animals were administrated doses of 5 mg praziquantel with small amounts of dry food and also in canned food after seven days. The palatability of the administrate formulations was assessed based on the food uptake by the cats. The results showed 100% food intake for all animals both with dry and canned food. The dissolution enhancing effect of small extrudate diameters and the addition of PEG did not have a negative influence on the taste-masking effect. The mechanism of taste masking by solid lipid extrusion was attributed to the coverage of the extrudate surfaces. During extrusion, the suspended drug particles evade the resistance of the die plate walls and a thin layer composed of softened lipid is formed on the surface of the extrudates.

9.5 Summary

HME extrusion processing can be successfully implemented as an alternative taste-masking approach to bitter active substances. The selection of the appropriate masking agent at the appropriate drug–excipient ratio is a prerequisite for the successful taste masking. In addition, processing parameters such as screw speed, zone temperatures and die diameter can lead to further masking optimization. HME has been used to develop various pharmaceutical dosage forms such as granules or pellets with high drug loading. Electronic tongues are taste-sensing systems that can overcome the drawbacks of conventional taste evaluation techniques and facilitate the development of pharmaceutical dosage forms by providing reproducible results, low detection limits and high sensitivity.

References

(1) FDA (2008) *Guidance for Industry: Orally Disintegrating Tablets* (Rockville, MD, Dec. 2008), http://www.fda.gov/downloads/Drugs/GuidanceComplianceRegulatoryInformation/Guidances/ucm070578.pdf.

(2) Douroumis, D. (2007) Practical approaches of taste masking technologies in oral solid forms. *Expert Opinion on Drug Delivery*, **4**, 417–426.

(3) Douroumis, D. (2010) Orally disintegrating dosage forms and taste-masking technologies. *Expert Opinion on Drug Delivery*, **8**, 665–675.

(4) Davies, H.E. and Tuleu, C. (2008) Medicines for children: a matter of taste. *Journal of Pediatrics*, **153**, 599–604.

(5) Cram, A., Breitkreutz, J., Desset-Brèthes, S., Nunn, T. and Tuleu, C. (2009) European Paediatric Formulation Initiative (EuPFI). Challenges of developing palatable oral paediatric formulations. *International Journal of Pharmaceutics*, **365**, 1–3.

(6) Doty, R.L., Shah, M. and Bromley, S.M. (2008) Drug-induced taste disorders. *Drug Safety*, **31**, 199–215.

(7) Schiffman, S.S. (2007) Critical illness and changes in sensory perception. *Proceedings of the Nutrition Society*, **66**, 331–345.

(8) EMEA (2008) Ethical considerations for clinical trials on medicinal products conducted with the paediatric population. http://ec.europa.eu/health/files/eudralex/vol-10/ethical_considerations_en.pdf.

(9) Strickley, R.G., Iwata, Q., Wu, S. and Dahl, T.C. (2008) Pediatric drugs: a review of commercially available oral formulations. *Journal of Pharmaceutical Sciences*, **97**, 1731–1774.

(10) Woertz, K., Tissen, C., Kleinebudde, P. and Breitkreutz, J. (2011) A comparative study on two electronic tongues for pharmaceutical formulation development. *Journal of Pharmaceutical & Biomedical Analysis*, **55**, 272–281.

(11) Woertz, K., Tissen, C., Kleinebudde, P. and Breitkreutz, J. (2010) Performance qualification of an electronic tongue based on ICH guideline Q2. *Journal of Pharmaceutical & Biomedical Analysis*, **51**, 497–506.
(12) Sbarbati, A. and Osculati, F. (2005) The taste cell-related diffuse chemosensory system. *Progress in Neurobiology*, **75**, 295–307.
(13) Breslin, P.A. and Huang, L. (2006) Human taste: peripheral anatomy, taste transduction, and coding. *Advances in Otor-hino-laryngology*, **63**, 152–190.
(14) Tortora, G.J. and Derrickson, B.H. (eds) (2006) The special senses (Chapter 17). In *Principles of Anatomy and Physiology*, Atlas and Registration Card. John Wiley & Sons.
(15) Northcutt, R.G. (2004) Taste buds: development and evolution. *Brain, Behavior & Evolution*, **64**, 198–206.
(16) Goel, H., Vora, N. and Rana, V. (2008) A novel approach to optimize and formulate fast disintegrating tablets for nausea and vomiting. *AAPS PharmSciTech*, **9**, 774–781.
(17) Douroumis, D.D., Gryczke, A. and Schminke, S. (2011) Development and evaluation of cetirizine HCl taste-masked oral disintegrating tablets. *AAPS PharmSciTech* **12**(1), 141–151.
(18) Woertz, K., Tissen, C., Kleinebudde, P. and Breitkreutz, J. (2010) Rational development of taste masked oral liquids guided by an electronic tongue. *International Journal of Pharmaceutics*, **400**(1–2), 114–123.
(19) Woertz, K., Tissen, C., Kleinebudde, P. and Breitkreutz, J. (2011) A comparative study on two electronic tongues for pharmaceutical formulation development. *Journal of Pharmaceutical & Biomedical Analysis*, **55**, 272–281.
(20) Harada, T., Uchida, T., Yoshida, M., Kobayashi, Y., Narazaki, R. and Ohwaki, T. (2010) A new method for evaluating the bitterness of medicines in development using a taste sensor and a disintegration testing apparatus. *Chemical & Pharmaceutical Bulletin*, **58**, 1009–1014.
(21) Woertz, K., Tissen, C., Kleinebudde, P. and Breitkreutz, J. (2010) Performance qualification of an electronic tongue based on ICH guideline Q2. *Journal of Pharmaceutical & Biomedical Analysis*, **51**, 497–506.
(22) Petereit, H.U., Meier, C. and Gryczke, A. (2003) Melt extrusion consisting of salts of active ingredients. WO03/072083 A2.
(23) Liu, H., Wang, P., Zhang, X., Shen, F. and Gogos, C.G. (2010) Effects of extrusion process parameters on the dissolution behaviour of indomethacin in Eudragit E PO solid dispersions. *International Journal of Pharmaceutics*, **383**, 161–169.
(24) Petereit, H.U., Meier, C. and Gryczke, A. (2006) Method for producing an immediately decomposing oral form of administration which releases active ingredients. US 2006/0051412 A1.
(25) Gryczke, A., Schminke, S., Maniruzzaman, M., Beck, J. and Douroumis, D. (2011) Development and evaluation of orally disintegrating tablets (ODTs) containing Ibuprofen granules prepared by hot melt extrusion. *Colloids Surf B Biointerfaces*, **86**, 275–284.
(26) De Brabander, C., Van Den Mooter, G., Vervaet, C. and Remon, J.P. (2002) Characterization of ibuprofen as a nontraditional plasticizer of ethyl cellulose. *Journal of Pharmaceutical Sciences*, **91**, 1678–1685.

(27) Qi, S., Gryczke, A., Belton, P. and Craig, D.Q. (2008) Characterisation of solid dispersions of paracetamol and EUDRAGIT E prepared by hot-melt extrusion using thermal, microthermal and spectroscopic analysis. *International Journal of Pharmaceutics*, **354**, 158–167.
(28) Saerens, L., Dierickx, L., Lenain, B., Vervaet, C., Remon, J.P. and De Beer, T. (2011) Raman spectroscopy for the in-line polymer-drug quantification and solid state characterization during a pharmaceutical hot-melt extrusion process. *European Journal of Pharmaceutics & Biopharmaceutics*, **7**, 158–163.
(29) Maniruzzaman, M., Boateng, J.S., Bonnefille, M., Aranyos, A. and Douroumis, D. (2011) Taste masking of paracetamol by Hot Melt Extrusion (HME): An Electronic Tongue Evaluation. 38th Annual Meeting & Exposition of the Controlled Release Society (CRS), July 30–August 3.
(30) Barthelemy, P., Laforêt, J.P., Farah, N. and Joachim, J. (1999) Compritol 888 ATO: an innovative hot-melt coating agent for prolonged-release drug formulations. *European Journal of Pharmaceutics & Biopharmaceutics*, **47**, 87–90.
(31) Faham, A., Prinderre, P., Farah, N., Eichler, K.D., Kalantzis, G. and Joachim, J. (2000) Hot-melt coating technology. I. Influence of Compritol 888 Ato and granule size on theophylline release. *Drug Development & Industrial Pharmacy*, **26**, 167–176.
(32) Suzuki, H., Onishi, H., Takahashi, Y., Iwata, M. and Machida, Y. (2003) Development of oral acetaminophen chewable tablets with inhibited bitter taste. *International Journal of Pharmaceutics*, **251**, 123–132.
(33) Breitkreutz, J., Bornhöft, M., Wöll, F. and Kleinebudde, P. (2003) Pediatric drug formulations of sodium benzoate: I. Coated granules with a hydrophilic binder. *European Journal of Pharmaceutics & Biopharmaceutics*, **56**, 247–253.
(34) Breitkreutz, J., El-Saleh, F., Kiera, C., Kleinebudde, P. and Wiedey, W. (2003) Pediatric drug formulations of sodium benzoate: II. Coated granules with a lipophilic binder. *European Journal of Pharmaceutics & Biopharmaceutics*, **56**, 255–260.
(35) Krause, J., Thommes, M. and Breitkreutz, J. (2009) Immediate release pellets with lipid binders obtained by solvent-free cold extrusion. *European Journal of Pharmaceutics & Biopharmaceutics*, **71**, 138–144.
(36) Michalk, A., Kanikanti, V.R., Hamann, H.J. and Kleinebudde, P. (2008) Controlled release of active as a consequence of the die diameter in solid lipid extrusion. *Journal of Controlled Release*, **132**, 35–41.
(37) Reitz, C. (2007) Extrudierte Fettmatrices mit retardierter Wirkstofffreigabe. PhD thesis, University of Düsseldorf.
(38) Witzleb, R., Kanikanti, V.R., Hamann, H.J. and Kleinebudde, P. (2011) Solid lipid extrusion with small die diameters–electrostatic charging, taste masking and continuous production. *European Journal of Pharmaceutics & Biopharmaceutics*, **77**, 170–177.
(39) Ishicawa, T., Wakabayashi, T., Matsuki, M., Kusunose, T. and Nobeoka, M. (1975) Modified polyamide compositions containing a polyethylene glycol derivative and a fatty acid or fatty acid salt. US Patent 356,068.
(40) Kuang, M., Zhou, S., Jingxin, L. and Qiman, L. (2008) Low environmental sensitive antistatic material based on poly(vinyl chloride)/quaternary ammonium salt by blending with poly(ethylene oxide). *Journal of Applied Polymer Science*, **109**, 3887–3891.

10

Clinical and Preclinical Studies, Bioavailability and Pharmacokinetics of Hot-melt Extruded Products

Sandra Guns and Guy Van den Mooter
Laboratory for Pharmacotechnology and Biopharmacy, Catholic University of Leuven Campus Gasthuisberg O&N2, Belgium

10.1 Introduction to Oral Absorption

Drug delivery via the oral route is still by far the most popular and patient-friendly way of drug administration. Bioavailability following oral administration is a complex process governed by the interplay between drug physicochemical properties, absorption and efflux potential and systemic and pre-systemic metabolism. From a physiological point of view, the gastrointestinal (GI) tract is mainly an absorptive and digestive organ. It allows the body to take up the necessary nutrients, electrolytes, etc. and it makes no difference between food and drugs. However, for drugs to be absorbed, certain physicochemical prerequisites need to be fulfilled. Solubility and dissolution rate in the fluids of the GI tract are very important. Drug compounds need to dissolve within a specific time frame (a few minutes up to ca. 6 hours; this is mainly driven by the stomach to distal ileum transit time) to allow adequate absorption.

One of the implications of combinatorial chemistry and *in silico* drug design used during the drug discovery phase is the increased molecular complexity leading to potent and selective compounds, often with aqueous solubility that is too low to allow the development to a marketable drug. The number of pharmacologically active molecules with poor physicochemical and biopharmaceutical properties has increased steadily over the past

Hot-melt Extrusion: Pharmaceutical Applications, First Edition. Edited by Dennis Douroumis.

15–20 years. More than half of all new drug candidates currently entering the development pipeline fail because of non-optimal physicochemical and biopharmaceutical properties. It is therefore crucial to evaluate the potential of a compound to be developed into a marketable formulation with adequate and reproducible systemic exposure as soon as possible in the drug discovery/development process. Fortunately, physicochemical profiling has currently become a part of the activities during lead optimization and candidate selection in many pharmaceutical companies [1].

Besides solubility and dissolution rate, lipophilicity will also contribute to absorption. Penetration of drug molecules through lipid bilayers of enterocytes requires a certain degree of lipophilicity. It is generally accepted that $\log P$ should be less than 5 in order to prevent excessive binding to cell membranes and reduced transport to the hydrophilic cytosolic environment. The size of drugs is another factor to consider with respect to absorption driven by passive diffusion. According to Lipinski's rule of five, molecules with a molar mass above 500 g/mol are generally not well absorbed. Finally, the charge of the drug molecules will also influence permeability as it is recognized that passive diffusion is more favorable for uncharged species [2].

Drug absorption from the gastrointestinal tract occurs mainly by transcellular (across cells) and paracellular (between cells) transport mechanisms. The contribution of the paracellular pathway is rather limited because of the presence of tight junctions and is therefore only important for ions and small hydrophilic molecules (molar mass up to 300 g/mol) [3]. Sugars and even amino acids can be transported paracellularly in case of saturation of their typical carrier systems. Passive diffusion is by far the most important mechanism for transcellular drug absorption. According to Fick's first law, dissolved molecules are transported through the lipidic membranes from a region of high concentration in the intestinal lumen to a region of low concentration in the enterocytes and then to the blood. Absorbed drug molecules are further transported by the blood flow, thus maintaining the necessary concentration gradient. Several so-called carrier-mediated systems are present in the gastrointestinal tract. Although their primary function lies in the absorption of nutrients, they also play a significant role in transcellular transport of many drugs.

It has become evident that many drugs possess structural properties suitable for carrier-mediated transport [4]. This type of transport is saturable and requires a specific interaction between the drug molecule and certain carriers (proteins) which are located in the apical cell membrane. The carrier binds to the drug molecule and translocates it through the membrane. This process needs energy (provided by hydrolysis of adenosine triphosphate or ATP), is temperature dependent and transports molecules against a concentration gradient. Several carrier-mediated systems have been described in the past for transport of peptides, sugars, amino acids, nucleosides and vitamins. Nucleoside analogs can also be transported through a carrier-mediated mechanism called facilitated diffusion. This is an energy-independent saturable system that needs a concentration gradient.

Molecules can also undergo efflux into the intestinal lumen, hence counteracting the absorptive process and thus reducing the oral bioavailibility [4]. Several efflux systems are currently known such as P-glycoproteins and multi-drug resistance-associated proteins. Besides carrier proteins involved in transporting molecules in and out of enterocytes, the intestinal membrane contains enzymatic proteins. Today, it is generally accepted that membrane-associated metabolic enzymes belonging to cytochrome P450, esterases (lipases), proteases or sulfotransferases play an important role in oral drug absorption.

Moreover, luminal (proteases, lipases) and microbial (reducing, hydrolytic) enzymes also contribute to the overall drug bioavailability. Finally, once drug molecules are absorbed, they are prone to hepatic metabolism which can be divided into phase I (mainly transformation of the molecule to make it more reactive) and phase II (production of more water-soluble metabolites to facilitate their elimination) reactions.

10.2 In Vivo Evaluation of Hot-melt Extruded Solid Dispersions

10.2.1 Oral Immediate Release

There exists a significant number of scientific papers on solid dispersions which are prepared by a variety of manufacturing techniques such as spray drying, rotary evaporation or hot-melt extrusion. Next to the necessary physical chemical characterization to understand the structure of the systems, *in vitro* dissolution is carried out to study the pharmaceutical performance, i.e. to verify the potential (rate and extent) of the solid dispersions to deliver the active pharmaceutical ingredient (API) in the dissolved state in the dissolution medium. It is generally accepted that *in vitro* data are not sufficient to appreciate the solubilizing power of a dosage form of a poorly soluble drug to generate the necessary driving force for absorption, despite the use of *in vitro* dissolution media which are claimed to be biorelevant.

Preclinical and finally clinical testing is absolutely mandatory. The value of preclinical testing is that it is relatively cheap and provides information to what extent a dosage form of a poorly soluble API is able to generate supersaturation in the GI tract and increase and stabilize the systemic exposure in a living species.

Although the general morphology of the mammalian GI tract exhibits some basic structural similarities, it differs considerably among species. In addition to metabolic differences, the anatomical, physiological and biochemical differences in the GI tract of the human and common laboratory animals can cause significant variation in drug absorption from the oral route [5]. Among the physiological factors, pH, bile, pancreatic juice, mucus and fluid volume and content can modify dissolution rates, solubility, transit times and membrane transport of drug molecules. The transit time of dosage forms can be significantly different between species due to different dimensions and propulsive activities of the GI tract. The lipid and protein composition of the enterocyte membrane along the GI tract can alter binding and passive, active and carrier-mediated transport of drugs. While small animals such as rats, mice, guinea pigs and rabbits are suitable for determining the mechanism of drug absorption and bioavailability values from powder or solution formulations, larger animals such as dogs, pigs and monkeys may be more suited to assess absorption from formulations such as tablets or capsules.

A representative example of the poor predictability of *in vitro* dissolution for *in vivo* performance has been reported by Zheng and co-workers [6] in a study of the bioavailability in rats of nimodipine from solid dispersions with hydroxypropylmethylcellulose (HPMC), Eudragit® EPO and polyvinylpyrrolidone-co-vinylacetate 64 (PVPVA). The results revealed that the commercial product Nimotop® gave the highest C_{max} followed by Eudragit® EPO solid dispersion, although the difference was statistically not significant. The mean $AUC_{0\text{–}12hr}$ after administration of Nimotop® and the Eudragit® EPO solid dispersion formulation were comparable while the other preparations exhibited lower values. Although *in vitro* dissolution studies showed that the rank order of drug dissolution rate was PVPVA > Eudragit® EPO > HPMC solid dispersion, *in vivo* investigations found that HPMC and PVPVA resulted in lower mean AUC $_{0\text{–}12\,\text{hr}}$ than Eudragit® EPO.

The authors speculated that the contradiction between *in vivo* and *in vitro* data may have resulted from the pH-dependence of Eudragit® EPO. *In vitro* dissolution in 0.1 N hydrochloric acid containing 0.05% (w/v) sodiumdodecyl sulfate (SDS) showed that the solid dispersion with Eudragit® EPO dissolved significantly faster than all formulations except Nimotop®, supporting the high bioavailability observed with the Eudragit® EPO solid dispersion. The authors also suggested that the increased bioavailability of the Eudragit® EPO solid dispersion might have resulted from rapid dissolution in the stomach that saturated the efflux pump and CYP3A4 in the intestinal mucosa, thus increasing drug bioavailability.

It has been reported that the API is metabolized by CYP3A both in the liver and small intestine [7–9] and that absorption of the drug in the intestinal mucosa is inhibited by a P-glycoprotein efflux pump [10]. Lack of *in vitro*/*in vivo* correlation has also been reported by Six et al [11]. In a double-blind single-dose cross-over study in 8 human volunteers, the performance of three solid dispersion formulations of itraconazole in comparison with Sporanox®, the marketed form, was investigated. Solid dispersions made up of itraconazole (40%, w/w) and HPMC 2910, Eudragit® E100 or a mixture of Eudragit® E100-PVPVA were manufactured by hot-melt extrusion and filled in gelatin capsules. The mean AUC values (0–72 h) after administration of Sporanox® and the HPMC formulation were comparable while they were lower in the case of the Eudragit® E100 and Eudragit® E100-PVPVA formulations (Figure 10.1). A significant difference was found between Sporanox® and the Eudragit® E100-PVPVA solid dispersion (1365.5 ± 619.9 ng h/ml versus 928.9 ± 355.7 ng h/ml), although there was no significant difference between the other formulations.

A large inter-individual difference in pharmacokinetic behavior of itraconazole in humans and laboratory animals has been reported in literature [12–15], and this may partially explain the high variability in the mean AUC. The *in vitro* release behavior of the solid dispersions showed the opposite of what was observed *in vivo* since the formulations based on Eudragit® E100 and Eudragit® E100-PVPVA showed the lowest mean AUC and C_{max} (Figure 10.2). The fact that fast *in vitro* dissolution behavior resulted in the lowest AUC, while a slower dissolution rate was observed for capsules showing the highest AUC seems to point to the importance of the polymer dissolution properties. The authors suggested that since, in case of the HPMC solid dispersion and Sporanox®, the polymer dissolution is much slower compared to Eudragit® E100 or the mixture of Eudragit® E100 and PVPVA. Hence a microenvironment was hypothesized to exist in which adequate drug solubility can be maintained for a longer period of time. If the polymer diffuses too rapidly to the bulk phase, the drug is poorly protected and is more prone to precipitation.

In an earlier study Baert and co-workers [16] described the performance of tablets consisting of itraconazole which had been melt-extruded with HPMC 2910 5mPas (the authors call this mixture 'Triaset') in the same ratio as described by Six et al. [11]. The Triaset was mixed with other typical tablet ingredients and compressed to biconvex tablets. In a limited human trial (5 healthy volunteers), the 200 mg melt-extruded tablet was compared to two 100 mg coated cores-capsules of the commercial formulation (Sporanox®) in fasted conditions. Contrary to the results reported by Six et al. [11], the AUC of itraconazole was 2.3 times higher after administration of the melt-extruded tablet.

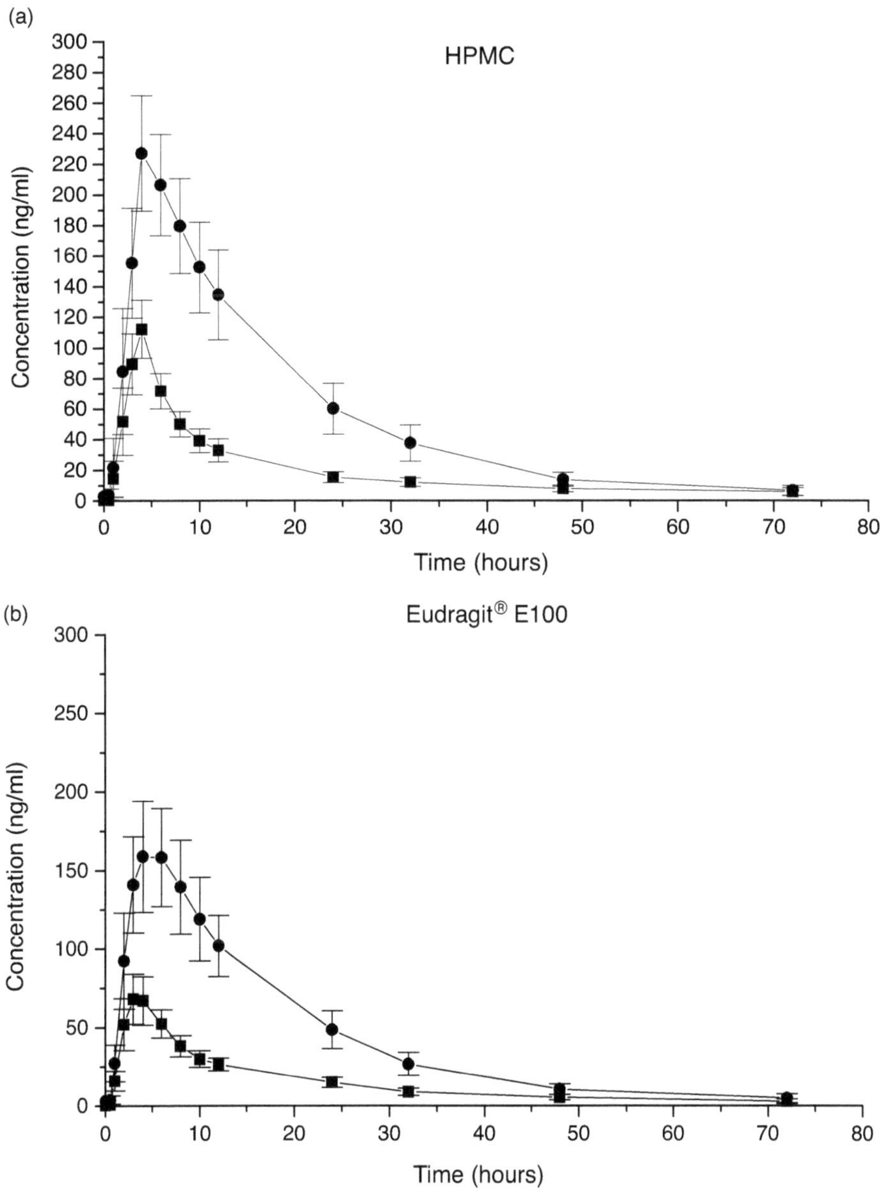

Figure 10.1 *Average plasma concentration versus time profiles of itraconazole (■) and hydroxy-itraconazole(•) after oral administration (100 mg itraconazole doses) of (a) HPMC formulation; (b) Eudragit® E100 formulation; (c) Eudragit® E100/PVPVA64 formulation; and (d) Sporanox® in healthy volunteers (n = 8). Error bars indicate the standard error of the mean. K. Six, et al. 2005, reproduced with permission from Elsevier.*

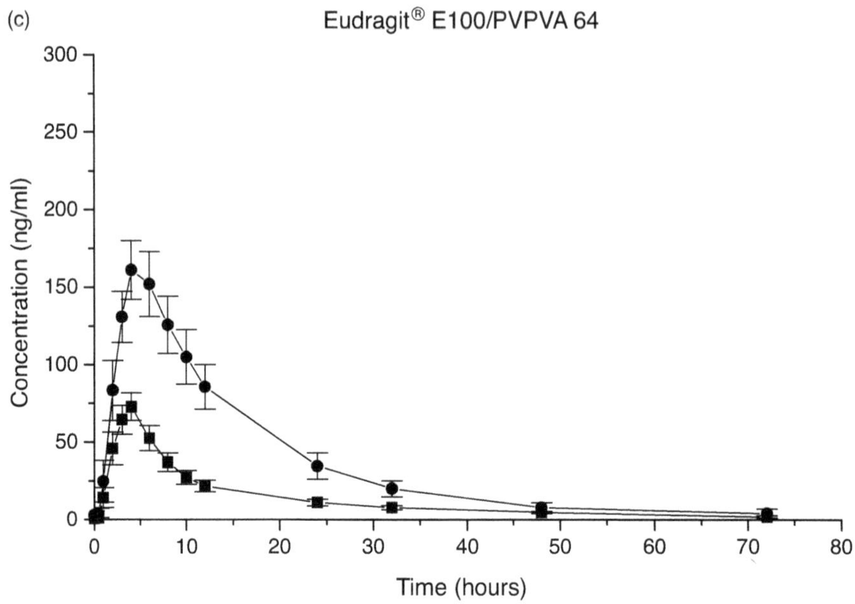

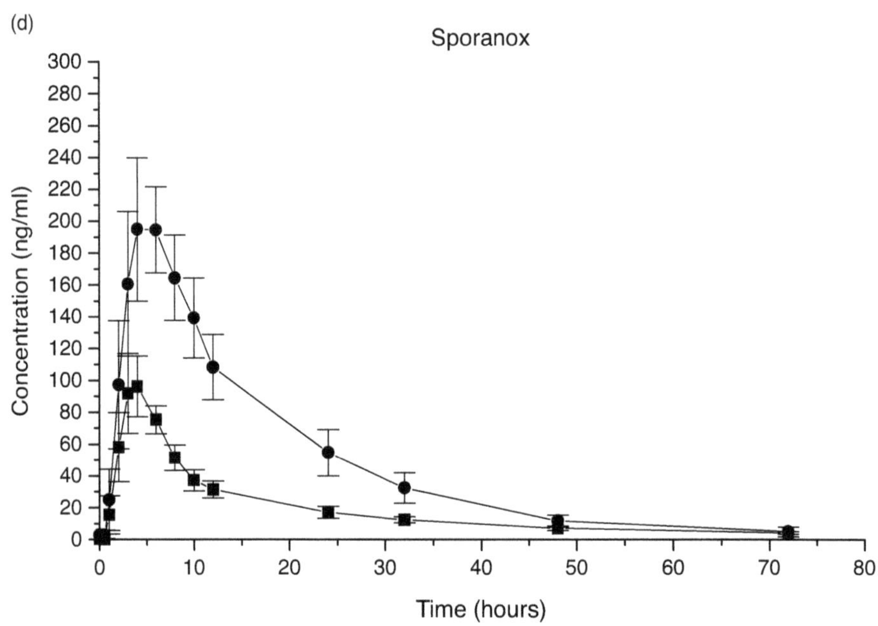

Figure 10.1 *(Continued)*

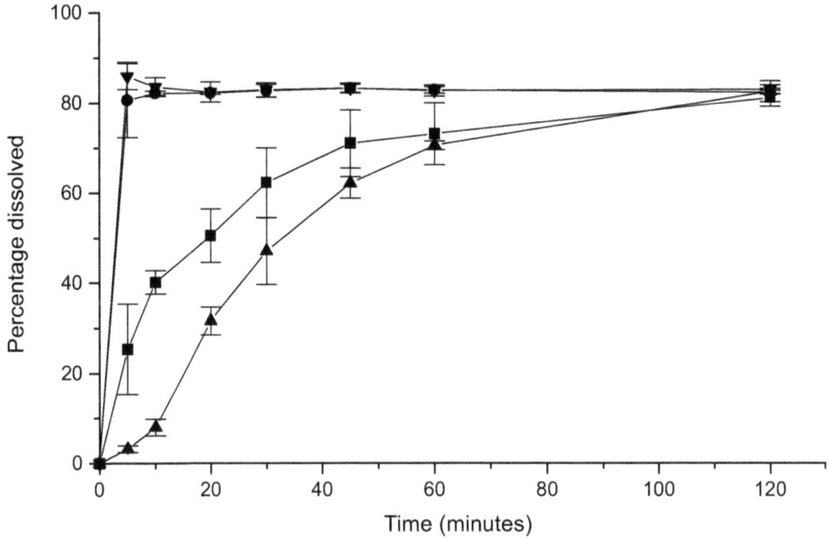

Figure 10.2 *Dissolution profiles of Sporanox® (▲), HPMC extrudate (■), Eudragit® E100 extrudate (▼) and Eudragit® E100/PVPVA 64 extrudate (•) in capsules containing 100 mg itraconazole in SGF_{sp}. The dissolution (paddle method) was carried out in 500 ml SGF_{sp} at 37°C and 100 rpm. Error bars indicate the standard deviation. K. Six, et al. 2005, reproduced with permission from Elsevier.*

Miller et al. [17] suggested that the optimum formulation approach for itraconazole is to control drug release in order to retard precipitation as pH is increased and extend the absorption window in the small intestine. The authors studied itraconazole exposure in rats after oral dosing of the API with melt-extruded binary systems made up of microparticles of itraconazole and either polyvinylpyrrolidone K25 (PVPK25) and HPMC (Methocel E3 premium LV). Primary to melt extrusion, the microparticles were mixed with poloxamer 407 and polyethyleneoxide 200M. The itraconazole-HPMC extrudate formulation exhibited a slightly delayed T_{max} and a greater C_{max} than the API-PVPK25 extrudate formulation; however, no statistical difference was seen between the AUC values for the two formulations. The authors reported that rapid precipitation of itraconazole occurred upon entrance into the more neutral pH environment of the small intestine, and this resulted in a brief opportunity for absorption.

The authors stressed the importance of supersaturation in another *in vivo* absorption study of itraconazole in rats from hot-melt-extruded solid dispersions using Methocel™ E50 and Eudragit® L 100-55 as polymeric carriers [18]. In this case, the superior behavior of the Methocel formulation correlated with the *in vitro* release profiles, as this was found to be a superior stabilizer for the supersaturated levels of the API in neutral pH than Eudragit®. A peak-trough-peak shape in the plasma curves was seen with both formulations.

The similarity with a previous study by Hardin and co-workers suggested that this observation could be attributed to enterohepatic recirculation of unmetabolized itraconazole [19]. Compared to an itraconazole-Methocel formulation prepared by aerosol solvent extraction, a significant improvement in total AUC was found for the melt-extruded formulation [20].

This could possibly be the result of poorer maintenance of the supersaturated state of the API by the low-molecular-weight HPMC used in the aerosol solvent evaporation system compared to the HPMC used in the hot-melt-extruded formulations.

Interestingly, the Eudragit® L 100-55 formulation exhibited the potential for prolonging the itraconazole absorption possibly by extending the API release from the polymeric matrix along the small intestine. The variability of the Eudragit® L 100-55 based formulation was higher than that of the Methocel formulation and this was attributed to poor stabilization of the itraconazole supersaturation at neutral pH. In order to improve the stability of supersaturated itraconazole in neutral media, Carbopol® 974P, an acidic polymer, was added to the Eudragit® 100-55 carrier [21]. *In vitro* dissolution showed that the addition of Carbopol® prolonged the release of supersaturated levels of itraconazole from the Eudragit® L 100-55 matrix following an acidic-to-neutral pH transition. *In vivo* evaluation of the absorption of itraconazole demonstrated that the addition of Carbopol® reduced the absorption variability observed with the Eudragit® L 100-55 carrier system.

Lakshman and co-workers [22] compared different formulation strategies for a poorly soluble Novartis compound by studying its oral bioavailability in dogs. A 20% drug-loaded melt extrudate using PVP-K30 provided a 7-fold greater bioavailability than the control formulation made up of crystalline API triturated with poloxamer 188. The solvent-evaporated solid dispersion in PVP-K30 that was processed using a rotary evaporator provided the second-best improvement in exposure. Among the other formulations tested, PVP-K30-based spray-dried spray-granulated and the PVP-K90-based solid dispersion produced relatively lower bioavailability enhancement of approximately 2- to 3-fold. PVP-K30 was clearly better than PVP-K90. The use of SDS with PVP-K90 improved the bioavailability further. Unfortunately, no studies were conducted to understand the differences in bioavailability among the different formulations. The authors hypothesized that spray-dried and spray-granulated formulations were not intimately or molecularly dispersed, in contrast to the melt extrudates that were exposed to high shear mixing. Further, the rate of dispersion and the degree of supersaturation of drug substance in aqueous media could also be different for samples prepared by different methods.

Interestingly, the company considered preparing dosage forms using various solvent evaporation methods such as spray-drying, spray-granulation and rotary evaporation to lead to manufacturing and scale-up issues due to the need for large volumes of environmentally unfriendly organic solvents to dissolve the drug substance. Manufacturing feasibility must be evaluated as soon as the preclinical formulation activities start to anticipate large-scale formulation and processing problems.

The glass thermoplastic system (GTS) developed by researchers from Janssen Pharmaceutica is an ingenious drug delivery system for poorly soluble drugs; however, it might be difficult to scale up. The system is made up of hydroxyl propyl beta cyclodextrin (HPβCD), citric acid and HPMC and probably combines inclusion complexation, solid dispersion and acidification solubilization principles [23]. In a study with 18 healthy human volunteers, four formulations of a water-insoluble microsomal triglyceride transfer protein inhibitor under various fed/fasted conditions were investigated. Besides the GTS system, an oral solution of the compound with HPβCD, one solid dispersion formulation of the compound in HPMC sprayed onto beads (the same principle as that for the oral delivery of the structural analogue itraconazole in the commercial product Sporanox®) and one melt-extrudate tablet based on HPMC as carrier were studied.

Plasma levels were obtained after single oral administration either after a standard breakfast or in fasted individuals. All formulations showed high oral bioavailability as well as a trend for positive food effects, consistent with the BCS class II nature of the drug. In the fasted state, the GTS capsule provided a significant increase of oral bioavailability compared to the bead capsule and a trend toward an improved AUC in the case of the melt-extruded system. The GTS was significantly more bioavailable in the fed state relative to both the bead and melt extruded formulations. Compared to the oral solution, the relative bioavailability for the capsule with the beads is about 27%, for the melt extrudate tablet 75% and for the GTS capsule 97% in the fasting state. In the fed state, AUC values increased for all three formulations (70% for the bead formulation, 12% for the melt extruded systems and 55% for the GTS) relative to the fasted conditions although only the difference for the GTS was significant.

The authors suggested that the tendency for the melt-extrudate tablet to give a higher relative bioavailability compared to the beads capsule may be to some extent related to the higher polymer concentration in the solid dispersion prepared by melt extrusion, suggestive of the stabilizing effect of HPMC. The best performance of the GTS system was explained in terms of the combined effect of the citric acid, the complex formation by HPβCD and the stabilizing effect of HPMC.

Another new processing technology (at least from a pharmaceutical perspective) is the KinetiSol Dispersing technology, a high-energy mixing process for the production of amorphous pharmaceutical solids. This technology utilizes a series of rapidly rotating blades to process the drug and polymeric carrier through a combination of kinetic and thermal energy without the aid of external heating sources.

Dinunzio et al. [24] compared the *in vivo* performance of solid dispersions of itraconazole-HPMC produced using KinetiSol Dispersing and hot-melt extrusion in Sprague–Dawley rats. Examination of the pharmacokinetic profile of itraconazole revealed that both formulations exhibited similar behavior showing an increase in plasma concentration over the initial 4 hour, followed by a brief plateau which has been previously attributed to enterohepatic recirculation [21] and then a final elimination phase. Both solid dispersion processes were also able to produce amorphous compositions that provided similar *in vivo* behavior to the currently marketed product Sporanox, which also contains HPMC E5 as the primary stabilizing polymer for the solid dispersion.

The formulation strategy for nimodipine used by Yunzhe et al. [25] is also complicated with respect to large-scale manufacturability. A solid dispersion consisting of nimodipine, Plasdone-S630 and Eudragit®-E100 is prepared by hot-melt extrusion. The solid dispersion is then suspended in a semi-solid vehicle comprising PEG400, Plasdone-S630 and PEG6000 and, finally, the semi-solid mixture needs to be transferred to hard HPMC capsules. The oral bioavailability of this system was evaluated in beagle dogs and compared to the commercially available tablets (Nimotop®). No statistical difference between the $AUC_{0-\infty}$ values and C_{max} of the two dosage forms was observed. However, the solid dispersion displayed a significantly lower T_{max} than the reference formulation. Interestingly, two peaks were found in the plasma concentration versus time profile which, according to the authors, was caused by the biphasic absorption process of the high dose of the lipophilic API. The second peak was higher in the reference formulation, whereas the maximum was found in the first peak for the solid dispersion formulation. This most likely is a result of differences in drug release behavior between the two formulations in the stomach fluid.

At a lower dosage, the bioavailability of the solid dispersion formulation may possibly be higher than that of Nimotop®.

Hot-melt extrusion can offer opportunities to reduce the number of dosage units administered per day, as pointed out in a study by Enders-Klein et al. [26] with lopinavir. This HIV protease inhibitor was coformulated with ritonavir to enhance its bioavailability and pharmacokinetics. The original solid formulation is a soft gel capsule from which 6 need to be taken daily, administered with food to maximize the lopinavir bioavailability. Three studies in healthy human volunteers were conducted to assess the bioavailability of tablets of lopinavir/ritonavir at 800/200 mg or 400/100 mg dose based on hot-melt extrusion under different food conditions, compared to equal doses of the original soft gel capsules. The tablet was bioequivalent to the capsule after a moderate fat meal with respect to lopinavir and ritonavir AUC values. Compared to the capsules, the tablet formulation resulted in more consistent lopinavir and ritonavir exposure within and across studies and across meal conditions. The authors explained the diminished food effect and decreased variability of the tablet by more consistent lopinavir and ritonavir exposures, thereby minimizing the likelihood of extreme high or low values compared to the soft gel capsules.

10.2.2 Oral Controlled Release

In addition to immediate release dosage form development, hot-melt extrusion has also been applied to manufacture modified release systems. High-dose modified-release formulations of imatinib mesylate (the drug load was ca. 90%) were prepared by melt granulation below the API melting point but above the glass transition temperature of the polymer using a twin-screw extruder [27]. By carefully selecting polymers based on their physicochemical properties, the release rate could be modified between desired times of 4 to more than 10 hours for the total drug release.

Administration to human volunteers of the marketed form of imatinib mesylate, Gleevec® 400 mg IR tablet, b.i.d. (separated by 12 hours) resulted in two plasma peaks. The first occurred at about 4 hours after the first dose and the second at ca. 18 hours, with a trough at 12 hours just before administration of the second dose. Administration of a single dose (800 mg) tablet of imatinib with hydroxypropyl cellulose (HPC) resulted in a peak concentration that was approximately twice that of the first peak and 1.5-fold of the second peak obtained after administration of Gleevec®. The peak maximum occurred approximately 2 hour later than the first peak with Gleevec® 400 mg. Tablets made up of the API and ethylcellulose demonstrated a similar pharmacokinetic profile to those with HPC but with a lower peak concentration, approximately 1.3-fold that of the first peak with Gleevec®, and very similar to the second peak. The maximum in the plasma concentration also occurred approximately 2 hours later than the first peak with the Gleevec® tablet.

These results showed that both test formulations had characteristics of a modified-release formulation with the peak concentration delayed for approximately 2 hours. The different C_{max} values for the two melt-extruded formulations pointed to the faster *in vivo* release rate of the HPC formulation. Interestingly, this agreed with *in vitro* dissolution. The authors hypothesized that the same plasma peak at ca. 6 hours for both of the modified-release formulations possibly reflected the existence of an absorption window or transit time difference of the modified-release formulation in the small intestine. Drug molecules released after 6 hours were possibly not well absorbed due to low absorption from the large intestine. *In vitro* data indeed showed that the EC formulation did not finish the API release until 8–10 hours.

Mini-matrices for the sustained release of ibuprofen were reported by Verhoeven et al. [28]. The systems were made up of an ibuprofen/ethylcellulose mixture and manufactured by hot-melt extrusion. Xanthan gum was added to tailor drug release. The influence of the concentration as well as the particle size of xanthan gum on the *in vitro* characteristics of the mini-matrices was investigated. The *in vivo* performance of these experimental formulations was evaluated in dogs and compared to an equivalent dose of a sustained-release ibuprofen matrix tablet (Ibu-Slow® 600 mg). A slower absorption was observed during the initial phase of the plasma concentration–time profiles for the mini-matrices containing 20% xanthan gum. A constant drug absorption pattern during 36 hours was observed for the 20% xanthan gum mini-matrices compared to the 30% xanthan gum mini-matrices and the reference formulation. The similar pharmacokinetic parameters of the formulations supported the hypothesis that the 30% xanthan gum mini-matrices behaved *in vivo* as a single-unit dosage form instead of multi-particulates, due to the immediate swelling upon contact with the GI-fluids and the formation of a plug. Although swelling was also noticed for the 20% xanthan gum mini-tablets, they performed *in vivo* as a multi-particulate dosage form since they differed in C_{max} and T_{max} values from Ibu-Slow® 600 mg.

Similar systems were investigated in dogs for the controlled delivery of metoprolol tartrate [29]. As well as matrices made up of xanthan gum/ethylcellulose, high-molecular-weight polyethylene oxide /ethylcellulose matrices were also investigated. An increasing xanthan gum concentration enhanced drug release, which was reflected in the higher $AUC_{0-36\,h}$ and relative bioavailability. Similar observations were made for the poly-ethylene oxide mini-matrices. The mini-matrices showed no statistical significant different exposure of metoprolol tartrate in terms of AUC compared to the sustained-release reference formulation Slow-Lopresor® 200 Divitabs, although the AUC values tended to increase at higher hydrophilic polymer concentration. None of the formulations showed a strong sustained-release effect either. In contrast to the behavior of the ibuprofen mini-matrices, a significant difference in pharmacokinetic parameters and plasma concentration–time profiles between 10% and 20% xanthan gum formulations was observed. This was attributed to the fact that metoprolol tartrate is water soluble while ibuprofen is not. An alternative explanation was the different drug/xanthan gum, drug/ethylcellulose and ethylcellulose/xanthan gum ratios for formulations with ibuprofen and metoprolol tartrate. Since the latter dissolves faster, it was suggested that it might therefore be less influenced by the rapid xanthan gum swelling and consequently release rate from the xanthan gum matrix.

10.2.3 Implants

Some API cannot be successfully delivered via the oral route due to stability or absorption-related problems, whereas in some cases it is desirable to have a prolonged drug release up to a few days or weeks. Vapreotide is a somatostatin analogue used for the therapy of hormone-dependent tumors and endocrine disorders. Like other peptides, it cannot be administered by the oral route and its plasma half-life is relatively short after parenteral administration. For these reasons, its use would be greatly enhanced by a sustained delivery system capable of maintaining controlled plasma levels of the peptide over an extended period of time. Vapreotide pamoate was therefore formulated as an implant using pol(lactide)-co-glycolide

(PLGA) as biodegradable carrier for the controlled release of the drug [30]. The implants were prepared by an extrusion method and the drug release was evaluated *in vivo* in rats. It was found that drug loading, polymer molecular weight, copolymer composition and end group modifications were critical factors affecting the *in vivo* release properties.

Lemmouchi et al. [31] studied the release of isometamidium chloride and ethidium bromide from copolymer rods prepared by extrusion and made up of poly-ε-caprolactone and L-lactide (P(CL-LLA)). The results obtained from the *in vivo* tests using the rods and the classical treatment (intramuscular injection) showed that the prophylactic period was significantly enhanced in the case of the administration of the polymer rods as compared to intramuscular injection. The *in vivo* experiments carried out in the laboratory on rabbits challenged with *Trypanosoma congolense* using P(CL-LLA) (ratio 74:26) implants containing either drug and the intramuscular injections of the same drugs showed that the prophylactic period in the case of isometadimidium rods is much longer than for the ethidium bromide rods.

Amman et al. [32] were able to demonstrate *in vitro*/*in vivo* correlation in rats for the delivery of the antipsychotic drug risperidone delivered as a biodegradable implant made up of 60% of PLGA. The implants were prepared by hot-melt extrusion using a single-screw system. This delivery strategy will contribute to adherence to antipsychotic therapy, which is still a major obstacle preventing optimal outcomes for patients with schizophrenia. Praziquantel (PZQ) loaded implants based on polyethylene glycol/polyεcaprolactone (PEG/PCL) blends of different ratios were fabricated by a combination of twin-screw mixing and hot-melt extrusion [33]. The *in vivo* performance of the implants was evaluated in rats. Interestingly, PEG release from all implants is far faster than PZQ release; complete PEG release occurs in 72 hours. After implantation, drug release becomes more moderate compared with *in vitro* drug release, and it tends to follow zero-order kinetics in the later stage. These results suggest that changing the composition of the PEG/PCL blends is an effective tool to adjust *in vitro*/*in vivo* drug release from the implants.

10.3 Conclusion

Hot-melt extrusion is a valuable process technology for manufacturing API-carrier solid dispersions, both on a laboratory and commercial scale. It provides efficient mixing capability of API and carrier, leading to enhanced drug dissolution properties and increased oral absorption and bioavailability. Moreover, the versatility of the technology creates opportunities for oral prolonged drug release as well as drug delivery by alternative routes, e.g. biodegradable implants.

References

(1) Balbach, S. and Korn, C. (2004) Pharmaceutical evaluation of early development candidates: the 100mg approach. *International Journal of Pharmaceutics*, **275**, 1–12.
(2) Lipinski, C.A., Lombardo, F., Dominy, B.W. and Feeney. P.J. (2001) Experimental and computational approaches to estimate solubility and permeability in drug discovery and development settings. *Advanced Drug Delivery Reviews*, **46**, 3–26.

(3) Söderlind, E. and Dressman, J. (2010) Physiological factors affecting drug release and absorption in the gastrointestinal tract. In Dressman, J. and Reppas, C. (eds.) *Oral Drug Absorption, Prediction and Assessement*. Drugs and the Pharmaceutical Sciences 193, Informa Healthcare, NY.
(4) Ungell, A.-L. (2010) Drug transport mechanisms across the intestinal epithelium. In Dressman, J. and Reppas, C. (eds.) *Oral Drug Absorption, Prediction and Assessement*. Drugs and the Pharmaceutical Sciences 193, Informa Healthcare, NY.
(5) Kararli, T. (1995) Comparison of the gastrointestinal anatomy, physiology and biochemistry of humans and commonly used laboratory animals. *Biopharmaceutics & Drug Disposition*, **16**, 351–380.
(6) Zheng, X., Yang, R., Zhang, Y., Wang, Z., Tang X. and Zheng, L. (2007) Part II: Bioavailability in beagle dogs of nimodipine solid dispersions prepared by hot-melt extrusion. *Drug Development & Industrial Pharmacy*, **33**, 783–789.
(7) Guengerich, F.P., Brian, W.R., Iwasaki, M., Sari, M., Baarnhielm, C. and Bernts-som, P. (1991) Oxidation of dihydropyridine calcium channel blockers and analogues by human liver cytochrom P4503A4. *Journal of Medicinal Chemistry*, **34**, 1938–1844.
(8) Ramsch, K.D., Ahr, G., Tettenborn, D. and Auer, L.M. (1985) Overview on pharmacokinetics of nimodipine in healthy volunteers and in patients with subarachnoid hemorraghe. *Neurochirurgia*, **28**, 74–78.
(9) Scherling, D., Buhner, K., Krause, H.P., Karl, W. and Wunsche, C. (1991) Biotransformation of nimodipine in rat, dog and monkey. *Arzneimittelforschung*, **41**, 392–398.
(10) Saeki, T., Ueda, K., Tanigawara, Y., Hori, R. and Komano, T. (1993) P-glycoprotein-mediated transcellular transport of MDR reversing agents. *FEBS Letters*, **324**, 99–102.
(11) Six, K., Daems, T., De Hoon, J., Van Hecken, A., Depre, M., Bouche, M.P., Prinsen, P., Verreck, G., Peeters, J., Brewster, M.E. and Van den Mooter, G. (2005) Clinical study of itraconazole prepared by hot-stage extrusion. *European Journal of Pharmaceutical Sciences*, **24**, 179–186.
(12) Yoo, S., Kang, E., Jun, H., Shin, B., Lee, K. and Lee, K. (2000) Absorption, first pass metabolism and disposition of itraconazole in rats. *Chemical & Pharmaceutical Bulletin*, **48**, 798–801.
(13) Bogaerts, M., Maertens, J., Van der Geest, R., Bosly, A., Michaux, J.L., Van Hoof, A., Cleeren, M., Woestenborghs, R. and De Beule, K. (2001) Pharmacokinetics and safety of a 7-day administration of intravenous itraconazole followed by a 14-day administration of itraconazole oral solution in patients with hematologic malignancy. *Antimicrobial Agents & Chemotherapy*, **45**, 981–985.
(14) Reynes, J., Bazin, C., Ajana, F., Datry, F., le Moing, J.P., Chwetzoff, E. and Levron, J. (1997) Pharmacokinetics of itraconazole (oral solution) in two groups of human immunodeficiency virus-infected adults with oral candidiasis. *Antimicrobial Agents & Chemotherapy*, **41**, 2554–2558.
(15) Suarez-Kurtz, G., Bozza, F., Vicente, F., Ponte, F. and Struchiner, C. (1999) Limited sampling strategy models for itraconazole and hydroxyitraconazole based on data from a bioequivalence study. *Antimicrobial Agents & Chemotherapy*, **43**, 134–140.
(16) Baert, L., Verreck, G. and Thone, D. (1997) Antifungal compositions with improved bioavailability, WO 97/44014.

(17) Miller, D., McConville, J., Yang, W., Williams III, R. and McGinity, J. (2007) Hot-melt extrusion for enhanced delivery of drug particles. *Journal of Pharmaceutical Sciences*, **96**, 361–376.
(18) Miller, D., Dinunzio, J., Yang, W., McGinity, J. and Williams III, R. (2008) Enhanced in vivo absorption of itraconazole via stabilization of supersaturation following acid to neutral pH transition. *Drug Development & Industrial Pharmacy*, **34**, 890–902.
(19) Hardin, T., Graybill, J., Fetchick, R., Woestenborghs, R., Rinaldi, M and Kuhn, J. (1988) Pharmacokinetics of itraconazole following oral administration to normal volunteers, *Antimicrobial Agents & Chemotherapy*, **32**, 1310–1313.
(20) Lee, S., Nam, K., Kim, M., Jun, S., Park, J., Woo, J. and Hwang, S. (2005) Preparation and characterization of solid dispersions of itraconazole by using aerosol solvent extraction system for improvement in drug solubility and bioavailability. *Pharmaceutical Research*, **28**, 866–874.
(21) Miller, D., Dinunzio, J., Yang, W., McGinity, J. and Williams III, R. (2008) Targeted intestinal delivery of supersaturated itraconazole for improved oral absorption. *Pharmaceutical Research*, **25**, 1450–1459.
(22) Lakshman, J., Cao, Y., Kowalski J. and Serajuddin, A. (2008) Application of melt extrusion in the development of a physically and chemically stable high energy amorphous solid dispersion of a poorly water soluble drug. *Molecular Pharmacology*, **5**, 994–1002.
(23) Verreck, G., Vandecruys, R., De Conde, V., Baert, L., Peeters, J. and Brewster, M. (2004) The use of three different solid dispersion formulations – Melt extrusion, film coated beads and a glass thermoplastic system – to improve the bioavailability of a novel microsomal triglyceride transfer protein inhibitor. *Journal of Pharmaceutical Sciences*, **93**, 1217–1228.
(24) Dinunzio, J., Brough, C., Miller, D., Williams, R. and McGinity, J. (2010) Fusion processing of itraconazole solid dispersions by Kinetisol dispersing: a comparative study to hot melt extrusion. *Journal of Pharmaceutical Sciences*, **99**, 1239–1253.
(25) Yunzhe, S., Rui, Y., Wenliang, Z. and Xing, T. (2008) Nimodipine semi-solid capsules containing solid dispersion for improving dissolution, *International Journal of Pharmaceutics*, **359**, 144–149.
(26) Klein, C.E., Chiu, Y-L., Awni, W., Zhu, T., Heuser, R., Doan, T., Breitenbach, J., Morris, J., Brun, S. and Hanna, G. (2007) The tablet formulation of lopinavir/ritonavir provides similar bioavailability of the soft gelatin capsule formulation with less pharmacokinetic variability and diminished food effect. *Journal of Acquired Immune Deficiency Syndrome*, **44**, 401–410.
(27) Vasanthavada, M., Wang, Y., Haefele, T., Lakshman, J., Mone, M., Tong, W., Joshi, Y. and Serajuddin, A. (2011) Application of melt granulation technology using twin-screw extruder in the development of high-dose modified-release tablet formulation. *Journal of Pharmaceutical Sciences*, **100**, 1923–1934.
(28) Verhoeven, E., Vervaet, C. and Remon, J.P. (2006) Xanthan gum to tailor drug release of sustained release ethylcellulose mini-matrices prepared via hot-melt extrusion: in vitro and in vivo evaluation. *European Journal of Pharmaceutics & Biopharmaceutics*, **63**, 320–330.
(29) Verhoeven, E., De Beer, T., Schacht, E., Van den Mooter, G., Remon, J.P. and Vervaet, C. (2009) Influence of polyethylene glycol / polyethylene oxide on the release

characteristics of sustained release ethylcellulose mini-matrices produced by hot-melt extrusion: in vitro and in vivo evaluations. *European Journal of Pharmaceutical Sciences*, **72**, 463–470.
(30) Rothen-Weinhold, A., Besseghir, K. and Gurny, R. (1997) Analysis of the influence of the polymer characteristics and core loading on the in vivo release of a somatostain analogue. *European Journal of Pharmaceutical Sciences*, **5**, 303–313.
(31) Lemmouchi, Y., Schacht, E., Kageruka, P., De Deken, R., Diarra, B., Diall, O. and Geerts, S. (1997) Biodegradable polyesters for controlled release of trypanocidal drugs: in vitro and in vivo studies. *Biomaterials*, **19**, 1827–1837.
(32) Amann, L., Gandal, M., Lin, R., Liang, Y. and Siegel, S. (2010) In vitro-in vivo correlations of scalable PLGA-risperidone implants for the treatment of schizophrenia. *Pharmaceutical Research*, **27**, 1730–1737.
(33) Cheng, L., Lei, L. and Guo, S. (2010) In vitro and in vivo evaluation of praziquantel loaded implants based on PEG/PCL blends. *International Journal of Pharmaceutics*, **387**, 129–138.

11

Injection Molding and Hot-melt Extrusion Processing for Pharmaceutical Materials

Pernille Høyrup Hemmingsen and Martin Rex Olsen
Egalet, Værløse, Denmark

11.1 Introduction

Hot-melt extrusion has proved to be widely applicable in the pharmaceutical industry throughout the latest decades [1–3]. The unique properties of materials processed by elevated temperatures have been used in drug products that would have posed extremely complex challenges in traditional pharmaceutical processing and, in some cases, even ensured prolongation of products in the market that would otherwise have been terminated. In particular, the reduction of unit operations that is offered by extrusion and injection molding, by integrating e.g. mixing, melting, forming and cooling in a single step, has inherent advantages. The implementation of process analytical technologies (PAT) and quality by design (QbD) approaches in pharmaceutical processes is readily applicable to extrusion and/or injection molding because of the careful control already employed. For example, this is achieved by means of temperature, pressure, speed and other critical process parameters which can be monitored and controlled in a semi-continuous manner.

Most of the excipients used in melt processes are compounds that are widely used for other pharmaceutical solid dosage forms. However, the process does call for specific properties of the composition. Primarily, the compounds need to be sufficiently stable at elevated temperature to withstand degradation throughout the manufacturing process. In the case of injection molding, at least one of the constituents of the composition needs

Hot-melt Extrusion: Pharmaceutical Applications, First Edition. Edited by Dennis Douroumis.

to be thermoplastic, i.e. posses the ability to change shape at elevated temperatures. Finally, the compatibility of the compounds needs to be acceptable not only at room temperature and under accelerated conditions, but also at the elevated temperatures during the process.

This chapter will introduce both hot-melt extrusion and injection molding. In particular, the injection molding process will be discussed as it is a relatively new manufacturing process within the pharmaceutical industry. Key process steps and the influence on product quality are treated. In addition, a few illustrative examples will elucidate the range of applications that are viable for products manufactured by means of hot-melt extrusion and/or injection molding.

11.2 Hot-melt Extrusion in Brief

Extrusion is a continuous process which creates objects of fixed geometry by means of a die. Extrusion was initially developed to create objects of fixed cross-sectional area such as cylinders, sheets, tubes etc. In principle, any material can be extruded as long as the material can be transported to and through the die. Accordingly, extrusion may be applied to both hot and cold materials such as metals, polymers, ceramics, concrete and foodstuffs and, of course, pharmaceutical products [4, 5]. Hot-melt extrusion has for example been employed in producing tablets for gastro-retentive controlled release systems [6] and implants [7]. This chapter is focused on the hot-melt process on polymers for the pharmaceutical industry.

Extruder systems come in many sizes and shapes. They can all, however, be described by three distinct parts: a hopper for feeding the material, a screw and barrel for heating and blending the material and, finally, a die section in which the material is shaped, cooled and cut (see Figure 11.1). The process is controlled by temperature and screw speed, which can easily be monitored by e.g. thermocouples and drive amperage.

The screw serves three purposes: feeding, melting/compression/blending and metering for the die. Accordingly, it is common to have three zones in the screw of the extruder. The feeding zone feeds the material into the extruder. In the melting zone, in which the material is melted and compressed and (in some cases) also mixed, the screw core has an increasing radius thus compressing the material. Finally, the screw has a metering system in which the

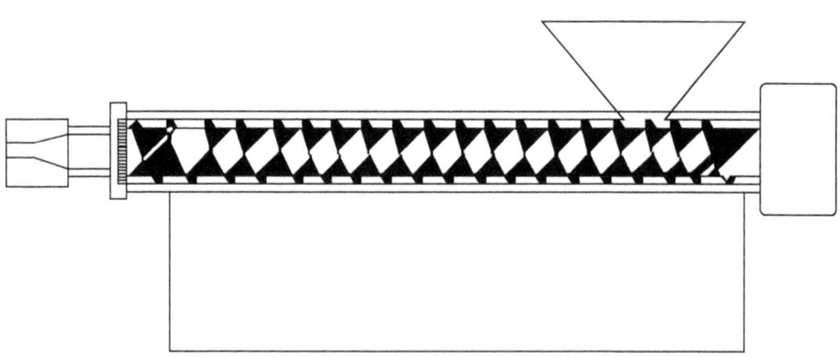

Figure 11.1 Sketch of an extruder system showing hopper, screw and die.

melt is conveyed into the die. The die shapes the material into the final geometry which is subsequently cooled by means of e.g. a cooling conveyer band.

Polymers generally exhibit poor thermal conductivity and therefore melt slowly. Accordingly, most polymer manufacturing processes involve rotating screw devices that ensure both uniform dosing and heating for the duration of the heating period [8]. This obviously poses the same challenge during the cooling step and it is commonly a durable process to cool the material sufficiently.

As described, the geometries that can be formed by extrusion are numerous; tubes are commonly produced by extrusion. Examples include drinking straws, medical tubing and various food and candy. An extension is co-extrusion in which two or more layers are extruded simultaneously. The die controls the shape and sizes of the different layers. Thermoforming is a common post-extrusion process for plastic sheet stock, where the sheet is heated until it is softened and then formed via a mold.

11.3 Injection Molding

Hot-melt extrusion and injection molding belong to the same class of processes in which material is formed at elevated temperatures. The key difference is based on the fact that injection molding shapes the material in three dimensions and therefore offers the ability to produce final pharmaceutical material in a single process step. As for an extruder, injection molding consists of a hopper, a screw and a forming unit (a mold). The 3D shaping enables the manufacturing of very diverse products encompassing stents, hearing aids and entire body panels of cars. The process is illustrated in Figure 11.2, which shows the main features of an injection-molding machine.

The polymer blend that includes the active pharmaceutical ingredient (API) falls, typically under gravity, from a hopper into a cylinder where it is propelled along by a rotating screw into an electrically heated section. As the material is heated, it softens and flows. When the cylinder contains enough material to fill the mold, the screw action is stopped. In the final stage, the screw moves axially, acting as a ram, injecting the material through a small nozzle and down channels (runners) into the shaped cavity within a cooled mold. Simplistically, the injection molding cycle consists of six steps: (1) mold closing; (2) injection; (3) holding; (4) cooling; (5) mold opening; and (6) final part ejection.

The mold-closing part of the cycle is split into at least three different steps. The first is the movement from open mold to the position where the mold can be damaged from i.e. parts

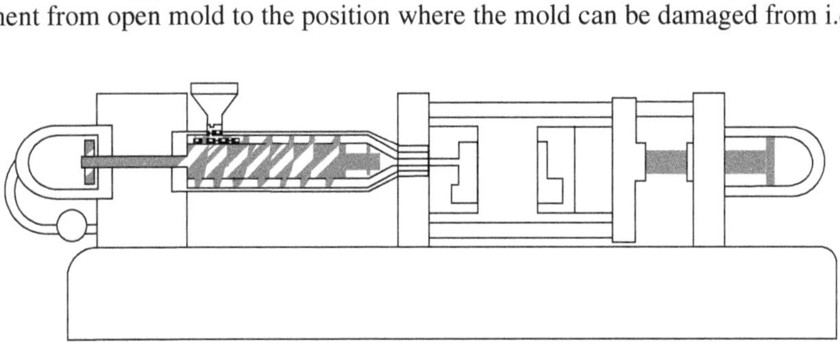

Figure 11.2 *Injection-molding machine sketched to show hopper, screw and die.*

or moldings stuck inside the mold; this step is controlled by velocity which is normally fast. To prevent any damage to the mold, the movement from the critical position to the mold-closed position is controlled by pressure, which should be kept just high enough to close the mold. Normally this step is monitored either by pressure or time. When the mold is completely closed, the building of clamping force is engaged.

The injection step is normally controlled by speed or volume/time. On modern machines, the injection speed can possibly be increased or decreased during the injection sequence. The choice of injection speed depends on several parameters and combinations of these. Parameters that influence the setting of injection speed include: material, wall thickness, flow length, part size, part shape and the part surface properties. Usually this step is set to 98% filling before switching to holding.

Holding is a second injection step controlled by pressure rather than by speed. The polymer melt is kept under constant pressure as long as the part gains weight, or until the gate seals off. This process step compensates for the shrinkage of the polymer during cooling, which is the main reason for the high consistency of the injection-molding process compared to the extrusion process. The choice of holding pressure and holding time depends on the required properties of the part or molding with regards to dimensions, weight, crystallinity, physical properties, solubility or combinations of these.

During the cooling time, the screw rotates and prepares the material for the next cycle. Meanwhile, the molded part cools off to a temperature low enough to be able to remove the part from the mold. Mold opening is controlled by speed, and it opens enough to make space for the part to fall out.

Finally, in the ejection step the machine pushes the part out of the mold and the part falls out.

11.4 Critical Parameters

11.4.1 Melt Temperature

The actual temperature on the molten polymer inside the barrel is the melt temperature at the point where the mold is ready for injection. This parameter is mainly influenced by the barrel temperature, but also by the back pressure and the screw speed. Some injection-molding machines have a built-in feature that measures the actual melting temperature; otherwise, the measures are performed by the operator, which introduces possible variations. The temperature of the polymer impacts the viscosity on the melt, meaning the higher the temperature the lower the viscosity or the further away from the solidification temperature of the polymer.

For technical applications, cost and efficiency are very important. Normal operation is to set the melt temperature as low as possible, while still being able to meet the requirements for the specific part (i.e. more energy added to the polymer is more energy to be removed again). For pharmaceutical controlled-release applications, the melt temperature has impact on the solubility of the matrix and hence the drug release. Furthermore, the melt temperature could have an impact on the product stability if the degradation of the polymer is initiated during the processing. The choice of melting temperature is based on process clarification and formal process development. The process development may conveniently be included in a QbD approach according to the latest guidelines from both the US Food & Drug Administration (FDA) and the European Medicines Agency (EMA). A robust melting temperature may therefore be found within the design space. Because of the impact from the barrel temperature, back pressure, screw speed and the possible variation in measurement, the process development may reveal settings for barrel temperature, back pressure and screw speed instead of a measured melting temperature.

11.4.2 Barrel Temperature

The machine microprocessor controls barrel temperature as measured by thermocouples placed within the steel. The heat is added from electrical heat bands, which can be inductive. The parameter has a high impact on the melting temperature and is therefore often used as the control parameter instead of the melt temperature. For process development, see Section 11.6.3.

11.4.3 Cooling Temperature

Mold temperature is also known as cooling temperature. The parameter is in most cases controlled by water circulation through water channels inside the mold. The temperature can also be controlled by either oil circulation or electrical heating in extreme cases, where high temperatures above 100°C are required. Measurement can be performed either directly inside the mold by thermocouples, or by a cooling water temperature measurement combined with a flow measurement. In industry, the choice of temperature for technical applications is based on cost efficiency: a temperature as low as possible while still being able to meet the physical, dimensional or visual requirements for the specific part. For pharmaceutical controlled-release applications the cooling temperature has some impact on the release properties and can have an impact on the crystallinity of the compound. For this reason it should be investigated in a QbD investigation set-up.

11.4.4 Holding Pressure

The holding pressure is the parameter compensating for the polymer shrinkage during cooling. After the speed-controlled injection movement, the material is kept under high pressure to fill the mold cavity to as near to 100% as possible.

11.4.5 Holding Time

The duration in which holding pressure is applied during material cooling is called the holding time. Solidification of the material in the gate sets the optimal holding time. Too short a holding time will lead to underfilling of the mold, while too long a holding time will eventually lead to flow of material back though the gate.

A feasible holding time is easily found by means of a gate seal experiment in which subjects molded at different holding times are weighed, alloeing a gate seal curve to be constructed. Once constant weight is obtained, the minimum holding time is found (see Figure 11.3).

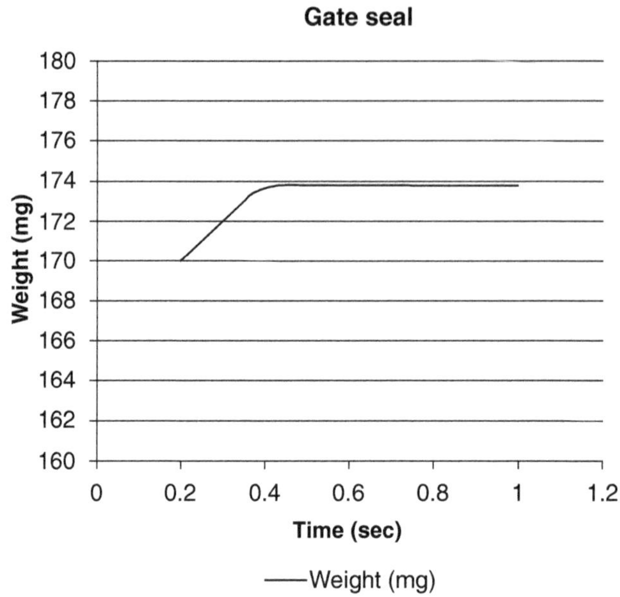

Figure 11.3 *Holding time is set to a duration that ensures constant weight and thereby process stability.*

11.4.6 Back Pressure

Back pressure is the pressure applied to the material during screw recovery after the holding step. By increasing back pressure, mixing and plastification are improved.

11.4.7 Injection Speed

Injection molding imposes high shear-flow rates on the polymer as it is squirted at high pressure into the die. This tends to align the long polymeric molecules and increase the fluidity of the polymer substantially. The shear is proportional to the flow speed or injection speed. This shear thinning of the molten polymer is essential to injection molding and can only be achieved if high injection pressures (and thereby speeds) are used. It is important to note that, in some ranges, the viscosity is non-Newtonian; small changes in injection speed will lead to large changes in viscosity. Accordingly, the injection speed should be set sufficiently high to ensure a high degree of polymer alignment and a small variation in viscosity (see Figure 11.4).

11.4.8 Cooling Time/Cycle Time

The mold is set at a temperature to ensure that the molten material solidifies almost as soon as the mold is filled. However, as discussed above, polymers have a relatively long cooling duration and, accordingly, the cooling time most accommodate this. In some cases the cooling time can influence the solid-state properties of the final product such as crystallinity, physical stability or other critical attributes of the product.

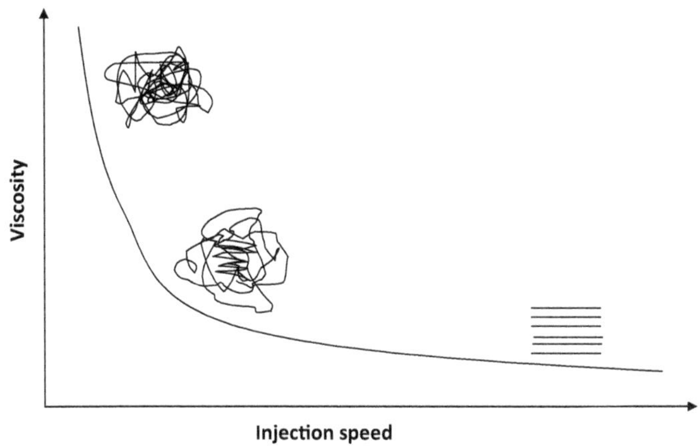

Figure 11.4 Injection speed influences viscosity by means of polymer alignment.

11.5 Example: Comparison of Extruded and Injection-molded Material

In some cases extrusion and injection-molding technology may both be applied to the same system [9]. In the current example, two tubular geometries, having the same theoretical release rate and using the same chemical composition, was directly compared.

The outer biodegradable and water-impermeable coat was composed of ethylcellulose and cetostearyl alcohol. The inner matrix was based on a poly(ethylene oxide) (PEO) carrier with paracetamol as model API. The chain length and API load was selected such that the inner matrix was eroded during dissolution.

For extrusion, a 12–mm-diameter 24-cm-length extruder designed for powder feeding was employed and fitted with a three-layer die-head (1 layer for the shell and 2 layers for the matrix core). A puller 250 × 25 mm, cool air-gun (–10°C) and cooling cylinder was fitted to the extruder. The shell cylinder heat zones were set at 100–125°C, pressure 160 bar and 25 rpm. The matrix cylinder heat zones were set at 60–85°C, pressure 140 bar and 9.6 rpm. The co-extrudate ran at 5–10 m/min.

For injection molding, a two-component Arburg allrounder injection machine was fitted with a customized mold, molding both shell and matrix. The matrix volume was 150 mm^3. The barrel temperature was set to 130–170°C and the matrix barrel was set to 60–70°C. The mold cooling temperature was 8°C. Holding pressure was 2000 bar for 5 seconds and cycle time was 26.4 seconds.

The results showed that both injection-molding and co-extrusion manufacturing produces tablets that are homogeneous and fully formed. The extruded tablet exhibited a matrix visibly shrinked as the matrix was partly loosened from the shell. The tablets were tested in a USP2 dissolution apparatus in pH 6.8 phosphate-buffered media. The time to complete dissolution ranged from 8 to 12 hours (see Figure 11.5).

The results of the tests as displayed in Figure 11.5 emphasize that manufacturing can have a significant influence on the key quality attributes of a product. Interestingly, the release properties change significantly upon shifting manufacturing method. It can be speculated that the origin of this change could be based on the differences in the pressure applied in

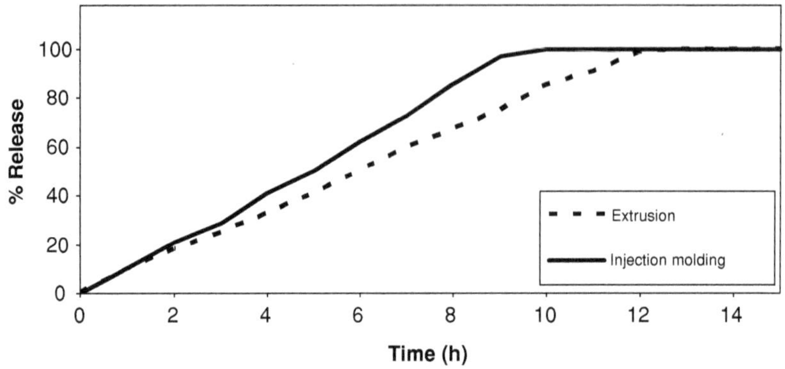

Figure 11.5 Comparison of dissolution properties of injection-molded (solid line) and extruded materials (dashed line).

co-extrusion and injection molding. The lack of holding pressure in the extrusion process allows for the material to anneal and therefore leads to a lower energy state. In turn, this leads to a higher energy barrier upon dissolution which (in this case) is detected by longer dissolution time.

11.6 Development of Products for Injection Molding

11.6.1 Excipients

Carriers selected for injection molding and/or extrusion of pharmaceutical products must possess thermoplastic properties: i.e. the carrier material (typically a polymer) must be able to deform easily at elevated temperatures and then solidify upon cooling. The material chosen should be thermally stable at the process temperature. This can in some cases be obtained by means of a plasticizer, which by definition lowers the softening temperature of the composition in which it is included. In some cases the API will function as a plasticizer by considerably lowering the processing temperature of the polymer [10–12].

Additives such as colorants, stabilizers, antioxidants, UV inhibitors and other functional excipients may also be included in the formulation. The list of suitable carriers includes a wide range of polymers from both natural and synthetic sources; a short list of carriers is included in Table 11.1.

Properties of material produced by means of hot-melt extrusion or injection molding highly depend on the excipients that are used in the composition [13]. Often, but not always, the thermoplastic carrier in the composition is also the release-modifying constituent.

The release kinetics in, for example, a controlled-release tablet can be controlled by varying the physicochemical properties of the polymer system.

By addition of more soluble excipients, the release rate of the matrix system will increase; for more hydrophobic excipients, the release rate will decrease. Fast solubility of the constituents leads to more rapid polymer disentanglement, giving rise to faster dissolution [14, 15]. Obviously this is equally true if the excipient is a thermoplastic material and

Table 11.1 *Examples of carriers used for hot-melt extrusion and injection molding.*

Chemical name	Trade name or trivial name
Ammonio methyacrylate copolymer	Eudragit RS/RL
Pectin	
Poly(ethylene glycol)	Carbowax
Poly(ethylene oxide)	Polyox
Poly(propylene glycol) and poly(ethylene glycol)	Lutrol, Poloxamer
Poly(methacrylic acid-co-mehyl methacrylate)	Eudragit S/L
Starch	
Poly(vinyl acetate)	PVA
PVB	
Poly(vinyl pyrrolidone)	PVP
Hydroxypropyl Methylcellulose	Methocel
Ethyl cellulose	Ethocel
Poly(lactide-co-glycolide)	PLGA

a carrier for the formulation. An example of controlling release properties by means of composition of hydrophilic and hydrophobic constituents is depicted in Figure 11.6.

The diffusion layer that often exists in a pharmaceutical formulation during dissolution is a hydrophilic barrier that can control water penetration and drug diffusion. Initially, the polymer becomes hydrated and swells. Here, the polymer chains are strongly entangled and the gel layer is highly resistant. Moving away from this swelling position, the diffusion layer becomes progressively hydrated however and, when sufficient water has accumulated, the chains disentangle and the polymer dissolves. The duration of the disentanglement depends, among other things, on the polymeric chain length. Accordingly it is possible to control

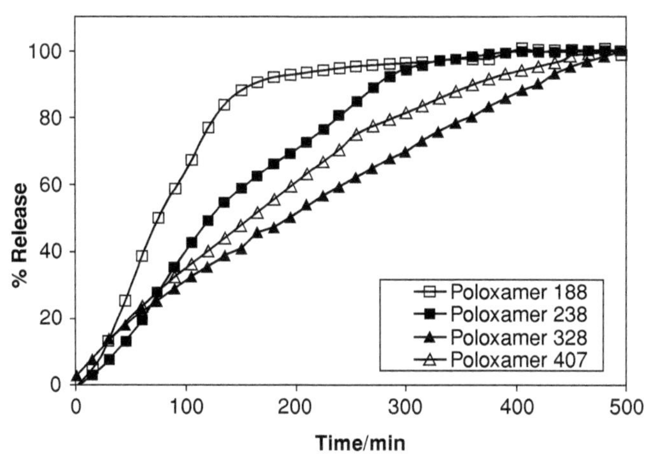

Figure 11.6 *Changing the hydrophobicity of excipients in a given formulation alters the release characteristics of a pharmaceutical composition.*

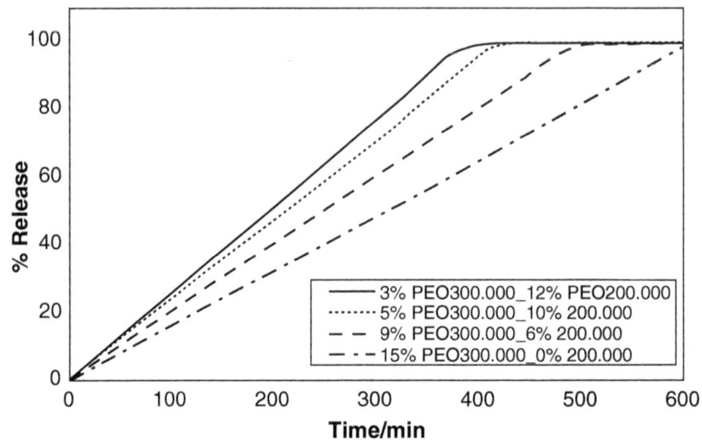

Figure 11.7 *Controlling release rate by means of polymeric chain length.*

the release rate of controlled-release systems by changing polymeric chain length. This is illustrated in Figure 11.7.

11.6.2 Stability

Products that are manufactured from molten material display distinct properties in regards to stability. Obviously normal attention should be paid to e.g. chemical degradation and compatibility between excipients and API. Often, however, control of physical stability is of key importance because of the conformational changes that are inherent in the cooling step of the extrusion and/or injection-molding process.

The heating involved in the melting of the material calls for special attention [16]. As most processes follow (or can be approximated by) Arrhenius kinetics, chemical degradation can be limited by decreasing the time at which the material is exposed to elevated temperature. In injection-molding processes the material transit time is often less than 10 minutes. Furthermore, it is important to note that the process is dry. The absence of water reduces the potential degradation pathways. It is not uncommon to observe surprisingly high chemical stability in injection-molded products compared to traditional tablets manufactured by, for example, wet granulation. This effect (i.e. the lack of chemical degradation) can be assigned to processing under dry conditions.

11.6.3 Process Development

A key to understanding properties of injection-molded materials is the fact that the product is formed under heating followed by cooling at high pressure. This enables an extremely high degree of control of the settling of the polymers which, in turn, gives rise to the desired properties of e.g. tablets.

The temperature necessary for processing by means of injection molding is coupled to the softening of the formulation. Typically the softening is overall governed by the physico-chemical properties of the polymeric carrier. The tensile modulus of a polymer will decrease

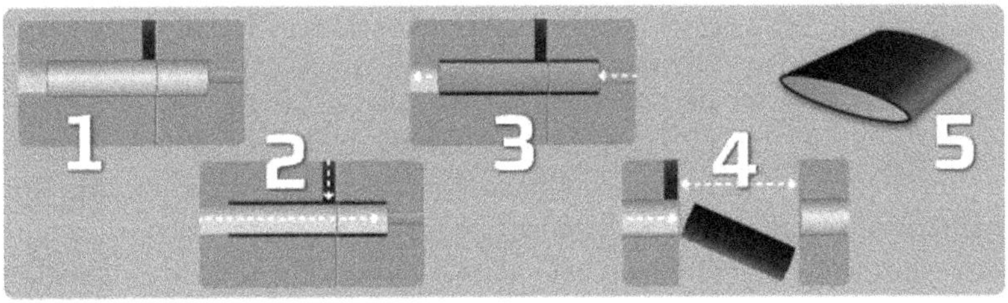

Figure 11.8 Manufacturing process of Egalet®: (1) cavity is empty; (2) piston moves forward, coat material is injected; (3) coat material hardens, piston recedes while matrix material (containing drug) is injected into the cavity; (4) matrix hardens, piston moves forward ejecting the finished Egalet® tablet; and (5) finished Egalet® tablet. For a better understanding of the figure, please refer to the color section.

above the glass transition temperature (T_g) and, further, above the melting temperature in the case of crystalline or semi-crystalline polymers. Most APIs degrade at elevated temperatures; the selection of carrier system is therefore important for maintaining a process that leaves the API in an acceptable state. The temperature necessary for processing may be lowered by including a plasticizer.

An example of a tablet manufacturing process is the development of Egalet® tablets [17, 18]. The process includes a conventional two-component injection-molding process in which an outer shell is initially molded and the matrix containing the active component is molten inside the shell (see Figure 11.8).

The manufacturing process that forms the basis of tablet properties thus becomes fairly simple consisting of mixing, molding, (optionally) coating and storage (see Figure 11.9). The controls in each step of the process include temperature, time, pressure and room conditions. The influence of each parameter on injection molding, in particular, was discussed in

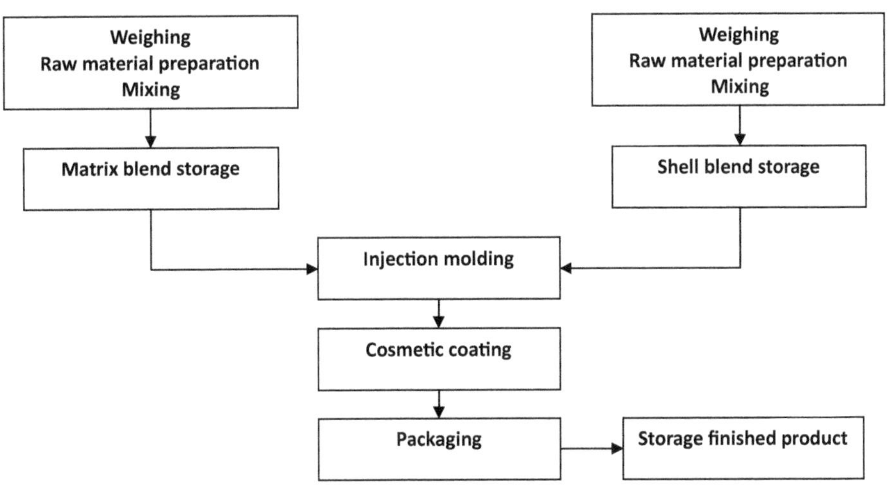

Figure 11.9 Process flow diagram for the manufacturing of Egalet® tablets.

detail in Section 11.4. Critical parameters in conventional tablet manufacturing technology are powder handling and powder flow properties. This is less critical in injection-molding and extrusion technology because the dosing step is at a time where the material is melted. In injection molding the tablet is defined by the mold; uniform tablet size and mass can thereby be obtained. Flow properties do influence the process robustness as the powder enters the barrel and screw cavity from a hopper. However, even very poorly flowing powder blends can be handled without influencing product quality. For large-scale manufacturing, hoppers with built-in powder-dosing solutions are commercially available.

A process that inherently possesses temperature cycling obviously influences the polymorphic behavior of the end product. Both API and excipients may display a multitude of conformations after injection molding. By controlling the rate of cooling, this may be employed actively to enforce a specific polymorphic conformation or in special cases of amorphous formulations. In particular, it is known from early studies that the rate of cooling influences the crystallinity of the polymer carrier [19]. The degree of crystallinity in turn influences the properties of the product. For example the release during dissolution may be influenced by crystallinity because the water diffusion, disentanglement of polymers and drug diffusion depends on the mobility in the polymer matrix.

The dynamics of polymer matrices are generally highly influenced by the conformational entropy that polymers possess. This means that it takes time to enter into the stable conformation. Usually the dynamics can be described by stretched exponentials. Time-dependent behavior, which originates from entropic and conformational processes, includes annealing phenomena, shrinkage, 'sweating', dissolution time variability etc. For example, it is standard procedure within the plastic industry to account for up to 1% shrinkage when designing molds (e.g. for cell phones).

The dynamics of crystallization may be studied for example by means of differential scanning calorimetry (DSC), X-ray or other solid-state characterization methods.

A way to stabilize the dissolution time and release properties is by the inclusion of an excipient which is capable of 'freezing' the structure at baseline. In that way, no further crystallization or de-crystallization can take place and dissolution changes should not occur. Such stabilizers include sugars and salts, for example.

Injection molding of polymeric materials includes the application of pressure during injection and cooling by means of holding pressure. Injection pressure is the pressure applied during injection of the melted material into the mold and is the key parameter that controls the rate at which the material enters the mold. The material starts to cool immediately, partly because the mold is cooled and partly because of the pressure drop after leaving the die. Accordingly, the rate of injection may influence the structure of the final product. For example, low injection pressure may lead to layering of the material in the final composition.

The filling of the mold is followed by a period of cooling before the mold is opened and the product expelled from the mold. During the cooling period, holding pressure is applied. The holding pressure governs the packing during cooling and thereby directly influences the final density of the product. The density may also influence important properties of a pharmaceutical product such as release rate, uniformity of content, etc.

The shrinkage that is also coupled to the dynamics of polymeric materials is particularly important for materials in which the properties of the product are coupled to geometry. For example, Egalet® ADPREM technology as described in the following section depends on both physical and chemical properties of injection-molded materials in order to maintain anti-tamper, dissolution and assay/impurity properties throughout the shelf life of the product.

11.7 Properties of Injection-molded Materials

11.7.1 Egalet® Technology

Egalet® technology is founded on an erosion-based drug-release system rather than the more conventional diffusion-based systems. The feasibility of erosion-based systems is being explored by a number of companies in the pharmaceutical business, including Egalet Ltd.

The vast majority of controlled-release formulations are based on delivery systems, in which the drug delivery relies on aqueous diffusion through a matrix or membrane to cause drug release.

The Egalet® technology differentiates from the conventional diffusion-based matrix since erosion of a polymer-based matrix is the controlling factor in drug release. The simplest Egalet® formulation consists of two components: shell and matrix (Figure 11.10). The active drug substance is distributed evenly throughout the matrix.

From a drug delivery point of view, the shell serves two purposes: to protect the matrix and maintain the integrity of the geometry in the formulation. This is important because the controlled-release functionality is highly dependent upon the constant surface area throughout the release of the drug in the GI tract [18, 20, 21]. Furthermore, the shell protects the matrix from being manipulated via crushing, grinding or chewing.

The drug release can be altered precisely by adjusting the polymeric composition of, for example, PEO carrier and poloxamer grade within the matrix. The active drug substance is blended into a single mixture and subsequently molded via injection molding.

The matrix gives a zero-order release profile of drug independent of pH and the presence of alcohol. The rate of release is determined by the geometry of Egalet® tablets in terms of surface area and length, and by the chemistry of the matrix.

It can be a lengthy and costly process to tailor the dose form to match the required in vivo release profile of the controlled-release drug. Predicting in vivo performance from in vitro dissolution is highly attractive, as it considerably aids the early development of new drug products and potentially decreases development time and cost. Several studies show

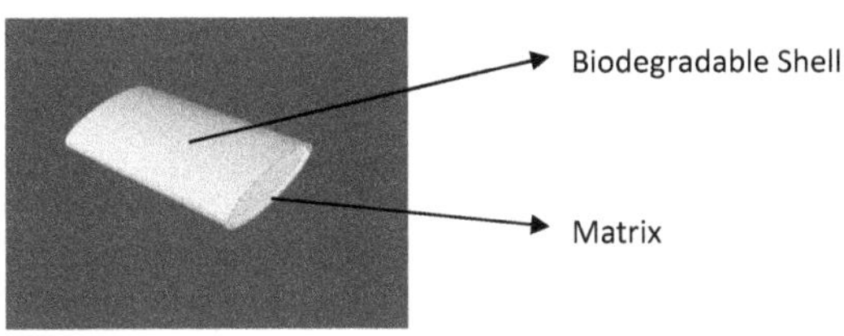

Figure 11.10 *Sketch of a simple Egalet® formulation showing the main features: a biodegradable shell and the matrix holding the active pharmaceutical ingredient.*

that this prediction is possible; however, just as many studies show that the prediction only holds for the formulation mechanism employed for the model. In vitro-in vivo correlation (IVIVC) has for example been performed for hydrophilic matrix systems [22], showing a correlation for diffusion systems.

The simple relationship of release time and erosion distance of the tablet has been discussed above. As a result, the erosion system becomes extremely predictable in vitro. It is therefore an obvious next step to investigate the predictability of such an erosion system in vivo. This was pursued for three Egalet® hydrocodone formulations, differing in length but not in their composition.

As expected, the dissolution time varies systematically with tablet length (see Figure 11.11). The dependence of release on tablet length was also found in vivo, in which the

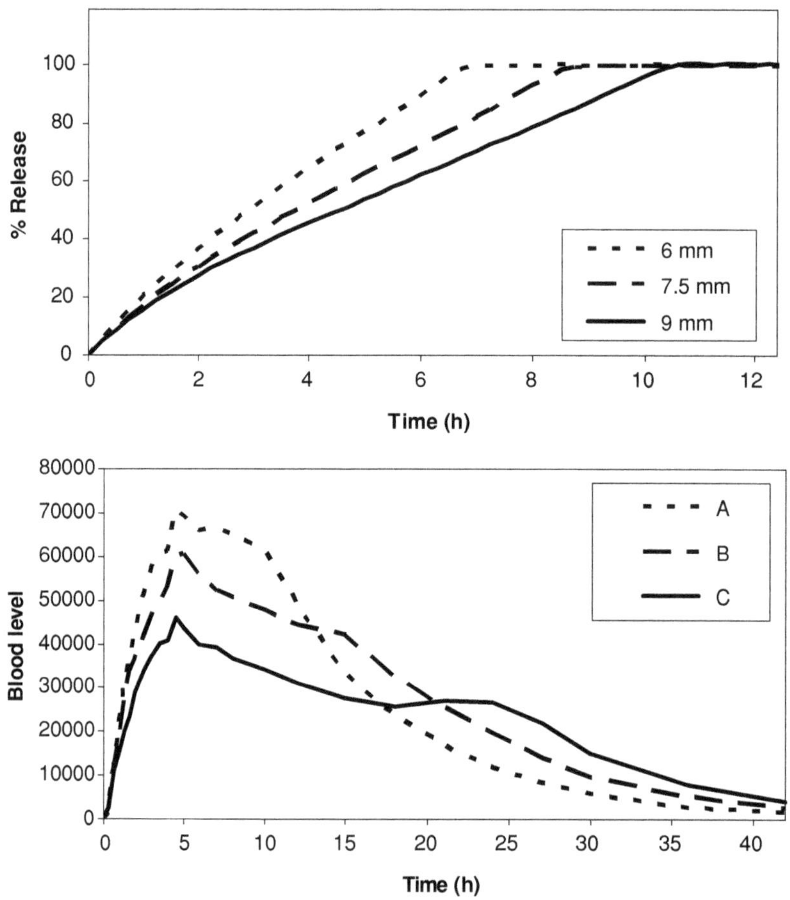

__Figure 11.11__ Figure showing main features of IVIVC. (Top) In vitro dissolution of the test formulations 6, 7.5 and 9 mm in USP2 pH 6.8 phosphate buffer, 50 rpm. The relative standard deviation is below 4% at all time points. The test formulations are fully dissolved (100%) after approximately 6.5, 8.5 and 10 hours, respectively. (Bottom) Plasma concentration up to 42 hours post-dose of the three test formulations (A, B and C: test tablet formulations).

C_{max} increased with release rate, C24 increased with tablet length and the total exposure (area under the curve) was independent of release rate. Accordingly, it was possible to obtain a point-to-point correlation between the in vitro release and the in vivo release: a level A IVIVC [23]. This correlation enables a predictability within the Egalet® abuse-deterrent prolonged-release erodible matrix (ADPREM) system that may be used beneficially to target specific release profiles in vivo.

It is important to note that the predictability only holds within formulations having the same release mechanism, i.e. Egalet® ADPREM is an erosion-based system in which the exposed area controls the release. Drug products with other release mechanisms would be expected to exhibit different correlations between in vitro and in vivo release [24].

11.7.2 Controlling Physical State by Means of Hot-melt Extrusion and Injection Molding

Materials produced by injection molding can be defined as being solid dispersions [25]. However, the actual physical state of the active pharmaceutical ingredient depends on the interaction with the carrier polymer. A multitude of possible scenarios exists for the physical state of compounds, either excipients or API. The compound may be either crystalline, amorphous or a combination of both. Furthermore, the compound may exist as discrete particles, as true solid solution or a combination of both. It is of outmost importance to control the properties in such a way that both chemical and physical properties are conserved throughout the shelf life of a pharmaceutical product.

Formulations composed of hydrophilic polymers such as PEO have been widely used as tablet constituents in conventional direct compression tabletting, but may also be employed as a carrier for both amorphous and crystalline APIs [26].

The properties of injection molding, in which the tablet is formed from melt under controlled pressure and cooling temperature, can be employed to develop amorphous formulations. In developing amorphous formulation from melt, it important to: (1) ensure the formulation and chemical composition supports the amorphous state of the drug compound and (2) carefully control the cooling process (preferably by means of quench cooling). The temperature and pressure control is an integrated part of standard injection-molding equipment. Often the drug candidates selected for developing amorphous formulations are compounds exhibiting poor aqueous solubility. This typically gives rise to yet another process control during manufacturing: humidity. Water acts in most formulations and, particularly for polymeric materials, as a plasticizer. Accordingly mobility is increased in the material if water has been taken up. Increased mobility leads to an increased likelihood of conformational changes, and hence to the possibility of crystallization of amorphous material. Furthermore, increasing the hydrophilicity of a formulation of a hydrophobic drug will increase the stability of the crystalline state.

A showcase of such a formulation of amorphous material is depicted in Figure 11.12, in which crystals were found after 6 weeks storage at 40°C (thereby decreasing dissolution rate). This was found to be fairly straightforward to resolve by means of controlling the manufacturing process: the cooling temperature (quench) was lowered and the process and process facilities were conducted at very low humidity. After these process changes, the product was shown to be stabile.

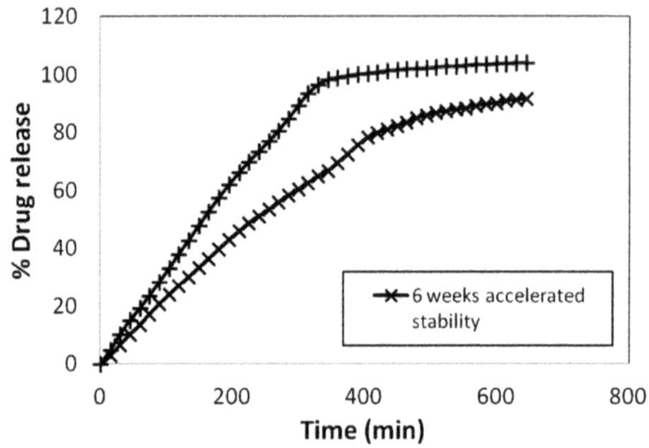

Figure 11.12 Formulation of amorphous drug, depicting a decreasing dissolution rate with storage time.

11.7.3 Anti-tamper Properties of Injection-molded Tablets

Controlling release rate or increasing bioavailability by means of manipulating manufacturing process is by no means a new discipline. In recent years, however, a new branch of quality attributes has been introduced. The success of controlled-release products has driven the introduction of products having higher drug content and thereby also a higher risk of unwanted effect and safety risks in the case of use outside of the prescription [27, 28]. For example, it is not uncommon for patients to chew tablets or otherwise tamper with controlled-release tablets in order to gain a faster relief of symptoms or simply because of difficulties with swallowing.

A severe problem that has been widely recognized by several authorities in the US is the fact that drug products may also be abused [29]. It is believed that at least 10% (and probably more) of all opioid products in the US are consumed during non-prescribed use, either in higher doses than prescribed (by defeating the release system by tampering) or by people other than the actual patient. This has formed the basis of a considerable drive from both the FDA and also from public and private insurance systems to develop products that are less prone to abuse.

The indications from the FDA are quite clear with respect to the commercial advantages of developing a truly abuse-resistant opioid analgesic product. In an advisory board meeting in November 2009, the FDA indicated that a company presenting proof of the fact that its product does lead to a clear reduction in drug abuse patterns will gain the advantage of the FDA removing other medications from the market. Several approaches to abuse deterrence are currently either under development or have been entered into the approval process (see Figure 11.13).

In combination with chemical formulation, injection molding provides the possibility to develop drugs with unique properties. The API is incorporated into a controlled-release matrix which is not easily crushed, grinded or in any other way mechanically manipulated to prepare the product for abuse (Table 11.2).

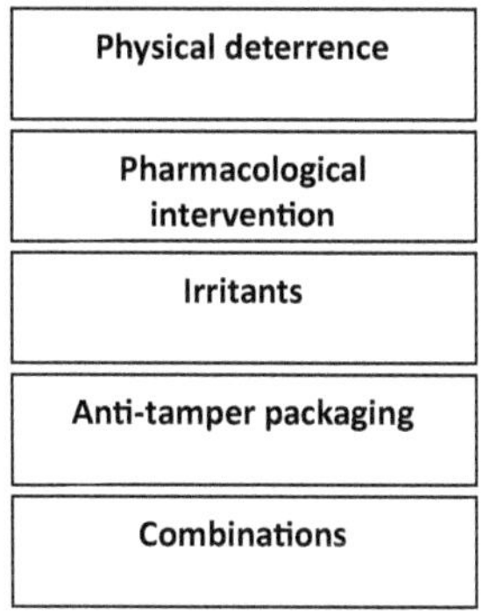

Figure 11.13 *Anti-abuse concepts.*

Focusing on tamper resistance, the physical and chemical robustness of new drug products requires development of standardized methods for testing tamper-resistance properties in vitro (see Figures 11.14 and 11.15). The Egalet® ADPREM technology provides a balanced approach to significant tamper-resistance properties combined with 'around-the-clock' pain management [29–31].

A key parameter in a drug product is that the drug works as intended. For controlled-release opioid products, that means simply that the patient experiences reduced pain balanced with a relatively low prevalence of unwanted adverse effects. Conceptually, the anti-tamper properties of tablets manufactured by means of injection molding comes as an inherent feature. Accordingly, anti-tamper products may be developed without compromising efficacy or safety. This might be seen as a balanced approach to abuse-deterrent opioid controlled-release product development.

In a clinical study in which Egalet® morphine once-daily dosing was compared to MST Continus twice-daily, Egalet® morphine was clearly as efficacious as MST Continus twice-daily [32]. Recent clinical data suggests that a 24-hour profile may also be achieved on oxycodone, which subsequently opens the possibility for a 24-hour profile for all opioid products currently in the pipeline.

Table 11.2 *The relative success of particle size reduction (1: a very easy, highly successful end-point and 10: a difficult, highly unsuccessful end-point).*

	Room temperature		Microwave	
Product	ADPREM	Comparator	ADPREM	Comparator
Mortar and pestle	10	1	8	2
Knife	10	1	5	1
Vise	8	2	6	3
Channel lock pliers	8	1	6	2
Krups coffee mill	10	1	10	1

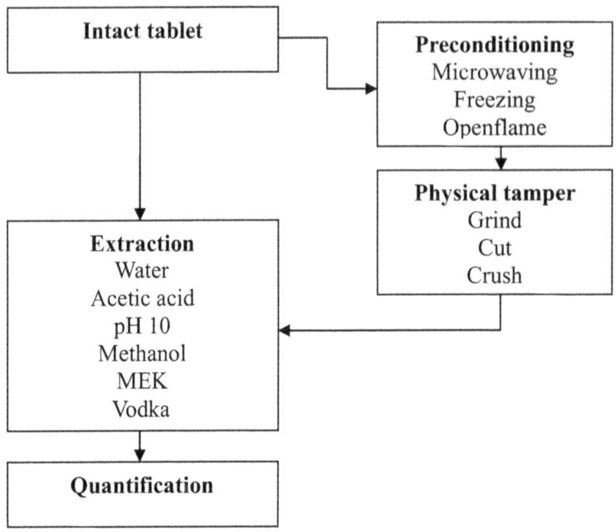

Figure 11.14 *Protocol for in vitro testing of tamper-resistance properties.*

Figure 11.15 *Tests of tamperablity. Top views: test of particle size reduction experiment in a coffee mill until either tablet or lid brakes. In this case the lid broke first. Bottom view: test of injectability. The tablet is dissolved in 2 ml water and forms a gel that is impossible to inject. For a better understanding of the figure, please refer to the color section.*

A primary advantage of the Egalet® formulation is that it is physically robust. Manufacturing by means of injection molding renders a very hard formulation which is not easily manipulated by conventional means such as chewing, crushing, grinding, milling or other means of mechanical manipulation. Due to the chemistry involved and the polymeric nature of the excipients, smoking or melting is not viable for abusing this drug product.

11.8 Concluding Remarks

The pharmaceutical industry faces the constant challenge of developing new and better products for patients worldwide. The technologies of extrusion and injection molding offer a method of preparing dosage forms that may contribute in the endeavor to innovate and formulate products that increase both safety and efficacy of pharmaceutical products. This is obtained by a high control of release properties of the dosage forms. Furthermore, the manufacturing process offers the possibility to integrate the approaches of QbD and PAT because critical process parameters may be tightly controlled. In principle, both injection molding and extrusion technologies may be applied to both crystalline and amorphous APIs; the concept is also well suited to develop dosage forms of poorly soluble compounds.

The increasing attention on anti-tamper products that has emerged within the last decade offers a new application of injection-molded tablets. In particular, Egalet® ADPREM has been developed specifically to address the need for new pain management products that resist deliberate and/or accidental tampering.

References

(1) Breitenbach, J. (2002) Melt extrusion: from process to drug delivery technology. *European Journal of Pharmaceutics & Biopharmaceutics*, **54**, 107–117.
(2) Bruce, L.D., Shah, N.H., Malick, A.W., Infeld M.H. and McGinity, J.W. (2005) Properties of hot-melt extruded tablet formulations for the colonic delivery of 5-aminosalicylic acid. *European Journal of Pharmaceutics & Biopharmaceutics*, **59**, 85–97.
(3) Follonier, N., Doelker, E. and Cole, E.T. (1994) Evaluation of hot-melt extrusion as a new technique for the production of polymer-based pellets for sustained release capsules containing high loadings of freely soluble drugs. *Drug Development & Industrial Pharmacy*, **20**, 132–1339.
(4) Repka, M.A., Repka, S.L. and McGinity, J.W. (2002) Biadhesive hot-melt extruded film for topical mucosal adhesion applications and drug delivery and process for preparation thereof. United States Patent 6375963.
(5) Zhang, F. and McGinity, J.W. (1999) Properties of sustained-release tablets prepared by hot-melt extrusion. *Pharmaceutical Development & Technolgy*, **4**, 241–250.
(6) Fukuda, M., Peppas, N.A. and McGinity, J.W. (2006) Floating hot-melt extruded tablets for gastroretentive controlled drug release system. *Journal of Controlled Release*, **115**, 121–129.

(7) Bhardwaj, R. and Blanchard, J. (1997) In vitro evaluation of Poly(D,L-lactide-co-glycolide) polymer-based implants containing the α-melanocyte stimulating hormone analog, Melanotan-I. *Journal of Controlled Release*, **45**, 49–55.
(8) Stevens, M.P. (1990) *Polymer Chemistry: An Introduction*, 2nd edition. Oxford University Press, Oxford.
(9) Rothen-Weinhold, A., Besseghir, K., Vuaridel, E., Sublet, E., Oudry, N., Kubel, F. and Gurny, R. (1999) Injection-molding versus extrusion as manufacturing technique for the preparation of biodegradable implants. *European Journal of Pharmaceutics & Biopharmaceutics*, **48**, 113–121.
(10) De Brabander, C., Van Den Mooter, G., Vervaet, C. and Remon J.P. (2002) Characterization of ibuprofen as a nontraditional plasticizer of ethyl cellulose. *Journal of Pharmaceutical Sciences*, **91**, 1678–1685.
(11) Forster, A., Hempensatall, J., Tucker, I. and Rades, T. (2001) Selection of excipients for melt extrusion with two poorly water-soluble drugs by solubility parameter calculation and thermal analysis. *International Journal of Pharmaceutics*, **226**, 147–161.
(12) Forster, A., Rades, T. and Hempenstall, J. (2002) Selection of suitable drug and excipient candidates to prepare glass solutions by melt extrusion for immediate release oral formulations. *Pharmaceutical Technology Europe*, **14**, 27–37.
(13) Henrist, D. and Remon, J.P. (1999) Influence of the formulation composition on the in vitro characteristics of hot stage extrudates. *International Journal of Pharmaceutics*, **188**, 111–119.
(14) Peppas, N.A., Wu, J.C. and von Meerwall, E.D. (1994) Mathematical modeling and experimental characterization of polymer dissolution. *Macromolecules*, **27**, 5626–5638.
(15) Siepmann, J., Streubel, A. and Peppas, N.A. (2002) Understanding and predicting drug delivery from hydrophilic matrix tablets using the "sequential layer" model. *Pharmaceutical Research*, **19**, 306–314.
(16) Crowley, M.M., Zhang, F., Koleng, J.J. and McGinity, J.W. (2002) Stability of polyethylene oxide in matrix tablets prepared by hot-melt extrusion. *Biomaterials*, **23**, 4241–4248.
(17) Bar-Shalom, D., Bukh, N. and Larsen, T.K. (1991) Egalet®, a Novel Controlled-Release System in: Temporal Control of Drug Delivery. *Annals of the New York Academy of Sciences*, **618**, 578–580.
(18) Bar-Shalom, D., Slot, L., Lee, W.W. and Wilson, C.G. (2003) Development of the Egalet technology. In Rathbone, M.J., Hadgraft, J. and Roberts, M.S. (eds), *Modified-release Drug Delivery Technology*. Marcel Dekker, New York.
(19) Buckley, C.P. and Kovacs, A.J. (1976) Melting behavior of low molecular weight poly (ethylene-oxide) frations. 2. Folded chain crystals. *Colloid & Polymer Science*, **254**, 695–715.
(20) Bar-Shalom, D., Andersen, C., Wilwon, C.G. and Washington, N. (2008) Development of the Egalet. In Rathbone, M.J., Hadgraft, J., Roberts, M.S. and Lane, E.S. (eds), *Modified-release Drug Delivery Technology*. Informa Healthcare, New York, London.
(21) Marvola, J., Kanerva, H., Slot, L., Lipponen, M., Kekki, T., Hietanen, H., Mykkänen, S., Ariniemi, K., Lindevall, K. and Marvola, M. (2004) Neutron activation-based gamma scintigraphy in pharmacoscintigraphic evaluation of an Egalet® constant-release drug delivery system. *International Journal of Pharmaceutics*, **281**, 3–10.
(22) Turner, S., Federici, C., Hite, M. and Fassihi, R. (2004) Formulation development and human in vitro-in vivo correlation for a novel, monlithic controlled-release matrix

system of high load and highly water-soluble drug niacin. *Drug Development & Industrial Pharmacy*, **30**, 797–807.

(23) Hemmingsen, P.H., Haahr, A.-M., Gunnergaard, C. and Cardot, J.-M. (2011) Development of a new type of prolonged release hydrocodone formulation based on Egalet® ADPREM technology using *in vivo–in vitro* correlation. *Pharmaceutics*, **3**, 73–87.

(24) Mehuys, E., Vervaet, C., Gielen, I., Van Bree, H. and Remon, J.P. (2004) In vitro and in vivo evaluation of a matrix-in-cylinder system for sustained drug delivery. *Journal of Controlled Release*, **96**, 261–271.

(25) Cuff, G. and Raouf, F. (1999) A preliminary evaluation of injection moulding as a tabletting technology. *Pharmaceutical Technology Europe*, **4**, 18–26.

(26) Maggi, L., Segale, L., Torre, M.L., Ochoa Machiste, E. and Conte, U. (2002) Dissolution behavior of hydrophilic matrix tablets containing two different polyethylene oxides (PEOs) for the controlled release of a water-soluble drug. Dimensionality study. *Biomaterials*, **23**, 1113–1119.

(27) Colemann, J.J., Schuster, C.R. and DuPont, R.L. (2010) Reducing the abuse potential of controlled substances. *Pharmaceutical Medicine*, **24**, 21–36.

(28) Haahr, A.-M., Lyhne-Iversen, L., Ridderberg, S., Nilssson, M. and Hemmingsen, P. (2007) Drug abuse resistant controlled release, using Egalet® dosage units. Proceedings of the 34th Annual Meeting and Exposition of the Controlled Release Society, poster 680.

(29) White, A.G., Birnbaum, H.G., Mareva, M., Daher, M., Vallow, S., Schein, J. and Katz, N. (2005) Direct costs of opioid abuse in an insured population in the United States. *Journal of Managed Care Pharmacy*, **11**, 469–479.

(30) Warner, M., Chen, L.H. and Makuc, D.M. (2009) Increase in fatal poisonings involving opioid analgesics in the United States, 1999-2006, NCFS Data Brief No. 22, Center for Disease Control and Prevention, US Department of Health and Human Services, National Center for Health Statistics.

(31) Hemmingsen, P.H., Haahr, A.-M. and Tygesen, P. (2010) Raising the bar for tamper resistance: Presenting a new protocol for in vitro testing of tamper resistance properties. *Proceedings of the annual meeting of AAPS*.

(32) Ridgway, D., Sopata, M., Burneckis, A., Jespersen, L. and Andersen, C. (2010) Clinical efficacy and safety of once-daily dosing of a novel, prolonged-release oral morphine tablet compared with twice-daily dosing of a standard controlled-release morphine tablet in patients with cancer pain: A randomized, double-blind exploratory crossover study. *Journal of Pain and Symptom Management*, **39**, 712–720.

12

Laminar Dispersive and Distributive Mixing with Dissolution and Applications to Hot-melt Extrusion

Costas G. Gogos and Huiju Liu
Department of Chemical, Biological, and Pharmaceutical Engineering, New Jersey Institute of Technology, Newark, NJ, USA
Peng Wang
Department of Chemical Engineering, University of Rhode Island, Kingston, RI, USA

12.1 Introduction

Pharmaceutical hot-melt extrusion (HME) has been explored and studied in the last few decades, by both industrial and academic investigators, because of its potential of rendering poorly water-soluble active pharmaceutical ingredients (APIs) readily bioavailable to patients through oral dosages. The HME field is currently being investigated even more intensively because of recent discoveries of large families of potent and promising, but essentially water-insoluble, APIs.

HME is a term that the pharmaceutical sector adopted to differentiate it from traditional oral dosage producing techniques, such as direct compression and tableting. It involves the use of single- or twin-rotor extruders for the processing of usually water-soluble polymeric excipients, mixing them while molten with APIs to affect partial or total API dissolution and pumping the homogeneous mixture through a die to form an extrudate, where the API exists in a totally or partially dissolved but (in both cases) stable form. Compared to the

Hot-melt Extrusion: Pharmaceutical Applications, First Edition. Edited by Dennis Douroumis.

traditional drug production processes, HME is a solvent-free continuous process and it may lead to fewer required processing steps.

However, degradation of the drug (API) and excipient may occur during HME due to the relatively high processing temperatures needed to melt the excipient and to laminar flow heating, due to viscous energy dissipation. This may limit universal application of HME for all excipient/API pairs. Informative accounts and information regarding equipment, formulation principles and process conditions and parameters used in HME can be found in several review articles and a recently edited book [1–6].

Extrusion processing has been used in the polymer and food industries for over a century, and a great wealth of knowledge has been generated and accumulated both in theory and practice. From a processing point of view, HME involves five elementary steps: handling of particulate solids, melting, mixing, devolatilization and stripping, pressurization and pumping [7]. These five steps are shown schematically in Figure 12.1 for the case of processing a polymer with solid particulate functional additive(s) to form compounded (or filled) plastic pellets or (with in-line compounding processes [7a]) compounded plastic products.

As noted, the two most important elementary steps for plastics compounding are melting and dispersive and distributive mixing of the additives in the polymer matrix. On the other hand, as shown conceptually in Figure 12.2, for HME pharmaceutical processing, dissolution of the API in the molten excipient is an additional and most important elementary step, along with melting which precedes it and mixing which assists and speeds up dissolution. Such elementary steps may interact with each other as well as occur simultaneously (i.e. be coupled). Although any of the elementary steps may be critically important to a particular HME process, this chapter will deal with laminar dispersive and distributive mixing of API particulates in the molten excipient as they occur simultaneously with the desired dissolution of the API. We will also examine the role these coupled elementary steps play in determining key product properties such as API release rate in aqueous media and shelf-life stability.

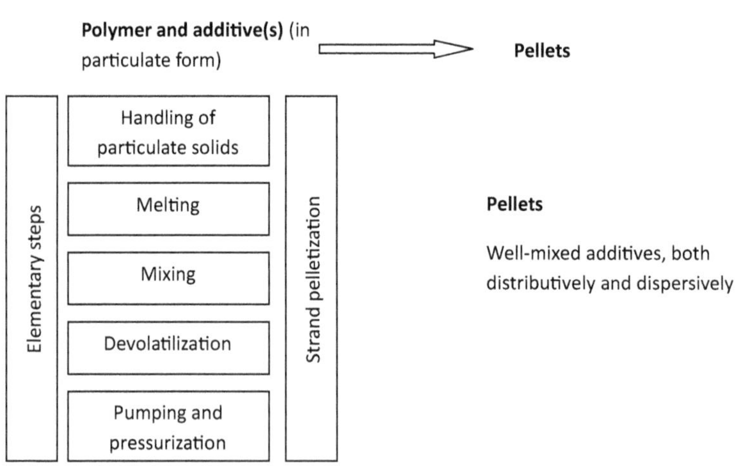

Figure 12.1 Conceptual structural breakdown of polymer-compounding processes.

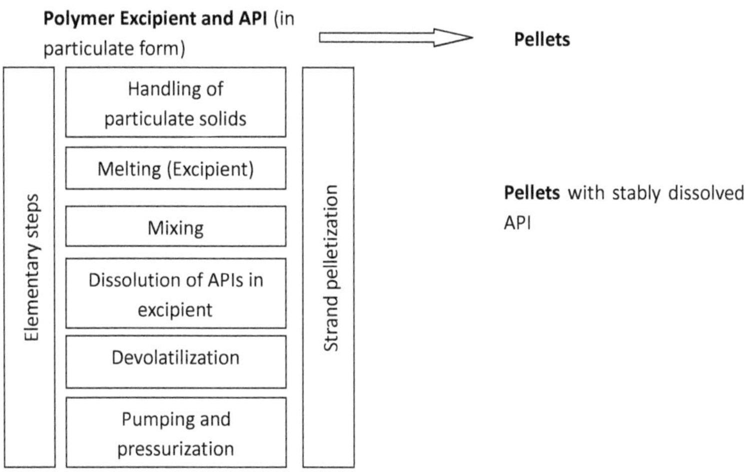

Figure 12.2 *Conceptual structural breakdown of pharmaceutical HME processes.*

12.2 Elementary Steps in HME

While briefly reviewing the features of the elementary steps which are relevant to HME processes and products, the following are worth considering.

12.2.1 Particulate Solids Handling (PSH)

Single-screw extruders (SSEs) are 'flood fed' through hoppers, the feed being a mixture of the excipient and API in particulate form. The fact that the barrel surface is rougher than that of the single screw allows for drag-induced packing of the particulates bed, as well as downstream movement and pressurization. In co-rotating twin-screw extruders (co-TSEs), which are commonly used in HME process development, the PS ingredients are fed gravimetrically or volumetrically controlled at constant rates. These rates are smaller than those needed to fully fill the parallel channels of the co-TSE, resulting in 'starve-fed' processing. PSH in co-TSEs may result in spatial particle segregation if the relative sizes or shapes of the API and the excipient are very different, due to different air resistive forces and different particle/wall kinematic friction coefficients. It is also worth noting that polymer excipients are commonly hygroscopic, so they may have to be dried prior to dry mixing with the API particulates [7b].

12.2.2 Melting

The physical mechanisms available for melting polymer systems in polymer processing equipment are as follows.

For SSEs, the mechanisms of *conductive melting* of the packed particulate bed surface by and next to the hot barrel surface and, after the thickness of the melted polymer layer exceeds that of the barrel-screw tip clearance, *viscous energy dissipation* during the drag flow causing the removal of the melt generated to the trailing end of the bed are important.

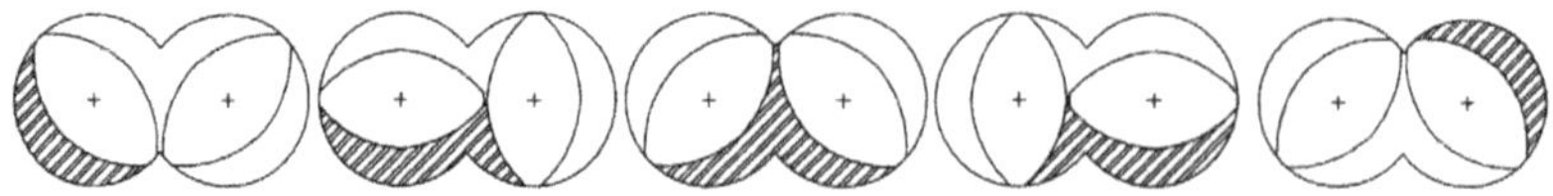

Figure 12.3 Snapshots of the repetitive expansion/contraction of each of the cross-sectional area pockets between a pair of kneading disks and the barrel of fully intermeshing co-rotating extruders. Z. Tadmor et al. 2006, reproduced with permission of John Wiley & Sons.

Melting is localized at the barrel surface and is gradual, requiring much of the extruder length to complete. Consequently, the age distribution of the melt generated in a typical SSE is of the order of its average residence time. This fact may result in adverse consequences for HME processing. First, since dissolution and mixing takes place only when the excipient is molten, there will be only limited dissolution of the API in that fraction of the excipient which melted late, potentially resulting in a wide distribution of the percentage of API dissolved. Second, the portion of the excipient which melted early will be more susceptible to thermal degradation [7c].

For co-TSEs, the available melting mechanisms are again *conductive melting* of the starve-fed loose particulates by the hot barrel. This mechanism is significant for the small co-TSEs used in HME development, where the surface-to-volume ratio is large. However, for larger-diameter co-TSEs, reverse-screw or reverse-kneading elements are used to create a filled section in which the packed particulates undergo repeated volume-wide deformations before exiting the fully filled region. During this process, the very powerful melting mechanism of *plastic energy dissipation* (*PED*) is important. It is also worth pointing out that the repeated large compressive deformations taking place in full kneading blocks induce particulate-to-particulate frictional heating and localized melting because of *frictional energy dissipation* (*FED*) [7d, 8].

Figure 12.3 shows five snapshots of a pair of co-rotating double-flight kneading elements. With a well-selected reverse-kneading element or screw-element sequence, PED is capable of completely melting the entire charge within an axial length of 1–3 diameters, giving rise to a melt stream which has almost the same 'age' which is very helpful for uniform API dissolution. Fully filled kneading elements, as discussed in Section 12.5.2, give rise to rapid and efficient chaotic laminar mixing comprising mostly extensional and 'folding' flows.

12.2.3 Devolatilization

This elementary step refers to the removal of low levels of volatiles of the order of 1000 ppm, dissolved in the molten matrix. Devolatilization is carried out in vented two-stage SSEs and co-TSEs in partially filled sections isolated from both the upstream and downstream sections by 'melt seals' so that vacuum can be applied. Under vacuum conditions, the dissolved molecules cause bubbles to be formed in the flowing melt stream (much like the bubbles formed by opening a carbonated refreshment container) which, when they reach the melt–vacuum interface, burst and are removed [7e]. Although there does not appear to be much work on devolatilizing HME extruders, the subject will receive attention because of FDA regulations.

12.2.4 Pumping and Pressurization

After the accomplishment of all the other elementary steps, extruders are required to pump ('meter') the molten charge through a die which shapes the exiting stream in operations such as pelletization and sheet, film, tube or profiled cross-sectioned products. *Drag-induced* pumping and pressurization is the mechanism enabling both co-TSEs and SSEs to be the pumps of choice for viscous fluids. SSEs can generate higher pressures under closed discharge conditions because their flight heights can be small in the metering section upstream of the die. On the other hand, because co-TSEs are fully intermeshing (and self-wiping, which is an advantage for HME operations) they are 'locked in' with wide channels which are incapable of generating high pumping pressures; in cases of very viscous extrudates, this limits the extrusion rate [7f]. Since excipients are water soluble care has to be taken in cooling strand extrudates, using chill rolls for conductive cooling and/or cold air for forced convective cooling.

12.3 Dispersive and Distributive Mixing

The mixing processes in single- and twin-screw extruders, the main topic of this chapter, are generally categorized into two types: dispersive mixing and distributive mixing [7g].

Dispersive mixing refers to the process involving the particle size reduction of cohesive components such as solid fillers (by de-agglomeration) or liquid droplets (by droplet deformation and break-up). *Distributive mixing* refers to distributing de-agglomerated particulates uniformly throughout space, or stretching the interfacial area between the components lacking a cohesive resistance and distributing them uniformly throughout the product volume. Dispersive mixing requires high flow stresses (either through high viscosity of high shear or elongational rates) in order to provide the dispersive forces to overcome the cohesive forces of the agglomerates or immiscible droplets; distributive mixing is dictated only by the flow-generated strain and does not require high stresses.

According to these definitions, the mixing of miscible liquids is regarded as distributive mixing; mixing of hard solid agglomerates, immiscible liquids and soft agglomerates is regarded as dispersive mixing [7g]. For illustrative purposes, the dispersive and distributive mixing of solid agglomerates is shown schematically in Figure 12.4.

12.4 HME Processes: Cases I and II

HME processes can be classified into two categories:

- Case I: in which the processing temperature is above the melting temperature (semi-crystalline polymer) or the softening temperature of an amorphous polymer ($T_g > 50$–$100°C$) but below the melting point of a crystalline API.
- Case II: in which the processing temperature is above both the melting or softening temperature of semi-crystalline or amorphous polymers, respectively, and above the melting point of the API.

'Processing temperature' refers to the melt temperature rather than barrel set temperature. Case I is more common, simply because it is carried out at temperatures which are safer

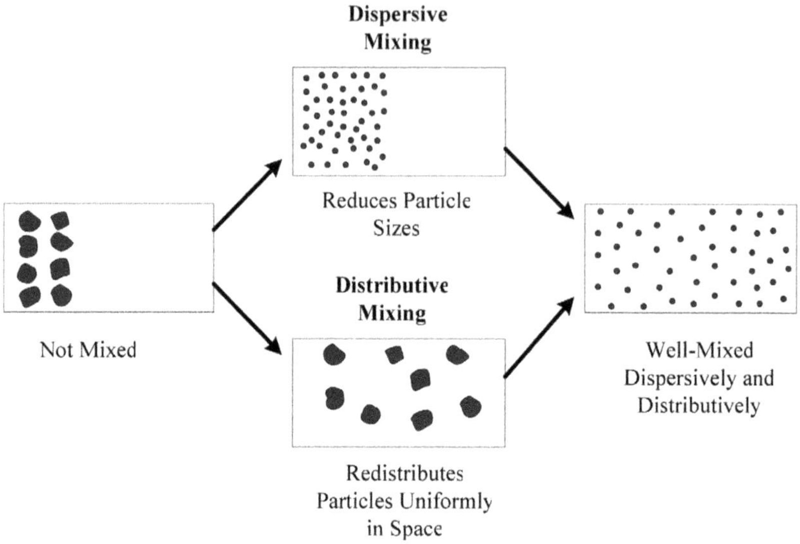

Figure 12.4 *Dispersive and distributive mixing of solid agglomerates and immiscible liquid droplets. Z. Tadmor et al. 2006, reproduced with permission of John Wiley & Sons.*

from an API degradation point of view (since that temperature is below its melting point). On the other hand, dissolution rates and solubility are expected to be larger if the process is carried out at higher temperatures.

12.4.1 Case I

The process represented by Case I provides a viable dissolution path which minimizes or circumvents the thermal degradation issue of drugs. The API is processed below its melting point and mixed with a polymer melt and the solid drug particles gradually dissolve into the polymer excipient melt, resulting in a desirable polymer–drug solid dispersion or solid solution. In this case, the solid API and the polymeric melt act as a solute and a highly viscous solvent, respectively, during HME. A physical model for Case I is schematically shown in Figure 12.5.

Firstly, the premixed drug (black) and polymer particles (white) are fed into the batch mixer or an extruder. The polymer particles then start melting due to the conductive heat from the mixer or extruder barrel and frictional and plastic energy dissipation for co-TSEs, leading to the solid drug particles being suspended in a continuous polymer melt matrix. While suspended at the processing temperature, which favors dissolution assuming intermolecular forces compatibility between the API and the excipient (i.e. miscibility), the drug molecules start dissolving and create a mass-transfer boundary layer around each drug particle. This layer is continuously wiped away and replaced by fresh polymer melt around each API particulate by the laminar distributive flow of the mixer. The same laminar mixing flow helps the drug molecules to diffuse and mix distributively into the molten excipient. The size of suspended drug particles diminishes as the diffusion continues until the particles

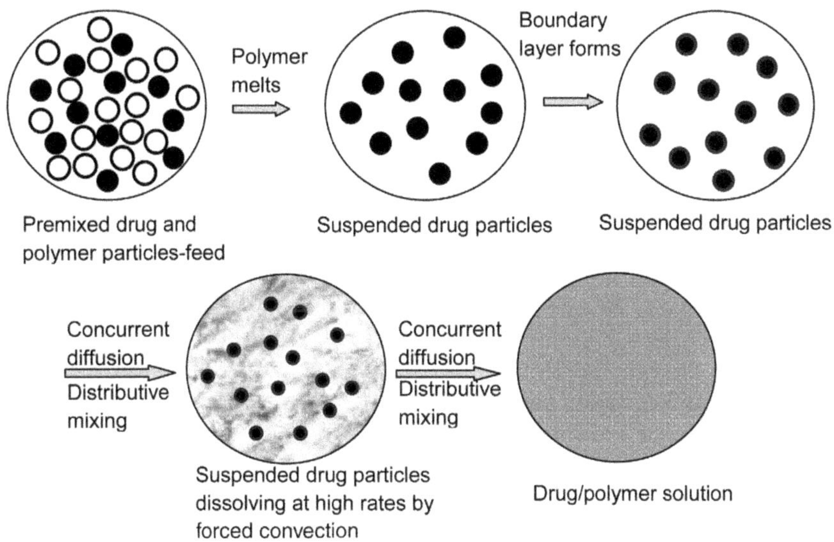

Figure 12.5 *Schematic representation of the morphological changes of the drug and polymer system in the solution formation process.*

disappear and a homogeneous solution is formed or until the limit of API solubility at the processing temperature is reached. In the latter case, they reach a minimum average size and remain suspended.

The dissolution of the drug in the polymer melt in an extruder is achieved by laminar forced convective mass transfer involving the dissolving and dissolved API molecules. Both dispersive and distributive mixing may play key roles on the dissolution of the drug into the polymeric excipient melt. Dispersive mixing may break up the drug agglomerates or even particulates due to high laminar flow forces generated by either material properties such as excipient viscosity, operating variables such as screw speed or extruder design variables such as the width of kneading elements in co-TSEs of Maddock 'barrier' mixing elements in SSEs. The total surface area of the drug particles exposed to the polymeric melt will therefore be increased and the dissolution rate will be increased. Distributive mixing can homogenize the drug concentration in the polymeric melt through shear or extensional flow or reorientation and bring more polymer melt into contact with the suspended drug particles, thus leading to dissolution rate enhancement.

Case I, the dissolution of APIs into polymer matrix, is of cardinal importance in practical HME because many APIs are heat sensitive, especially at temperatures above their melting point. In traditional polymer processing, only few examples involving dissolution of small molecule additives into molten polymer matrices can be found (e.g. physical blowing agents and certain process stabilizers). Additionally, these examples involve dissolving additives at much lower concentrations (>1%) than those used in typical API formulations. The main task of Case I is to completely dissolve drugs in polymeric melt within the shortest possible residence time without raising the processed stream melt temperature.

12.4.2 Case II

Case II, relevant at processing temperatures above the melting point of the crystalline API, involves liquid–liquid mixing between miscible or partially miscible components. The criterion that the solubility parameter difference $\Delta\delta$ between the drug and excipient be less than 7 $MPa^{1/2}$ is generally accepted during HME formulation screening, implying the need of partial miscibility [9, 10]. Such systems possess dynamic surface tension. The morphological evolution of mixing such partially miscible systems will therefore involve liquid phase break-up of the minor phase, and may follow the Scott/Macosko lacing/sheeting mechanism in the softening/melting region [11]. The minor phase break-up may actually be more complex, since the API viscosity is orders of magnitude smaller than that of the polymer excipient [7h].

Schematically, the morphology changes for Case II may follow the sequence shown in Figure 12.6. At the beginning, the premixed drug (black) and polymer (white) particles are fed into an extruder and conveyed by the conveying elements. The polymer particles melt first due to the energy input from the barrel and frictional and plastic dissipation. After the polymer particles totally or partially melt, the drug particles suspended in the molten polymer melt rapidly, and the drug droplets begin to be deformed by the mixing laminar flows of the polymer melt. After that, the drug liquid phase breaks up into much smaller droplets due to the competition of surface tension and flow stress. The small droplets are deformed along the shear direction. With numerous very small droplets, which have an enormous surface, diffusion between the droplets and the polymer predominates causing the droplets to disappear, creating drug–polymer solution. Diffusion also occurs during the break-up of the large drug droplets. Note that the 'characteristic diffusion time' t_D in Figure 12.6 is proportional to the square of the API phase droplet or ligament radius or the thin dimension of a sheet. Thus, for a molten excipient-API system with a diffusivity $D = 10\times10^{-11}$ m^2/s and where the API exists in 20 μm diameter droplets,

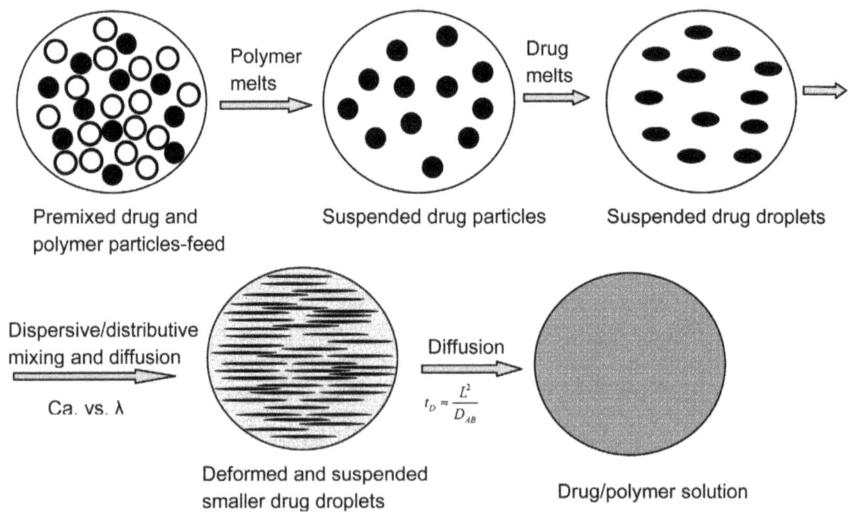

Figure 12.6 Morphological changes in drug/polymer system for Case II.

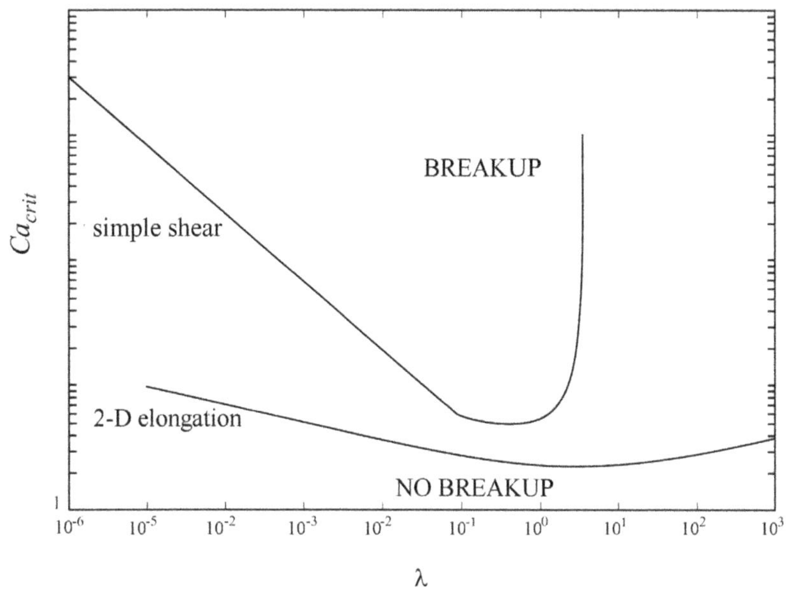

Figure 12.7 *Critical capillary number versus the viscosity ratio λ. H.P. Grace et al. 1982, reproduced with permission of Taylor & Francis.*

the diffusion characteristic time is of the order of 10 seconds. It is worth emphasizing that all of the steps described in Section 12.2, including polymer melting, drug melting, the break-up of drug droplets and even diffusion between drug droplets and polymer melt, may occur simultaneously over the span of several seconds (especially with co-TSEs) so that experimental evidence of the individual phenomena may not be easily obtainable.

Similarly to Case I, both dispersive and distributive mixing play key roles on the break-up of drug droplets in Case II. Droplet break-up for partially miscible systems during both simple shear and 2D extensional (elongational) flows is shown in Figure 12.7, where the critical capillary number Ca_{crit} is plotted against the viscosity ratio λ for Newtonian fluids. Ca_{crit} is defined as the ratio of viscous (dispersive) to the interfacial (cohesive) stresses [12]. The viscosity ratio is defined as the ratio of the dispersed phase to the continuous phase viscosities.

As seen in Figure 12.7, the 2D elongational (squeezing) flow is more efficient for droplet break-up than shear flow with much lower Ca_{crit} for the physically very broad region where $\lambda > 4$ and $\lambda < 10^{-3}$ (the latter being relevant to case II of HME). It should be noted that extensional flows are also more efficient than shear flows for distributive mixing because they are capable of increasing the resulting strain and interfacial area exponentially. In contrast, shear flows increase shear strain linearly and are therefore less efficient [7g]. We will see in the following section that fully filled co-TSE kneading element sections generate compressive extensional flows, much as they compressively deform particulate solid beds. Because of this capability, they are well suited to mixing and dispersing the rheologically mismatched fluids involved in HME.

12.5 Dissolution of Drug Particulates in Polymeric Melt

Similarly to the dissolution of drugs in an aqueous medium, the dissolution of drug particulates in molten polymeric excipients during HME can also be described by the Noyes–Whitney equation [13, 14]

$$\frac{dC}{dt} = \frac{D \times A \times (C_S - C)}{h \times V} \tag{12.1}$$

where D is the diffusion coefficient; A is the total surface area of the drug exposed to the dissolution media; C_s is the saturation solubility of the drug in the liquid which (for HME) is the excipient melt; C describes the concentration of the dissolved solid phase in the bulk at time t; h represents the diffusion boundary layer at the solid-liquid interface; and V is the volume of the dissolution medium.

The variables influencing the dissolution rate of drug particulates in the excipient melt can be grouped into three categories: process, equipment and material.

12.5.1 Process Variables

Process parameters have an important impact on the dissolution rate of drug particulates in polymeric melts. For co-TSEs, the most important process parameters are the barrel set temperature, the screw speed and the feeding rate. Screw speed and feeding rate can be used to calculate the characteristic channel shear rate, shear stress, specific mechanical energy [5, 15] and the mean residence time [16, 17]:

$$\dot{\gamma} = \frac{\pi \times D \times n}{h \times 60} \tag{12.2}$$

$$\tau = \dot{\gamma} \times \eta \tag{12.3}$$

$$\mathrm{SME} = \frac{\text{consumed motor power}}{Q} = \frac{n \times (\%\text{torque}) \times \text{motor rating} \times 0.97}{\text{max rpm} \times Q} \tag{12.4}$$

$$t_m = \frac{A}{Q} + \frac{B}{n} \tag{12.5}$$

where $\dot{\gamma}$ is shear rate in $\sec^{-1}$; D is the screw diameter in mm; n is the screw speed in rpm; h is the over-flight clearance in mm; τ is shear stress in kPa; η is the melt viscosity in Pa s; SME is the specific mechanical energy (kW h/kg); Q is the feeding rate in kg/h;% torque is the percentage used of the maximum allowable torque; the motor rating is in kW; 0.97 is the gear box efficiency; max rpm is the maximum number of attainable screws rotations per minute; t_m is the mean residence time in seconds; and A and B are constants.

Many studies have been published concerning the effect of process variables on the final dissolution characteristics of drugs in aqueous media [18, 19], and the significance of process variables has been widely recognized [20, 21]. For example, Shibata et al. studied the preparation of solid dispersions of indomethacin (melting point = 162°C) with crospovidone (T_g = 60°C) using a twin-screw extruder and a twin-screw kneader, which has very long and strong kneading sections. All the barrel zones were set at the narrow temperature range 125–150°C, which is below the melting point of the drug. They concluded that the residence time, screw speed and heating temperature are significant factors for the properties of the solid dispersions obtained [21].

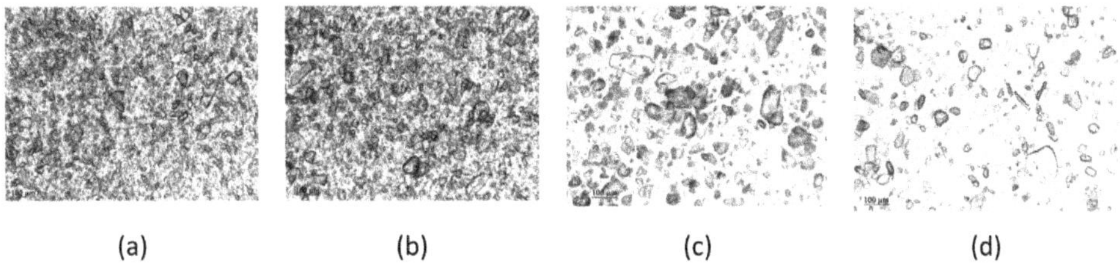

Figure 12.8 Optical micrographs of run at 100°C and 20 rpm: (a) 100; (b) 145; (c) 285; and (d) 420 sec.

In order to understand the dissolution mechanism from the polymer mixing perspective and gain insight on how to promote the dissolution rate, Liu et al. investigated the effects of extrusion process parameters on the dissolution behavior of indomethacin (melting point = 162°C) in Eudragit® E PO (T_g = 48°C) matrix using a batch mixer. The screw speeds can be controlled separately without altering the residence time in a batch mixer, which is more difficult to realize if an extruder is used [22]. The barrel temperature was set at 100, 110 and 140°C, the screw speed was set at 20 and 100 rpm for each temperature and samples were taken at 50, 95, 140, 280 and 420 seconds for each run. In all runs, the actual melt temperature was below the melting point of indomethacin. The weight ratio of Eudragit® E PO to indomethacin was kept at 70:30.

Figures 12.8 and 12.9 show optical micrographs of samples processed at 100°C, 20 rpm and 110°C, 100 rpm. The amount of the solid drug particles decreases with increasing mixing/processing time. For the run at 100°C, 20 rpm, both optical micrographs and scanning electron microscopy (SEM) pictures (not shown) show that there are still considerate amounts of drug particulates which were not dissolved after 420 sec of mixing. For the run at 110°C, 100 rpm however, both optical micrographs and SEM pictures (not shown) show that essentially no drug particulates can be found in the 285 sec sample. This morphological observation was also supported by differential scanning calorimetry (DSC) and X-ray diffraction (XRD) results.

The evolution of the specific enthalpy (integration DSC-obtained enthalpy of the broad peak/total drug mass) with mixing time, at different processing conditions, is depicted in Figure 12.10. All three processing variables used in this study, namely barrel set temperature, counter-rotating twin-rotor screw speed and residence time, are found to influence the dissolution of the indomethacin into the EPO melt. Given the same rotor speed of 20 rpm, all indomethacin particles are dissolved into the matrix within 3 min at the highest

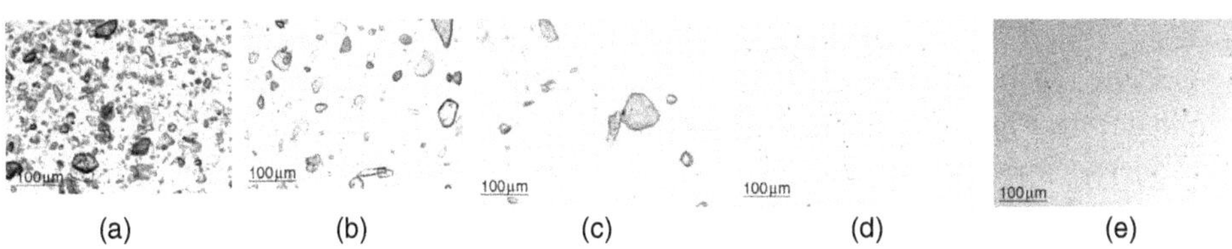

Figure 12.9 Optical micrographs of run at 110°C and 100 rpm: (a) 55; (b) 100; (c) 145; (d) 285; and (e) 420 sec.

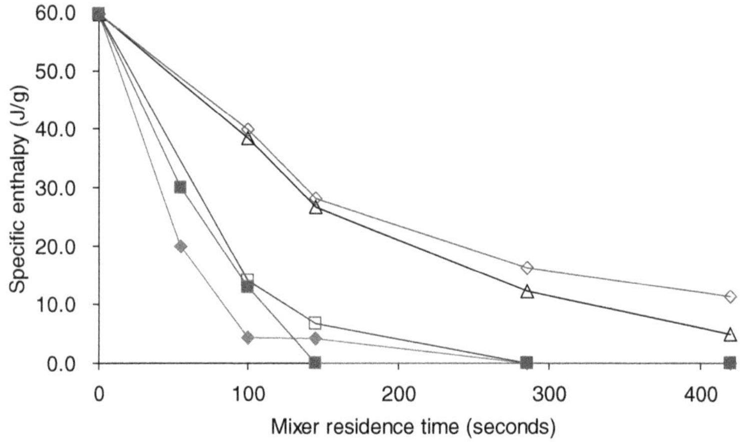

Figure 12.10 *The evolution of the specific enthalpy with residence time, screw speed and setting temperature: ◇ 100°C and 20 rpm; △ 110°C and 20rpm; □ 100°C and 100 rpm; ◆ 110°C and 100 rpm; and ■ 140°C and 20 rpm.*

set temperature of 140°C employed in this study. At the set temperatures of 100°C and 110°C, the drug particulates are not fully dissolved after 420 sec at the lowest rotor screw speed used; increasing the screw speed to 100 rpm allows a full dissolution of drug particulates within 300 sec. Obviously, both the barrel set temperature and screw speed can increase the dissolution rate appreciably.

The effects of barrel set temperature and rotor speed can be explained by the Noyes–Whitney equation. On one hand, if the mixer set temperature increases, the diffusion coefficient will increase due to the increased temperature and resultant decreased matrix viscosity; on the other hand, C_s also will increase. Both of these factors contribute to an increase of the API dissolution rate in the molten polymer excipient. When the mixer rotor speed increases, the distributive mixing is improved within the chamber and thus a higher concentration gradient in the region of the particulate surface is available. Moreover, the thickness of mass transfer boundary layer decreases as forced convective mass transfer prevails. Both effects lead to an increased dissolution rate.

As discussed in the introduction section and above, both dispersive mixing and distributive mixing may significantly enhance the dissolution rate. However, in this study there was no evidence that dispersive mixing was involved because the size reduction of drug particles was due to the diffusion of the drug molecules to the polymeric melt rather than shear forces [22]. Furthermore, the drug particles of the system used do not form agglomerates in the mixture based on the SEM pictures (not shown) of tumble-mixed solid before hot-melt processing. Thus, dispersive mixing is not needed in the system studied. Nevertheless it should be mentioned that, in other drug and polymer systems, dispersive mixing may break up existing drug agglomerates or even individual particles due to the high laminar flow forces. The total surface area of the drug particles exposed to the polymeric melt will then be increased, thus increasing the dissolution rate. Miller et al. demonstrated that HME processes can de-agglomerate and disperse 'engineered' drug particulates into an excipient matrix without altering their drug properties, and achieving enhanced dissolution properties [23].

The experimental results also lead to an important finding which has been overlooked before [22]: the times needed for the drug to dissolve inside the polymer melt and the typical residence time for an extrusion process have to be comparable under appropriate processing conditions. Depending on the barrel set temperature and the screw speed, the drug dissolution process may take from 1 minute up to a few minutes. The residence time of a typical continuous manufacturing extrusion process falls into the same range [16]. Hence, the batch internal mixer results obtained in this work are relevant to the expected performance of the API/excipient system in HME processes currently being used or under evaluation by the pharmaceutical industry. Furthermore, the HME process offers the possibility of further shortening the residence time to fully dissolve drug particulates into polymer melt through the liberal use of screw distributive or dispersive elements. On the other hand, one has to be careful when optimizing the HME process parameters such as screw speed, feeding rate and barrel temperature because changes in those parameters may lead to a change of the residence time of the processing stream, as Equation (12.5) shows.

It should be mentioned that, based on the Noyes–Whitney equation, the dissolution rate of drug particulates in polymer melt will increase if the total surface area of the drug particulates exposed to the dissolution media increases. Therefore, after drug loading is fixed in formulation development, the micronization of drug particulates is beneficial for enhancing the dissolution rate of drug particulates in polymer melt. Furthermore, the narrower the drug particle size distribution, the more uniform the total dissolution time distribution needed for complete dissolution of drugs in polymer melt will be. There are many commercialized mills available such as fluid energy mills (FEM) [24, 25].

12.5.2 Equipment Variables

Equipment or design variables concern mainly screw design. There are primarily three kinds of co-TSE screw elements: conveying screw elements, kneading elements and toothed screw elements. Comprehensive discussion can be found in several books [3, 7, 15]. Screw design plays an important role for both distributive mixing and dispersive mixing. In co-TSEs, for example, the wider the kneading blocks (KB), the more intensive will the dispersive mixing be; the narrower the KBs, the more distributive will the mixing be. Toothed screw elements, such as Coperion's SME (screw mixing element), TME (turbine mixing element) and ZME (Zahnmishelement), can generally offer more distributive mixing while inputting less mechanical viscous energy [15].

Although the importance of screw configuration has been mentioned in several review articles [1, 4], there are only a few publications specifically addressing the effect of screw configuration on preparing solid dispersions using twin-screw extruders.

Nakamichi et al. [18] reported that kneading blocks play a key role in transforming the crystalline nifedipine (melting point $= 175°C$) to an amorphous form inhydroxypropylmethylcellulose phthalate (HPMCP, $T_g \sim 160$–$170°C$). The barrel temperature in all experiments was set at 100°C.

Verhoeven et al. [26] studied the system of metoprolol tartrate (melting point $= 120°C$) and ethylcellulose ($T_g \sim 123°C$) mini-matrices using a twin-screw extruder. The barrel temperature in all experiments was set at 60°C. They found that the release rate of metoprolol

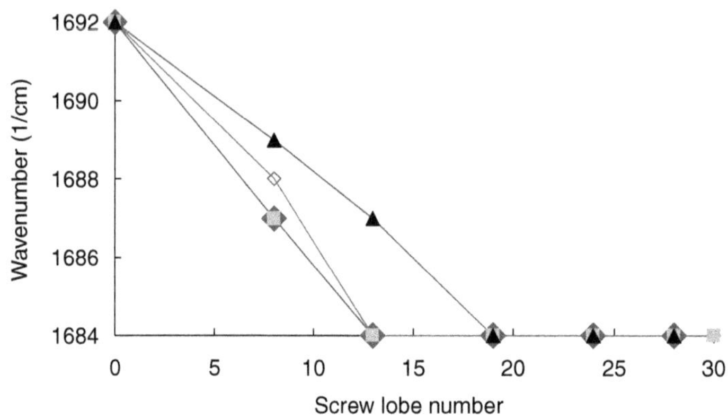

Figure 12.11 *Shift of the peak of the indomethacin benzoyl C=O stretch along the screw length: ■ screw with two kneading/mixing zones; ◇ screw with one kneading/mixing zone; ♦ screw with one stronger kneading/mixing zone; ▲ screw without kneading/mixing zone.*

tartrate in ethylcellulose and the homogeneity of the drug component were not affected by the number of kneading mixing zones or their position along the extruder barrel, as long as one kneading mixing zone was present [26]. Aside from reporting experimental findings, the mechanisms of melting and laminar mixing relevant to API dissolution into the excipient matrix by means of kneading blocks and other potentially beneficial screw elements were not discussed in these publications.

Liu et al. [27] examined the evolution of the extent of dissolution of indomethacin (melting point = 162°C) in Eudragit® E PO (T_g = 48°C) on four screw configurations processed in an APV 15-mm co-rotating twin-screw extruder. The first and second kneading blocks were set at lobe numbers 13 and 24, respectively. The ratio of barrel length to the screw diameter is 15. The barrel was set at 140°C. The screws were pulled out and quenched by water, which allowed for quick access to the processed stream carcass. The weight ratio of Eudragit® E PO to indomethacin was kept constant at 70:30.

Figure 12.11 shows the shifts of the indomethacin's benzoyl C=O stretch peak for the four screw configurations. The benzoyl C=O stretch peaks for the γ-form and amorphous indoemthacin are at 1692 and 1684 cm^{-1}, respectively [28]. The original indomethacin particulates fed into the hopper is in γ-form. Figure 12.11 indicates that the crystalline indomethacin did not totally transform to the amorphous state till the 19th lobe when the screw without kneading/mixing zone was used; for the other three screws which have at least one kneading/mixing zone, the transformation was complete at the 13th lobe. IR analysis therefore shows that the first kneading/mixing zone promotes the complete dissolution of indomethacin into Eudragit® E PO melt. The unique capability of fully filled kneading blocks in mixing and melting associated with the dissolution of API into polymeric excipient matrix needs further discussion.

Axial flow and back mixing take place due to the expansion and contraction [29] of each of the cross-sectional area pockets between a pair of kneading paddles and the barrel shown earlier in Figure 12.3, in connection with PED melting. The time sequence of the snapshots documents the evolution of one of the three cross-sectional area pockets (shaded area) A_S, which is available to be filled by the processed stream. As the pair of kneading disks co-rotate, A_S varies with time: it first expands from the minimum cross-section in the first snapshot to the maximum in the third snapshot and then contracts to the minimum in the right lobe of the barrel.

Back (axial) mixing plays a key role on the compositional uniformity of HME products since the multi-flight screw design splits the feeding material into different flights which cannot meet each other before the kneading blocks. Regarding to the back mixing, Brouwer et al. investigated the drag and pressure flow characteristics and residence time distribution in conveying elements, SME, ZME and TME, and suggested that SME's would be the best choice in melt sections of the extruder to dampen out feed variations with time [30].

Distributive and dispersive mixing can be induced by kneading blocks. Distributive mixing can be induced by the following two processes: (1) the separate kneading tips created by offsetting the individual kneading paddles relative to each other are split and recombine the flow; and (2) the expansion/contraction process causes time-varying extensional and folding chaotic flows. It is worth noting that extensional flows are more efficient than shear flows for distributive mixing because they are capable of increasing the resulting strain and interfacial area exponentially. Fully filled kneading blocks therefore create rapid (exponential) distributive mixing. Dispersive mixing can be induced in the high-shear gap between the intermeshing kneading paddles and the barrel wall, which is desired for agglomerate break-up.

It should be emphasized that the melting of polymer excipient and the dissolution of API into polymeric melt may simultaneously occur and be complete in a single kneading block, resulting in a much narrower 'melt age' distribution. One of the breakthroughs in the polymer processing field during the last two decades was polymer–polymer blend morphology evolution studies; it was found that a major reduction in phase domain size takes place in conjunction with the melting or softening of the components in a melting zone, usually consisting of kneading blocks [31]. Similarly to polymer–polymer blends preparation, the melting of polymer excipients and the dissolution of drugs can simultaneously take place and may be complete once passing through a kneading block, although no publications have explicitly presented this phenomenon to the date.

In Verhoeven's and Liu's work, the significance of the second kneading/mixing zone is not distinct [26, 27]. However, in the split feeding protocol, i.e. feeding drug particles from the downstream of extruders to shorten the time of heat exposure of the drug, the existence of the mixing zone after the drug feeding is necessary for promoting the dissolution within the limited residence time. It should be noted that the mixing zone located downstream of the feeding zone can also be composed of other toothed mixing elements such as SME, TME or ZME.

12.5.3 Material Variables

12.5.3.1 Possible Physical States of the Mixture

Ultimately, the drug release profile of a drug–polymer mixture is determined by the physical states of both components which, in turn, are decided by their miscibility. The mixing mechanism depends on the miscibility of the drug and polymer. Table 12.1 summarizes the possible physical states of the binary mixture. The complexity of the binary mixture

Table 12.1 *Possible physical states of a drug–polymer mixture.*

	One-phase solid solution		Two-phase solid mixture			
Drug	Molecularly dispersed	Molecularly dispersed	Crystalline	Crystalline	Amorphous	Amorphous
Polymer	Amorphous	Crystalline	Crystalline	Amorphous	Crystalline	Amorphous

is manifested by the six possible physical states. Very little is known about how drug molecules distribute in the two phases of the polymer after the HME-processed stream has been quenched to room temperature, especially for the cases where the polymer is semi-crystalline. All six physical states listed in the table are generally referred to as 'solid dispersions' in the literature.

Differential scanning calorimeters (DSC), X-ray diffractometers (XRD) and microscopes are the three most widely used analytical tools for the determination of the crystalline states of the drug and the polymer in a solid dispersion. Although the characterization may sound straightforward, care is required to avoid wrong conclusions. For example, the presence of a melting peak from the drug is commonly used as the indictor for crystalline drug. Our experimental results however show that, for certain binary mixtures, drug crystals can quickly dissolve into the polymer during the thermal scan. As a result, no drug melting peak will appear in the DSC thermogram, although the original sample does contain crystalline drug particles. It is also difficult to differentiate the drug from the polymer based only on the morphologies appearing in microscopic photos. That is why multiple characterization tools are often used to analyze a solid dispersion sample.

12.5.3.2 Drug–polymer Miscibility

From a thermodynamic aspect, mixing a drug with a polymer is not so much different from mixing a plasticizer with a polymer. Two strategies have been applied to predict/estimate the drug-polymer miscibility. The first strategy is based on a simple assumption: the solubility parameters of two miscible chemicals should be smaller than a critical value. The solubility parameter δ is defined as follows:

$$\delta = \sqrt{\frac{\Delta E}{v}} \tag{12.6}$$

where ΔE is the molar change in internal energy on vaporization and v is the molar volume of liquid. For a process that occurs at constant volume and constant pressure, the change in internal energy is equal to the change in enthalpy.

The research that follows this strategy often aims to answer two general questions: (1) what is the upper limit of the difference in the solubility parameters between the drug and the polymer for having a miscible system? and (2) how is the value of the solubility parameters of a drug and a polymer of interest theoretically and experimentally determined?

As mentioned Section 12.4.2, Forster et al. proposed an empirical criterion for miscibility prediction: a drug and a polymer can form a solution if their difference in solubility parameter is less than 7.0 $MPa^{1/2}$. If the difference is larger than 10 $MPa^{1/2}$, then the two are immiscible [9, 10]. This criterion has been widely applied for a rough estimation of drug–polymer miscibility. It should be pointed out that the polymer engineering and industry has been using 1.3–2.1 MPa$^{1/2}$ as the critical difference in solubility parameter to estimate the miscibility of two polymers [32]. Considering that the entropy of mixing is much smaller for a polymer–polymer system than that of a drug–polymer system, it is understandable that the critical solubility parameter difference is larger for a drug–polymer system.

The biggest challenge in using the solubility parameter criterion to determine the drug–polymer miscibility comes from the lack of accurate solubility parameter data. Hansen suggested four decades ago that the total change in internal energy on vaporization may be considered the sum of three individual contributions: one from dispersion forces ΔE_{d}, another from permanent dipolar interactions ΔE_{p} and the third from hydrogen bonds ΔE_{h}. As a result, the solubility parameter can be written

$$\delta^2 = \delta_{\mathrm{d}}^2 + \delta_{\mathrm{p}}^2 + \delta_{\mathrm{h}}^2 \tag{12.7}$$

where $\delta_i = (\Delta E_i/v)^{1/2}$ for i = d, p, h. Each solubility component can be further considered the sum of contributions from different groups, which can be found in the literature. This method has good practical utility, although it has been well recognized that the value thus obtained may not be very accurate.

The second strategy applied for miscibility predication is based on the calculation of Gibbs free energy of mixing. A polymer is regarded as a viscous solvent and the solubility of a solid drug in the polymer is described by a mixture phase equilibrium expression [33]:

$$\ln x_1\gamma_1 = -\frac{\Delta H_{\mathrm{fus}}}{RT}\left(1-\frac{T}{T_{\mathrm{m}}}\right) - \frac{1}{RT}\int_{T_{\mathrm{m}}}^{T}\Delta C_p dT + \frac{1}{R}\int_{T_{\mathrm{m}}}^{T}\frac{\Delta C_p}{T}dT \tag{12.8}$$

where x_1 is the saturation mole fraction of a solid drug in the polymer; γ_1 is the activity coefficient of the drug in the polymer at the solubility limit; T_{m} is the drug's melting temperature, T is the designated temperature of interest; ΔH_{fus} is drug's heat of fusion at the melting temperature; R is the universal gas constant; and ΔC_p is the heat capacity difference between solid and liquid drug ($\Delta C_p = C_p^L - C_p^S$) where liquid drug refers to the amorphous drug. ΔC_p is a function of temperature and the data can be determined experimentally. If ΔC_p does not change significantly in the temperature range of interest, the above equation can be rewritten as:

$$\ln x_1\gamma_1 = -\frac{\Delta H_{\mathrm{fus}}}{RT}\left(1-\frac{T}{T_{\mathrm{m}}}\right) - \frac{\Delta C_p}{R}\left[1-\frac{T_{\mathrm{m}}}{T}+\ln\left(\frac{T_{\mathrm{m}}}{T}\right)\right] \tag{12.9}$$

To calculate the mole solubility x_1, it is necessary to know the activity coefficient γ_1. Based on the classical Flory–Huggins lattice theory [34], the activity coefficient of APAP can be calculated.

$$\ln\gamma_1 = \ln\frac{\varphi_1}{x_1} + \left(1-\frac{1}{m}\right)\varphi_2 + \chi\varphi_2^2 \tag{12.10}$$

where ϕ_1 is the volume fraction of the drug; ϕ_2 is the volume fraction of the polymer; m is the volume ratio of a polymer molecule to a drug molecule; and χ is the Flory–Huggins interaction parameter. The relationship between ϕ_1, ϕ_2 and x_l is expressed:

$$\phi_1 = \frac{x_1}{x_1 + m(1 - x_1)} \quad \text{and} \quad \phi_2 = \frac{m(1 - x_1)}{x_1 + m(1 - x_1)} \tag{12.11}$$

If the value of χ is known, the drug's solubility at a specific temperature can be calculated using Equations (12.9)–(12.11). The interaction parameter χ can be calculated from the melting point depression method:

$$\frac{1}{T_m^{mix}} - \frac{1}{T_m} = \frac{-R}{\Delta H_{fus}} \left[\ln \varphi_1 + \left(1 - \frac{1}{m}\right) \varphi_2 + \chi \varphi_2^2 \right] \tag{12.12}$$

where T_m is the melting point of the pure drug while T_m^{mix} is the melting point of the drug mixed with a polymer (determined experimentally for different compositions). The interaction parameter χ can be obtained by fitting

$$\left(\frac{1}{T_m^{mix}} - \frac{1}{T_m} \right) \frac{\Delta H_{fus}}{-R} - \ln \varphi_1 - \left(1 - \frac{1}{m}\right) \varphi_2 \tag{12.13}$$

linearly with ϕ_2^2.

12.6 Case Study: Acetaminophen and Poly(ethylene oxide)

Characterization of the solid dispersion of poly(ethylene oxide) (PEO) and acetaminophen (APAP) is prepared by the batch melt mixing of PEO, a semi-crystalline polymer, which is often chosen as the excipient in HME operations because of its broad processing window. The model drug APAP, a pain reliever and fever reducer, has a solubility of 14.90 mg/g in water at 25°C. The difference in solubility parameter between PEO and APAP is 7.46 $MPa^{1/2}$. According to the criterion described in the literature [9, 10], PEO and APAP are likely to be immiscible.

A Brabender FE-2000 counter-rotating batch mixer (C.W. Brabender Instruments Inc., South Hackensack, NJ) was used to mix APAP and PEO. The processing (barrel) temperature was kept at 100°C and the rotor screw speed was controlled at 50 rpm by a computer program. The processing temperature is above the PEO's melting temperature, but lower than the melting temperature of APAP. The torque generally drops steadily after mixing due to gradual dissolution of APAP particles into molten PEO. Mixing was stopped after 6 min, when the torque did not drop any more.

The mixtures were cooled in air and then characterized using different analytical tools. Figure 12.12 shows SEM pictures of mixture samples of different drug loadings. It is evident that APAP recrystallizes and forms micron-sized particles, appearing predominately at the surface of each sample. The presence of crystalline APAP particles was further confirmed by the XRD spectra of the relevant samples (not shown here). The recrystallization of the API confirms that APAP and PEO are immiscible at ambient temperature.

It is interesting that the DSC thermograms of the mixtures show only one melting peak from PEO, as shown in Figure 12.13. Although the SEM pictures show APAP particles on the surface, no melting peak from APAP can be observed in the DSC thermograms. To understand the apparent contradiction, the mixture was heated on a hot stage and examined, in real time, using a polarized optical microscope. Figure 12.14 suggests that crystalline APAP particles, shown as bright yellow spots, start to disappear due to dissolution into

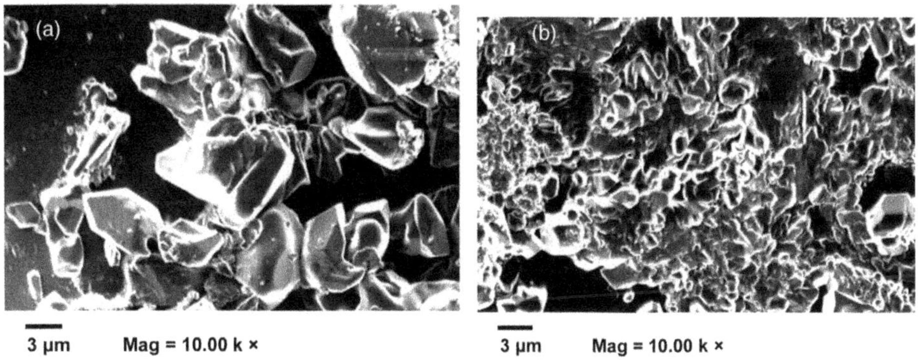

Figure 12.12 SEM images of APAP PEO mixtures after being stored for 18 days with drug loading of (a) 10 wt%; and (b) 20 wt%.

molten PEO even at temperatures below 100°C. The dissolution kinetics are very fast and the whole process is complete within minutes at 100°C. In other words, most crystalline APAP will dissolve into PEO during the thermal scan from room temperature to 171°C, the melting point of APAP. This finding suggests that the DSC characterization *by itself* could lead to wrong conclusions.

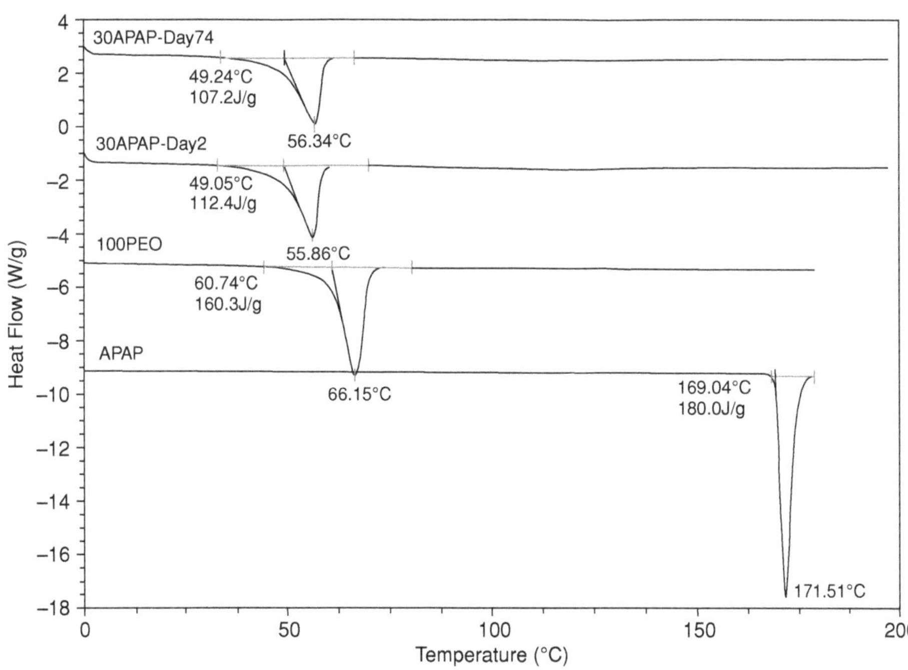

Figure 12.13 DSC thermograms of APAP powder, PEO, a mixture with 30 wt% drug loading after 2 day storage and a mixture with 30 wt% drug loading after 74 day storage (from bottom to top).

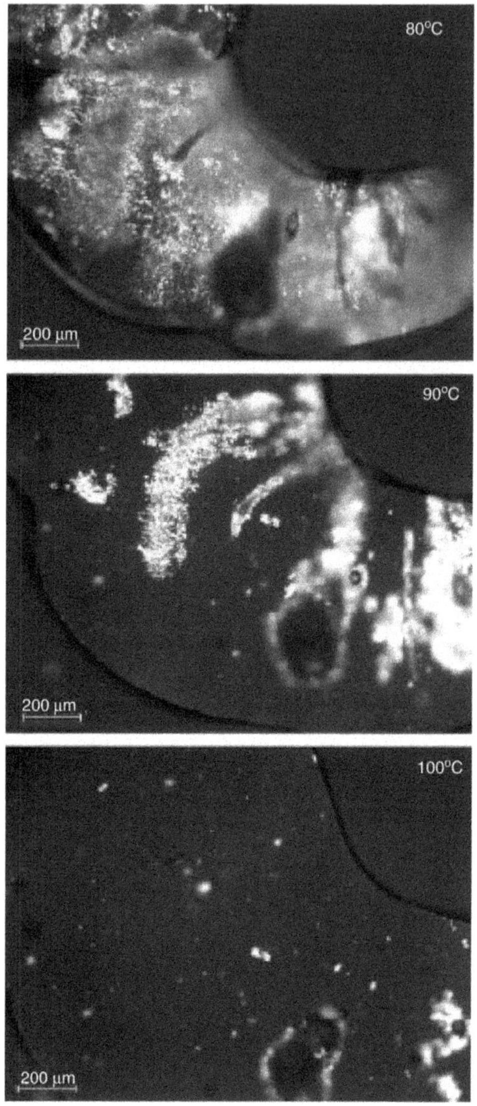

Figure 12.14 Polarized microscopic images of an APAP-PEO mixture on the hot stage at 80, 90 and 100°C. For a better understanding of the figure, please refer to the color section.

12.7 Determination of Solubility of APAP in PEO

It should be stated that the drug's solubility in a polymer is strongly dependent on the temperature, increasing with increasing temperature. As described in the previous part, APAP and PEO can not form one phase at room temperature, as evidenced by the recrystallized APAP particles on sample surfaces. However, at elevated processing temperatures, APAP can have significant solubility in the molten polymer. Determination of a drug's solubility in a polymeric excipient is a generally interesting topic because the solubility data can be a strong guide of process and formulation optimization.

Unfortunately, to the best of our knowledge, only a couple of studies have attempted to experimentally determine the solubility of a drug in a polymer [35, 36]. The method described in the literature involves heating the drug–polymer mixture at an extremely slow rate using a DSC. The temperature T at which the drug totally dissolves in the polymer was determined based on the endpoint of the endothermic dissolution peak. In other words, at temperature T the drug's solubility in the polymer is equal to the drug loading in the sample. The method can be applied to measure a drug's solubility in a polymer at a different temperature by determining the endpoint temperature of the dissolution peak at different drug loadings. Because the method requires heating a mixture at an extremely slow rate from the room temperature, it may very well cause thermal degradation of the drug, the polymer or both. For this reason, methods based on the rheological behavior and the glass transition temperature of the mixture were developed by our group to address this issue [37, 38]: unlike the method based on the dissolution process [35, 36], mixtures are directly equilibrated at the temperature of interest, followed by measuring the mixture's viscosity and the glass transition temperature directly. As a result, the experimental time required is dramatically shortened avoiding thermal degradation.

Figure 12.15 shows the relationship between the *reduced viscosity*, i.e. the viscosity of the APAP-PEO mixture divided by that of a pure PEO, and the drug loading. Each curve corresponds to one temperature. Four curves at different temperatures exhibit the same trend: the reduced viscosity value drops first with increasing drug loading, and then increases after reaching a concentration characteristic of each of the temperatures used. The initial decrease of viscosity indicates an increase of the mixture's polymer structure mobility, due to the drug dissolution. The dissolved drug acts as a plasticizer at small drug loadings, which leads to a decrease in the viscosity with an increase of drug concentration. On the other hand, the rise of the viscosity at higher drug concentrations occurs when

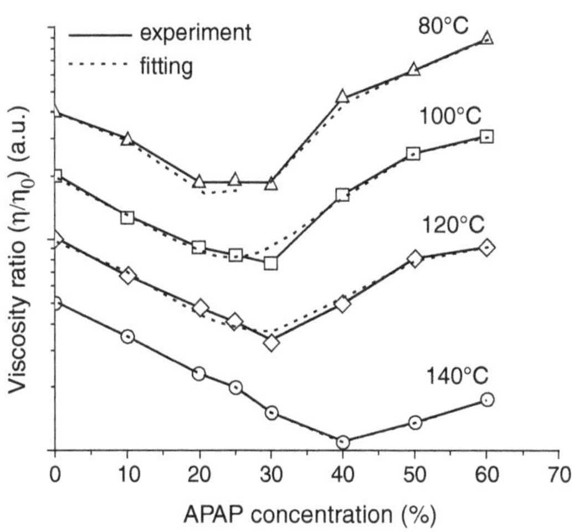

Figure 12.15 *Experimental and polynomial fitted reduced viscosity of the mixture with different drug loading at different temperature: △ 80°C; □ 100°C; ◇ 120°C; and ∘ 140°C.*

the drug solubility is exceeded and undissolved solid drug particles act as suspended filler particulates, increasing the mixture's viscosity. The APAP loading at the critical point, where the reduced viscosity reaches a minimum, gives us the APAP's solubility in PEO at that temperature. These data were further confirmed by results from measurements of the glass transition temperature and a direct observation using a polarized optical microscope.

In summary, APAP and PEO represent one category of drug–polymer mixtures which can form one phase during melt processing but where the drug will recrystallize when the temperature is lowered to room temperature. In the immiscible APAP-PEO case, recrystallization occurs almost instantly when the molten mixture is taken out of the hot batch mixer (as manifested by visible changes of the sample appearance from transparent to milky white). However, the entire crystallization process can last months. During this process, the crystallinity and morphology of the mixture undergoes slow changes which will alter the drug's release profiles, causing serious shelf life quality issues. This case study is presented in order to establish the fact that room temperature immiscibility is the cause for rejecting a given API-excipient system from being considered as a candidate for the HME process.

References

(1) Breitenbach, J. (2002) Melt extrusion: from process to drug delivery technology. *European Journal of Pharmaceutics & Biopharmaceutics*, **54**, 107–117.

(2) Crowley, M.M., Zhang, F., Repka, M.A., Thumma, S., Upadhye, S.B., Battu, S.K., McGinity, J.W. and Martin, C. (2007) Pharmaceutical applications of hot-melt extrusion: Part I. *Drug Development & Industrial Pharmacy*, **33**, 909–926.

(3) Ghebre-Sellassie, I. and Martin, C. (2007) *Pharmaceutical Extrusion Technology*. Informa Healthcare, New York.

(4) Repka, M.A., Battu, S.K., Upadhye, S.B., Thumma, S., Crowley, M.M., Zhang, F., Martin, C. and McGinity, J.W. (2007) Pharmaceutical applications of hot-melt extrusion: Part II. *Drug Development & Industrial Pharmacy*, **33**, 1043–1057.

(5) Martin, C. (2008) Continuous mixing of solid dosage forms via hot melt extrusion. *Pharmaceutical Technology*, **32**, 76–86.

(6) Liu, H., Wang, P., Xanthos, M., Zhu, L. and Gogos, C.G. (2009) Reveiew paper: processing aspects of hot melt extrusion in pharmaceutical applications. *Proceedings of 2009 SPE-ANTEC*, **4**, 2370–2375.

(7) Tadmor, Z. and Gogos, C.G. (2006) *Principles of Polymer Processing*. 2nd edition (a: p. 13; b: Chapter 4; c: Chapter 5 and Section 9.3; d: Sections 5.9 and 10.3; e: Chapter 8; f: Chapters 6, 9 and 10; g: chapter 7; h: p.667). Wiley-Interscience, New Jersey.

(8) Gogos, C.G., Tadmor, Z. and Kim, M.H. (1998) Melting phenomena and mechanisms in polymer processing equipment. *Advances in Polymer Technology*, **17**, 285–305.

(9) Greenhalgh, D.J., Williams, A.C., Timmins, P. and York, P. (1999) Solubility parameters as predictors of miscibility in solid dispersions. *Journal of Pharmaceutical Sciences*, **88**, 1182–1190.

(10) Forster, A., Hempenstall, J., Tucker, I. and Rades, T. (2001) Selection of excipients for melt extrusion with two poorly water-soluble drugs by solubility parameter calculation and thermal analysis. *International Journal of Pharmaceutics*, **226**, 147–161.

(11) Burch, H.E. and Scott, C.E. (2001) Evolution of structure in the softening/melting regime of miscible polymer mixing. *Polymer Engineering & Science*, **41**, 1038–1048.
(12) Grace, H.P. (1982) Dispersion phenomena in high viscosity immiscible fluid systems and application of static mixers as dispersion devices in such systems. *Chemical Engineering Communication*, **14**, 225–277.
(13) Noyes, A.A. and Whitney, W.R. (1897) The rate of solution of solid substances in their own solutions. *Journal of American Chemical Society*, **19**, 930–934.
(14) Wurster, D.E. and Taylor, P.W. (1965) Dissolution rates. *Journal of Pharmaceutical Sciences*, **54**, 169–175.
(15) Todd, D.B. (1998) *Plastics Compounding Equipment and Processing*. Hanser Publishing, New York.
(16) Gao, J., Walsh, G.C., Bigio, D., Briber, R.M. and Wetzel, M.D. (2000) Mean residence time analysis for twin screw extruders. *Polymer Engineering & Science*, **40**, 227–237.
(17) Gasner, G.E., Bigio, D. and Marks, C. (1999) A new approach to analyzing residence time and mixing in a co-rotating twin screw extruder. *Polymer Engineering & Science*, **39**, 286–298.
(18) Nakamichi, K., Nakano, T., Yasuura, H., Izumi, S. and Kawashima, Y. (2002) The role of the kneading paddle and the effect of screw revolution speed and water content on the preparation of solid dispersions using a twin-screw extruder. *International Journal of Pharmaceutics*, **241**, 203–211.
(19) Lyons, J.G., Blackie, P. and Higginbotham, C.L. (2008) The significance of variation in extrusion speeds and temperatures on a PEO/PCL blend based matrix for oral drug delivery. *International Journal of Pharmaceutics*, **351**, 201–208.
(20) Fujii, M., Okada, H., Shibata, Y., Teramachi, H., Kondoh, M. and Watanabe, Y. (2005) Preparation, characterization, and tableting of a solid dispersion of indomethacin with crospovidone. *International Journal of Pharmaceutics*, **293**, 145–153.
(21) Shibata, Y., Fujii, M., Sugamura, Y., Yoshikawa, R., Fujimoto, S., Nakanishi, S., Motosugi, Y., Koizumi, N., Yamada, M., Ouchi, K. and Watanabe, Y. (2009) The preparation of a solid dispersion powder of indomethacin with crospovidone using a twin-extruder or kneader. *International Journal of Pharmaceutics*, **265**, 53–60.
(22) Liu, H., Wang, P., Zhang, X., Shen, F. and Gogos, C.G. (2010) Effects of extrusion process parameters on the dissolution behavior of indomethacin in Eudragit® E PO solid dispersions. *International Journal of Pharmaceutics*, **383**, 161–169.
(23) Miller, D.A., McConville, J.T., Yang, W., Williams III, R.O. and McGinity, J.W. (2007) Hot-melt extrusion for enhanced delivery of drug particles. *Journal of Pharmaceutical Sciences*, **96**, 361–376.
(24) Teng, S., Wang, P., Zhu, L., Young, M.W. and Gogos, C.G. (2009) Experimental and numerical analysis of a lab-scale fluid energy mill. *Powder Technology*, **195**, 31–39.
(25) Wang, P., Zhu, L., Teng, S., Zhang, Q., Young, M.W. and Gogos, C.G. (2009) A novel process for simultaneous milling and coating of particulates. *Powder Technology*, **193**, 65–68.
(26) Verhoeven, E., De Beer, T.R.M., Van den Mooter, G., Remon, J.P. and Vervaet, C. (2008) Influence of formulation and process parameters on the release characteristics of ethylcellulose sustained-release mini-matrices produced by hot-melt extrusion. *European Journal of Pharmaceutics & Biopharmaceutics*, **69**, 158–166.

(27) Liu, H., Zhu, L., Wang, P., Zhang, X. and Gogos, C.G. (2011) Effects of screw configurations on the dissolution behavior of indomethacin in Eudragit® E PO solid dispersions. *Advances in Polymer Technology*, doi: 10.1002/adv.20256.
(28) Taylor, L.S. and Zografi, G. (1997) Spectroscopic characterization of interactions between PVP and indomethacin in amorphous molecular dispersions. *Pharmaceutical Research*, **14**, 1691–1698.
(29) Todd, D.B. (1993) Melting of plastics in kneading blocks. *International Polymer Processing*, **8**, 113–118.
(30) Brouwer, T., Todd, D.B. and Janssen, L.P.B.M. (2002) Flow characteristics of screws and special mixing enhancers in a co-rotating twin-screw extruder. *International Polymer Processing*, **17**, 26–32.
(31) Scott, C.E. and Macosko, C.W. (1995) Morphology development during the initial stages of polymer-polymer blending. *Polymer*, **36**, 461–470.
(32) Mason, J.A. and Sperling, L.H. (1976) *Polymer Blends and Composites*. New York Plenum Press, New York.
(33) Sandler, S.I. (2006) *Chemical, Biochemical, and Engineering Thermodynamics*. Hoboken, New Jersey, 658–702.
(34) Flory, P.J. (1953) *Principles of Polymer Chemistry*. Ithaca, New York.
(35) Tao, J., Sun, Y., Zhang, G. and Yu, L. (2009) Solubility of small-molecule crystals in polymers: d-mannitol in PVP, indomethacin in PVP/VA, and nifedipine in PVP/VA. *Pharmaceutical Research*, **26**, 855–864.
(36) Sun, Y., Tao, J., Zhang, G.G.Z. and Yu, L. (2010) Solubilities of crystalline drugs in polymers: An improved analytical method and comparison of solubilities of indomethacin and nifedipine in PVP, PVP/VA, and PVAc. *Journal of Pharmaceutical Sciences*, **99**, 4023–4031.
(37) Yang, M., Wang, P., Huang, C.Y., Ku, M.S., Liu, H. and Gogos, C.G. (2010) Solid dispersion of acetaminophen and poly(ethylene oxide) prepared by hot-melt mixing. *International Journal of Pharmaceutics*, **395**, 53–61.
(38) Yang, M., Wang, P., Suwardie, H. and Gogos, C.G. (2011) Determination of acetaminophen's solubility in poly(ethylene oxide) by rhelogical, thermal and microscopic methds. *International Journal of Pharmaceutics*, **403**, 83–89.

13

Technological Considerations Related to Scale-up of Hot-melt Extrusion Processes

Adam Dreiblatt
Century Extrusion

13.1 Introduction

The most common type of extruder being used for hot-melt extrusion (HME) is a fully-intermeshing co-rotating twin-screw extruder (TSE). The design criteria and operating principles for this equipment have been documented in the literature [1, 2] and are discussed in Chapter 2 of this text. The positive conveying mechanism of the TSE coupled with the self-wiping characteristic provides predictable scale-up from one machine size to another, as long as both machines are of this same geometry (i.e. both machines are co-rotating and fully-intermeshing). This is a significant advantage of the HME process compared to traditional pharmaceutical batch mixing applications such as wet granulation. Scale-up of these TSEs has been successfully demonstrated in the food and plastics industry since 1953 when the first patent was issued, resulting in thousands of large-scale production machines in operation today.

Transferring the HME process from one TSE to another (different in size, for example) requires characterization of the HME process to verify that the transferred process has not been changed. The characterization we describe here refers to measurable process parameters that identify a unique set of operating conditions which results in a unique product. It is critical to understand which variables must be held constant in order to reproduce the same process (and therefore the same product) on another TSE. As a

Hot-melt Extrusion: Pharmaceutical Applications, First Edition. Edited by Dennis Douroumis.

scale-up analogy, if we transfer the process of driving an automobile from one vehicle to another, the important parameter to be maintained is the actual ground speed of the vehicle (e.g. as measured on the speedometer in km/hr). Because the two vehicles are different, perhaps with different engine displacement, different transmission gear ratio, different size wheels, etc. the engine in the transferred vehicle may need to operate at a higher (or lower) speed in order to maintain the same ground speed. Both vehicles have a speedometer and a tachometer, but the important criteria in this case is the speedometer (the tachometer tells you that something is different, but is not critical to the outcome). In this example, the driver must understand (a) why the tachometer reading is different from the first vehicle and (b) that the tachometer reading is not critical to the final outcome.

Some of the data available from the HME process is obtained directly from the instrumentation (e.g. screw speed, barrel/die temperatures, motor load, melt pressure, etc.) while some of the information must be collected manually such as measurement of residence time and residence time distribution. When transferring the HME process from one extruder to another, it may not be critical to maintain the same screw speed on both machines for example, as long as the effect of screw speed (e.g. specific mechanical energy input and/or mixing quality) is maintained as a constant value.

As for the two vehicles described above, TSEs are not created equal; they differ slightly in geometry (e.g. diameter ratio, defined later in this section), torque density (available power compared to free volume) and the specific types of screw elements used for conveying, melting and mixing from one extruder manufacturer to another. Duplicating the HME process from a lab-scale extruder to a production-scale extruder therefore requires a detailed understanding of the HME process as well as TSE geometry. Figure 13.1 illustrates the interaction between extruder configuration, operating parameters and product properties for a given formulation:

- The interaction of screw speed, feed rate and barrel/die temperature settings on a specific extruder and screw configuration manifests as energy input to the raw materials through applied shear stress (i.e. frictional dissipation) and thermal heat transfer. This energy input per unit mass is referred to as specific energy and is a response variable that can be quantified from both mechanical (drive motor) and thermal (barrel heating) energy sources.

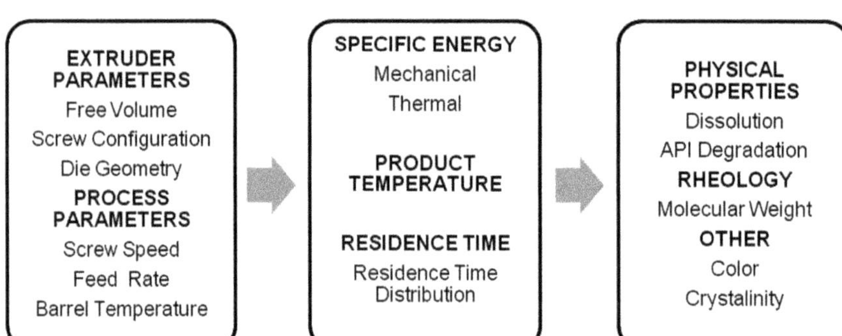

Figure 13.1 *Flow diagram representing the interaction of TSE geometry and operating conditions on the HME process.*

- The combined thermal and mechanical energy input occurs within the resulting residence time distribution; the product temperature increases as a consequence. When sufficient energy is applied, polymer(s) and/or active pharmaceutical ingredients (APIs) transform from solid to melt. Continued agitation of the molten material through mixing elements within the residence time further increases the temperature of the melt.
- The applied energy input and thermal history produce molecular changes to the polymer(s), excipient(s) and/or API(s). Too much mechanical energy input, for example, can result in loss of polymer molecular weight; too much temperature and/or residence time can lead to API degradation.

The response variables (specific energy, residence time distribution and product temperature) are therefore critical variables that characterize the HME process, since these variables are responsible for producing the desired morphology from the raw materials. These same response variables are therefore to be maintained constant when scaling-up the HME process from one extruder to another. Some of the difficulties in scale-up arise, for example, because it is not possible to *measure* the actual product temperature throughout the HME process (e.g. the temperature of the product exiting the extruder is usually measured and monitored, but not the actual product temperature throughout the extruder). Extruders are also not instrumented to measure residence time distribution (these measurements can be made, for example, in offline experiments to validate the HME process).

13.2 Scale-up Terminology

The concept of scale-up means different things to different people. The first step must therefore be to define what is meant when using this term. There are three possible scenarios where the term 'scale-up' can be used:

- Case 1: Increasing production output on a HME process by increasing the batch size (using the same extruder and or multiple extruders of the same size).
- Case 2: Increasing production output on a HME process by increasing the feed rate (with or without changes to other process parameters) on the same extruder.
- Case 3: Increasing production output on a HME process by increasing the feed rate (with or without changes to other process parameters) on a different extruder with larger dimensions.

All of the above cases result in increased output; the differences arise when the output per unit time is considered. Each case has a different consequence as far as changes to the HME process and resulting product(s) are concerned.

13.2.1 Scale-up: Batch Size

In this first case, the HME process remains unchanged. All process parameters relating to the extrusion process are maintained as a constant while the only variable here is the length of time over which the extruder is operated (as far as the Master Batch Record is concerned, it remains unchanged except for the length of time the extruder is operated). On an hourly basis, the process is no different; the production increase (i.e. increased batch size) results only from increased length of operation.

Extruders are continuous devices and are designed to operate in a continuous mode: continuous feeding of solids (e.g. excipients, API, etc.), continuous discharging of molten materials through a die/orifice with subsequent continuous cooling/solidification of the molten extrudate into particulates (e.g. pelletizing, milling, etc.). The time interval for starting/stopping of a batch is more a matter of convenience for regulatory purposes and is typically dictated by the batch size of upstream blending or downstream post-processing equipment. It is not within the scope of this chapter to discuss the regulatory aspects of HME; however, the only aspect of the HME process that changes in this first case is related to declaration of batch size for regulatory filing. As for the technical aspect of the extruder or extrusion process, there are no changes to the process or resulting product in this first case. The product performance and attributes (e.g. dissolution profile) remain unchanged as long as the extrusion process remains within the prescribed operating parameters.

A similar scenario exists when the production output is changed by increasing the number of extruders used for the HME process (e.g. installation of multiple extrusion lines). There is no technical (or regulatory) risk associated with increased output in this case, as long as all aspects of each extrusion line are identical and as long as each line is operated within the same prescribed operating parameters. The batch size per line (operating interval) can remain the same, or increase as described above.

When the words 'scale-up' are used to refer to this first case, the product and process remain unchanged in all aspects. There is no difference in the product emerging from the HME process with respect to time from the beginning to the end of the batch or from one extruder installation to the next.

13.2.2 Scale-up: Feed Rate

This use of the term 'scale-up' is also applicable when a product has been developed at one feed rate condition and the same product is now desired at a higher output on the same extruder. This can occur for an established product (i.e. an existing validated process) or during the development phase. The HME product development phase is usually limited to screening formulations at a reduced feed rate to conserve API; however, the impact of feed rate (as well as all other process variables) on the HME process must be clearly understood when the decision is made to proceed at a reduced feed rate for screening purposes.

- The resulting average residence time (i.e. thermal history) will be longer at a reduced feed rate, possibly contributing to increased API degradation.
- The resulting average shear rate could also be higher at a reduced feed rate, possibly contributing to increased degradation of polymer molecular weight.

In this case, the extrusion process *is* changed when the process is shifted to a higher feed rate on the same TSE; however, this change may or may not affect the product properties. It is not possible to run the same TSE at two different feed rate conditions and maintain all process parameters at the same values, since the free volume of the extruder is fixed. The magnitude of the effect of increased feed rate on the HME process (and resulting product) depends on many different factors and needs to be determined through experimentation. For thermally sensitive API's extruded at the relatively high temperatures necessary for

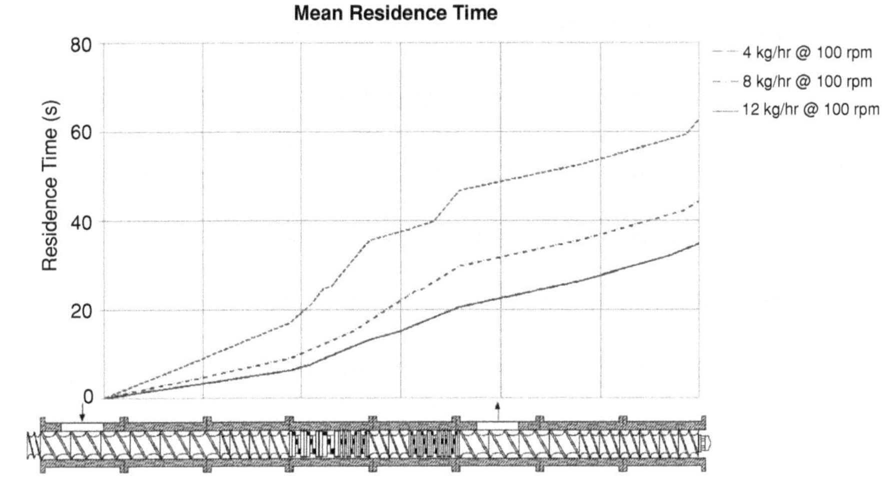

Figure 13.2 *Decrease in average residence time as a result of increasing feed rate at constant screw speed.*

polymer processability, any changes in average residence time could have a significant impact on the resulting product performance.

The variable with the largest impact on average residence time is feed rate; operating the same extruder at higher output will result in a lower average residence time as shown in Figures 13.2 and 13.3. The only possibility of maintaining an average residence time constant at two different feed rates would be to extend the length (i.e. free volume) of the TSE together with feed rate, although this is not a practical solution.

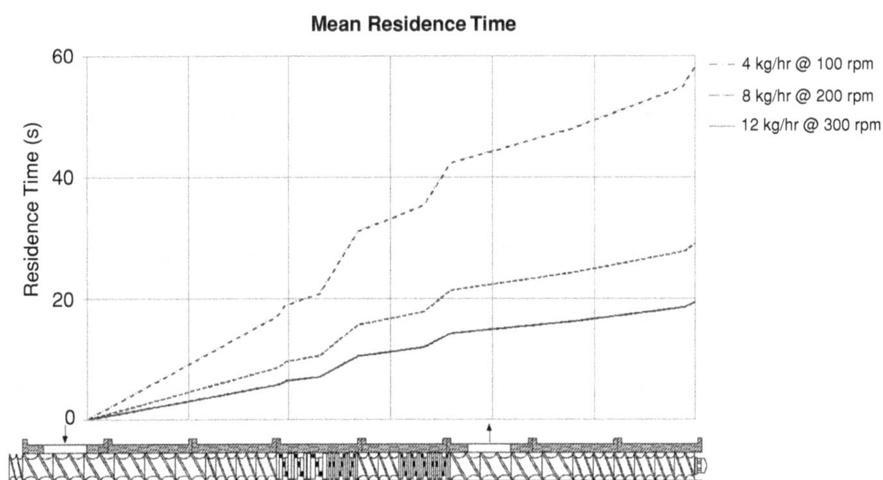

Figure 13.3 *Decrease in average residence time as a result of increasing feed rate with constant degree-of-fill (increase in screw speed in same proportion as feed rate).*

Scale-up by increasing feed rate on the same extruder therefore implies a difference in the HME process; the accompanying changes in the process may or may not have an effect on the resulting product attributes. It is possible to reproduce specific energy and product temperature at two different feed rates on the same extruder; characterization of a HME product produced at high and low feed rate is the only method available to determine the impact of residence time.

13.2.3 Scale-up: Extruder Diameter

In this case, the extrusion process *may* change during scale-up depending on the geometric differences between the two extruders and the specific operating conditions selected for the larger-scale extruder. Reproducing the HME process from one extruder to another requires understanding of TSE geometry as well as identifying the process limitation on the lab-scale extruder. Process limitations that can occur in HME are based on volume, power or heat transfer and are defined as the rate-limiting boundary above which an acceptable product can no longer be achieved. The method used to scale-up to a larger extruder will differ depending on which process limit is encountered on the smaller-scale extruder.

13.3 Volumetric Scale-up

A volumetric scale-up applies where the HME process is limited by the free volume of the TSE. This occurs when the feed section of the machine is full (can be in the main feed opening or in a downstream feeding section) as a result of the density of the raw materials and volumetric conveying capacity of the TSE. Increasing screw speed produces increased output up to the point where screw speed is at the machine limit.

A residence time limitation is also considered a volumetric limit since the residence time is a direct function of free volume. In this case, extending the machine length to increase free volume will also produce increased output (whereas increasing the machine length for a feed limitation does not result in any increase in output). Residence time limits occur when there is insufficient residence time, for example, for a reaction to occur.

Scale-up according to the volumetric ratio between a small and a large TSE will reproduce the average residence time on the larger extruder; if the two extruders are 'geometrically similar' (defined in the following sections), the degree-of-fill and average shear rate are also reproduced.

The primary parameter for sizing an extruder is the outer screw diameter (D_o) expressed in millimeters (mm). TSEs are produced in a range of sizes as small as 3 mm and as large as 450 mm. Each supplier of TSE provides their own range of sizes which varies from one supplier to another – there is no industry 'standard' to which all suppliers comply. As an example, supplier A produces machines with screw diameters of 18, 27, 40 and 50 mm while supplier B produces 18, 26, 32 and 45 mm sizes. The most common size of TSEs utilized for HME applications and typical production outputs are listed in Table 13.1.

The most important factor in scaling up to a larger diameter extruder is to match the desired output (in kg/hr) to the size of the extruder, since TSEs are made in discrete sizes. The dependent variable in this case will be the length of time the extruder must operate to complete a given batch size.

Table 13.1 *Extruder size versus output and application range.*

Screw diameter, D_o (mm)	Application range	Typical HME output (kg/hr)
< 10	Formulation screening	0.01–0.1
10–20	Process/product development	0.1–5
20–30	Small-scale production	5–20
30–50	Medium-scale production	20–100
50–70	Large-scale production	100–500

As an example, we take a product that has been developed at a feed rate of 4.2 kg/hr on an 18 mm extruder and scale-up this process volumetrically to a production-scale extruder which will operate at 25 kg/hr. We will assume the total batch size is 250 kg and the manufacturing plan is to complete the batch within one 12-hour shift. The 25 kg/hr production rate requires a machine with a free volume that is 5.95 times greater than the free volume of the 18 mm machine (25/4.2) to maintain the same average residence time. An extruder which matches this volume difference exactly would have an outer diameter of 32.6 mm; such a machine does not exist commercially, so we must select from the available machine sizes and see what happens to the process as a result.

- A TSE with 27 mm screw diameter is 3.375 times larger in free volume; operating this machine at 25 kg/hr feed rate will result in a significantly shorter residence time than the 18 mm process.
- A TSE with 32 mm screw diameter is 5.6 times larger in free volume; operating this machine at 25 kg/hr feed rate will result in a slightly shorter residence time than the 18 mm process.
- A TSE with 40 mm screw diameter is 10.95 times larger in free volume; operating this machine at 25 kg/hr feed rate will result in a significantly longer residence time than the 18 mm process.

Alternative possibilities include the following.

- Operate the 27 mm machine at the volumetrically scaled-up feed rate of 14.16 kg/hr (4.2 kg/hr × 3.375) to achieve the same average residence time as the 18 mm process; this machine now must operate for more than 17 hours to complete the specified 250 kg batch. This option does not meet the requirements for the manufacturing plan to complete the batch within one 12-hour shift.
- Operate the 40 mm machine at the volumetrically scaled-up feed rate of 45.99 kg/hr (4.2 kg/hr × 10.95) to achieve the same average residence time as the 18 mm process; this machine now can complete the specified 250 kg batch in less than 6 hours. While this option does meet the manufacturing plan guidelines, the upstream batch-mixing process or downstream processing of the extruded material (e.g. cooling or milling) could be a rate-limiting factor for such an increase in capacity.
- The 32 mm machine should be operated at the volumetrically scaled-up feed rate of 23.52 kg/hr to provide an exact replica of the 18 mm process; the process must operate for nearly 11 hours to complete the batch and will be within the specified manufacturing plan.

The free volume of a TSE is independent of screw design and expressed in cm^3/diameter or absolute volume (e.g. cm^3) when the length of the extruder is specified. Such data are available from the extruder manufacturer. When comparing the volumetric scale-up between two extruders, the difference in free volume can be estimated if the two machines have the same diameter ratio D_o/D_i *and* the same L/D (see the following section for a definition of notation).

The ratio of free volume is defined as $(D_{o1}/D_{o2})^3$, where D_{o1} and D_{o2} are the outer diameter of each extruder (expressed in millimeters). An extruder with twice the diameter of another, both having the same diameter ratio and L/D, has 8.0 times the free volume. If these two machines have different diameter ratio, the actual free volume of each machine must be obtained from the respective manufacturer to determine the ratio for volumetric scale-up.

13.3.1 Volumetric Scale-up: Length/Diameter (L/D)

The processing length of a TSE (measured in millimeters) is defined in relationship to the outer screw diameter (D_o) and is expressed as L/D. The required L/D for a TSE is determined by the HME process requirements:

- solids feeding;
- melting;
- downstream solids feeding (optional);
- mixing/homogenization;
- venting (optional); and
- die pressurization.

The barrel assembly of a TSE comprises modular barrel segments configured in a specific sequence to accomplish the desired processing tasks (e.g. as outlined above). The L/D assigned to each process task is a variable and optimized for each HME application.

As such, the overall processing length L/D of the TSE and the strategic position of each processing unit operation is to be maintained constant when scaling up from one extruder to another. As illustrated in Figure 13.4, a 20 mm extruder with $L/D = 40$ has a processing length of 800 mm.

To duplicate this HME process on a 40 mm extruder, the L/D should be maintained constant resulting in an extruder with a physical length of 1600 mm. These two machines (a 25 mm extruder with 800 mm length and a 40 mm extruder with 1600 mm length) would be considered 'geometrically similar' as far as the ratio L/D is concerned. If the 40 mm extruder is operated at the volumetrically scaled-up feed rate, the average residence time in this process will be identical to the 25 mm machine.

13.3.2 Volumetric Scale-up: Diameter Ratio

The diameter ratio of a TSE describes the relationship between the outer and inner screw diameters (D_o/D_i; Figure 13.5) and is a design parameter specific to each TSE manufacturer (i.e. the diameter ratio is constant for each machine size and can vary from one TSE manufacturer to another). Most TSEs today are manufactured with diameter ratios between 1.45 and 1.75. The diameter ratio controls the average shear rate since shallow screw

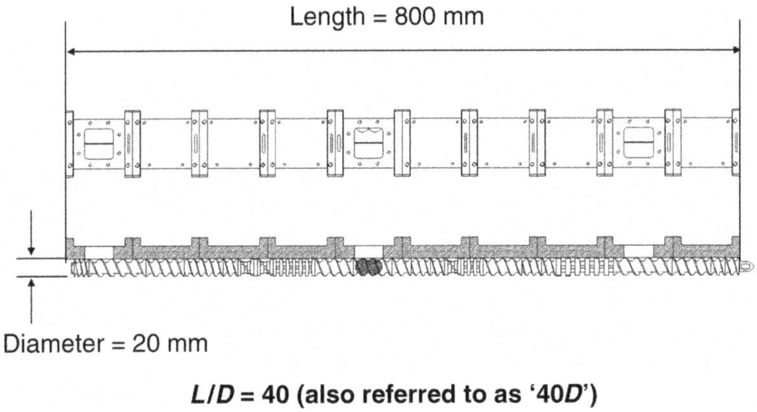

Figure 13.4 *The length of a TSE is determined by the process requirements and is specified as L/D (length/diameter).*

channels (i.e. lower diameter ratio) produce higher average shear rate at the same screw speed as compared to deeper screw channels (i.e. higher diameter ratio). Average shear rate is a critical parameter for scale-up in order to reproduce the same shear stress on the material.

The average shear rate and degree-of-fill (percent fill of each part of the screw along the axis of the extruder) will be identical for two extruders of different diameters if they have the same diameter ratio and are operated at the same screw speed and volumetrically scaled feed rate. Conversely, if these two extruders have different diameter ratios, the average shear rate will be different when operated at the same screw speed and volumetrically scaled feed rate. In the latter case, the operating screw speed for the larger extruder can be calculated

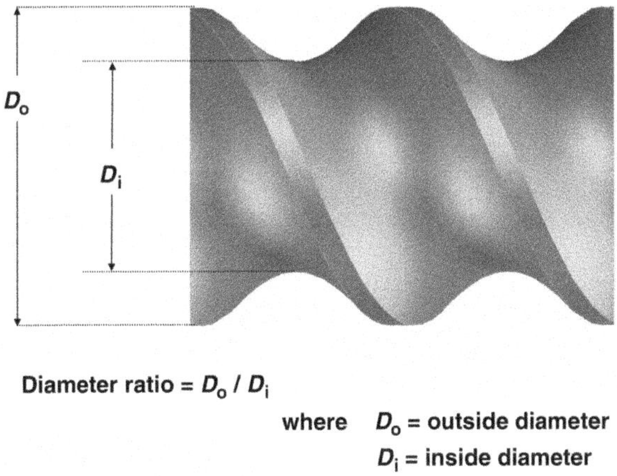

Figure 13.5 *The diameter ratio of a TSE is a critical design parameter for scale-up since it affects free volume, shear rate, degree-of-fill and average residence time.*

to reproduce average shear rate; however, the degree-of-fill between the two extruders will be different as a result. Two extruders are considered 'geometrically similar' if they have the same diameter ratio.

13.3.3 Volumetric Scale-up: Screw Design

The screws of the TSE, consisting of segmented conveying and mixing elements, are configured in a specific sequence in relationship to the processing steps described above (i.e. feeding, melting, mixing/homogenizing/venting/die pressurization) and interact with the operating variables to create the desired morphology. The characteristic function of each type of conveying and mixing element is documented in the literature [3] and discussed in Chapter 2 of this text; the critical aspect of screw design for scale-up involves transformation of the screw configuration from a small-scale extruder to a larger-scale extruder. The screw configuration for the larger-scale extruder is considered 'geometrically similar' if it reproduces the residence time distribution and specific energy.

While the diameter ratio and *L*/*D* of the TSE determine the free volume and average residence time, the screw configuration controls the residence time *distribution*. It is critical to reproduce this same effect on a larger screw diameter using elements with different geometries. The approach to scale-up of a screw configuration uses the extruder outer diameter (D_{O}) as a scaling factor.

TSE conveying elements are characterized by conveying direction (forward/reverse), pitch (in millimeters) and length (in millimeters). The conveying direction is to be maintained the same for both machines in all cases.

- Each conveying section of the screw configuration is transformed by relating the pitch and length to the outer diameter (D_{O}) of the TSE. For example, a 30 mm pitch conveying element which is 30 mm long on a 20 mm extruder can also be described as a 1.5*D* pitch element that is 1.5*D* in length; a 30 mm pitch conveying element which is 15 mm long on a 20 mm extruder can also be described as a 1.5*D* pitch element that is 0.75*D* in length (Figure 13.6).
- For this example, a geometrically equivalent screw configuration on a 40 mm extruder would then use 80 mm pitch conveying elements to reproduce the same degree-of-fill as the smaller-scale extruder if operated at the same screw speed as the 20 mm extruder and at the volumetrically scaled-up feed rate.
- TSE manufacturers do not always provide an exact scale-up of each type of element for every machine size (e.g. in this case, the larger extruder may use 75 mm elements instead of the desired 80 mm elements), making the task of scaling-up the screw configuration a bit more challenging.

Kneading and mixing elements are characterized by conveying direction (e.g. forward, neutral or reverse), offset angle (in degrees), number of kneading discs and length (in millimeters). The conveying direction is to be maintained the same for both machines in all cases.

For each kneading section of the screw configuration (e.g. differentiated by conveying direction), the kneading elements are transformed by relating the element length to the outer diameter (D_{O}) of the TSE where the number of kneading discs is the same on both TSEs. For example, a five-disc forward-conveying kneading element which is 30 mm long on a

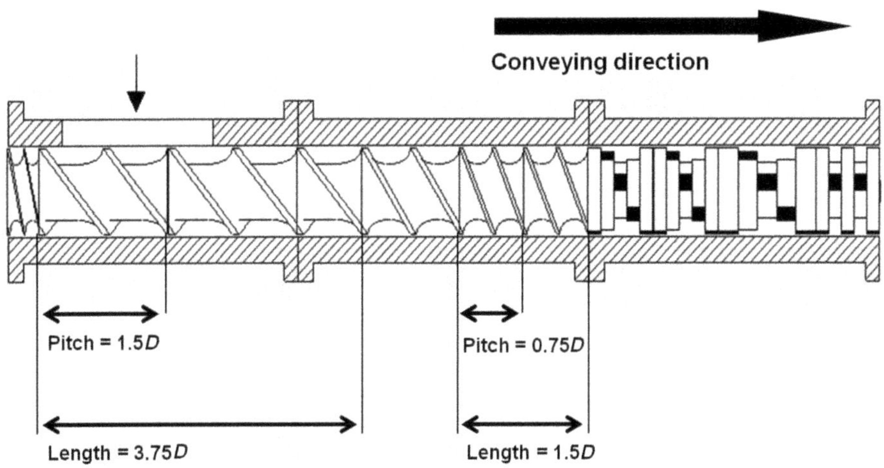

Figure 13.6 *Scale-up of conveying elements uses the ratio of pitch/diameter and an equivalent L/D for each pitch type.*

40 mm extruder can also be described as a 0.75*D* forward-kneading element; a five-disc neutral-kneading element which is 30 mm long on a 40 mm extruder can also be described as a 0.75*D* neutral kneading element (Figure 13.7). Where the number of kneading discs is dissimilar for the two TSEs, the individual disc width is related to the outer diameter in this same fashion to obtain a similar disc width on the larger-scale TSE.

The offset angle for each kneading element is kept constant if both TSEs share the same angles; where the angles are different between the two machines (e.g. 45 degree angles used on a lab-scale TSE and only 30 or 60 degree kneading elements are available for

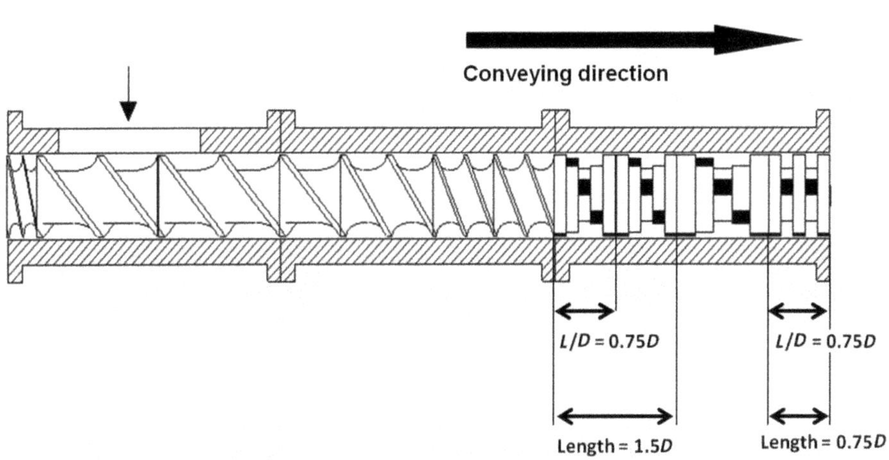

Figure 13.7 *Scale-up of kneading elements uses same ratio of length/diameter and an equivalent L/D for each kneading element type.*

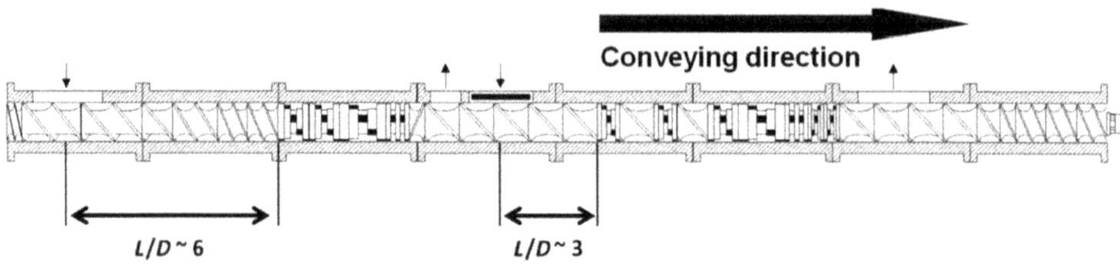

Figure 13.8 *Melting/mixing elements should be positioned with respect to feed and vent openings on the extruder (i.e. install these in the same relative positions using the same L/D on the larger scale extruder as installed on the smaller-scale extruder). Use the centerline of feed/vent openings for reference.*

the larger-scale TSE), the choice must be made to either decrease the mixing effect (e.g. transform a 45 degree kneading element to 30 degrees) or to increase the mixing effect (e.g. 45 degree kneading element transformed to a 60 degree kneading element).

When the conveying and kneading sections of the screw configuration are transformed as described, each part of the screw configuration should remain within the same relative position for both TSEs (refer to Figure 13.8). Slight adjustments may be necessary on the larger-scale TSE to reposition specific parts of the configuration with respect to the feed and vent openings, since extruder manufacturers do not always provide an *exact* replica of each type of kneading element for all machine sizes.

13.4 Power Scale-up

A second process limitation exists where the HME process consumes all of the available power (i.e. torque) for a given machine size. The rate-limiting factor then becomes the specific mechanical energy (SME, units of kWh/kg) input to the material which will be maintained constant when scaling from one size extruder to another:

$$\text{SME} = \frac{\text{Power (kWh/h)}}{\text{Throughput (kg/h)}}$$

SME represents the cumulative amount of mechanical energy from the extruder main drive motor that is required to melt, mix, convey and pressurize the extrusion die on a unit mass basis. The power is derived from TSE instrumentation or can be calculated from motor load, operating screw speed and extruder maximum available power (maximum power data is typically specified on the gearbox nameplate or in the equipment manual):

$$\text{Power (kW)} = \frac{\text{Operating screw speed (rpm)}}{\text{Maximum screw speed (rpm)}} \times \text{Maximum available power} \times \%\,\text{Motor load}$$

The power value for computing specific energy can also be calculated when the extruder maximum available power data is provided in Newton-meters (Nm):

$$\text{Power (kW)} = \frac{\left[\text{Extruder torque rating (Nm/shaft)} \times 2\right] \times \text{Screw speed (rpm)} \times \%\,\text{Torque}}{9550}$$

The parameter used to compare the available power for different TSEs is torque density, defined as torque/(centerline distance)^3 in units of Nm/cm^3. The torque density relates the available power relative to the free volume for a given machine size and can be used to compare TSEs of different sizes and/or manufacturers. The maximum production output can be estimated from the available power and the SME requirement:

$$\text{Capacity (kg/h)} = \text{Available Power (kW)/SME (kWh/kg)}$$

The critical aspect for power-limited scale-up is to evaluate the power in the above relationship at the intended operating speed, since available power increases with screw speed. As an example of a power-limited scale-up, the specific energy is computed for a lab-scale 18 mm TSE which operates at a feed rate of 4.2 kg/hr, 300 rpm and 90% torque. In order to calculate specific mechanical energy, the maximum screw speed and maximum available power for the 18 mm machine must be obtained from the gearbox nameplate. 18 mm TSE data implies a maximum power of 2.2 kW at 500 rpm (22 Nm/shaft). For a centerline distance of 15 mm and torque density of 6.5 Nm/cm, SME is calculated as $[(300/500) \times 2.2 \times 0.90]/4.2 = 0.283$ kWh/kg.

To determine the maximum output for a 40 mm TSE, the available power at the intended operating speed is obtained from the gearbox nameplate. 40 mm TSE data implies a maximum power of 53.4 kW at 600 rpm (425 Nm/shaft). For a centerline distance of 33.4 mm, torque density of 11.4 Nm/cm, we can calculate

$$\text{Available power} = 300/600 \times 53.4\ \text{kW} = 26.7\ \text{kW@100\%torque}$$

Since the TSE cannot operate at 100% motor load, the power is evaluated at 90% torque, i.e.

$$\text{Available power at 90\% torque} = (26.7 \times 0.9) = 24.03\ \text{kW}$$

The maximum capacity is then determined using the calculated SME. Power-limited scale-up capacity = available power/SME = 24.03/0.283 = 84.9 kg/hr.

Note that the volumetric difference between these two TSEs is 10.97 (assuming both machines have same diameter ratio); that is, the 40 mm machine has 10.97 times the free volume of the 18 mm machine although it has more than 20 times the available power. The torque density data reveals the difference in available power between these two example machines.

While the 40 mm machine has sufficient power to process 84.9 kg/hr at the same screw speed as the 18 mm machine, operating at this feed rate would result in much higher degree-of-fill, much lower average shear rate and much lower average residence time. Operating the 40 mm machine at the volumetric feed rate 46.074 kg/hr (4.2 kg/hr $\times$ 10.97 volumetric scale-up factor) would reproduce the 18 mm process conditions on the 40 mm machine. The resulting torque on the 40 mm machine would be:

$$\text{power requirement} = 46.074 \times 0.283 = 13.04\ \text{kW}$$
$$\text{available power at 300 rpm} = 26.7\ \text{kW}$$
$$\text{operating torque} = 13.04/26.7 = 49\%.$$

The disparity between the volumetric and power-limited scale-up factors illustrated in this example implies a TSE with higher torque density will operate at a lower motor load (i.e. percent torque) to reproduce the HME process. When scaling-up to a machine with lower torque density for a power-limited process, the larger-scale machine may operate at a feed rate that is less than the volumetric difference. The resulting process will have lower degree-of-fill, higher average shear rate and a longer average residence time compared to the smaller-scale TSE.

Most modern TSEs are manufactured with constant torque density across the range of machine diameters, that is, a TSE with twice the free volume has twice the available power, making scale-up according to volumetric relationship possible for machines with the same torque density. The deviation in torque density described in the above example occurs when scaling from older-generation TSEs with lower torque density to modern high-speed high-torque designs.

13.5 Heat Transfer Scale-up

The third process limitation for scale-up occurs when the maximum output for the HME process is limited by the extruder's ability to (1) remove heat from the material through the barrel cooling system or (2) introduce thermal energy into the material through the barrel heating system. In the first case, the product temperature within the extruder screws increases beyond a specified thermal limit when the feed rate exceeds the cooling capacity of the extruder. In the second case, the product within the extruder screws cannot reach a specified temperature as a result of insufficient thermal energy input.

Thermal heat transfer on TSE is based upon degree-of-fill (to ensure product contact with the inner barrel surface area), temperature gradient (between the product and the barrel), surface area (function of screw diameter and *L*/*D*) and residence time. Due to the circular geometry of extruders, scale-up to a larger screw diameter results in decreased thermal heat transfer because TSE volume (a function of D_o^3) increases faster than the inner barrel surface area (a function of D_o^2).

The scale-up factor used for a heat transfer limited process is based on the inner barrel surface area and can be estimated for machines with constant diameter ratio:

$$\text{Ratio of surface area} = (D_{o1}/D_{o2})^2$$

For TSEs with different diameter ratio, the absolute value of inner barrel surface area must be obtained from each extruder manufacturer to determine the scale-up factor. The above relationship demonstrates the challenges for scale-up based on heat transfer: a machine with twice the diameter has approximately eight times the free volume, but only four times the surface area.

Operating a larger-diameter extruder at a feed rate corresponding to the heat-transfer limit will result in a lower degree-of-fill as compared to the smaller-scale machine. The lower fill further reduces thermal heat transfer since the product is not in contact with the inner barrel surface (compared to the smaller-scale extruder). Lower fill at a constant screw speed also increases the average shear rate. Further, average residence time is much longer on the larger-scale extruder.

Reducing screw speed on the larger-scale extruder to increase the degree-of-fill, as an example, reduces the wiping of the melt film on the inner barrel surface which also negatively impacts thermal heat transfer. One option to improve this situation is to extend the length (*L*/*D*) of the barrel on the larger-scale extruder to provide additional heat transfer surface area.

A heat transfer limit is not commonly encountered in pharmaceutical applications; most HME processes use TSE to convert mechanical energy into product temperature through viscous heating and do not depend significantly on barrel heating or cooling to create the desired product properties.

13.6 Die Scale-up

When a shaping/forming die is part of the HME process, the critical parameter to maintain as a constant through scale-up is the die pressure. Die pressurization on a TSE corresponds to increased energy input (i.e. increased product temperature) and increased melt residence time in the filled screw section prior to the die (referred to as back-up length). Higher melt pressure requires more back-up length which results in increased product temperature and increased average residence time in this thermally sensitive part of the HME process.

In order to maintain the same die pressure on a larger-scale extruder, the die open area must be scaled-up at the same proportion as the feed rate. If the feed rate increases more than the die open area, the die pressure will be higher on the larger-scale extruder. The parameter to maintain constant through scale-up is mass flow per unit open area ([kg/h]/mm^2).

The dependent variable in this case will be the open area required on the larger-scale extruder. For the case of simple round extruded strands, the die hole diameter and number of die holes are manipulated for the larger-scale die to produce constant mass flow per unit open area (scale-up of die geometry is not within the scope of this chapter).

13.7 Conclusion

Scaling-up of batch size has no effect on the HME process or product where the increased batch size is achieved by increasing the operating time for the TSE or by installing multiple TSE lines. The assumption here is that the HME process operates within the prescribed operating conditions and that all extruders are identical when operating multiple lines.

Scale-up by increasing feed rate on the same TSE will reduce the average residence time in all cases; it is not possible to predict if this change to the HME process will influence the resulting product performance (e.g. dissolution behavior). Thus, screening experiments at low feed rate to conserve API will eventually require such experiments to be conducted at higher feed rates for candidate formulations.

Scaling-up the extruder dimensions to achieve higher production capacity can be accomplished using a factor based on free volume, available power or heat transfer.

Using a volumetric scale-up between two TSEs can duplicate the HME process exactly if the two machines are considered geometrically similar in terms of diameter ratio and *L*/*D* and if the larger-scale TSE is operated at the screw speed and feed rate corresponding to the volumetric ratio between the two machines. Any deviation from the volumetric feed rate or differences in geometry between the two machines (e.g. different diameter ratio or *L*/*D*) will result in changes to the HME process. Successful scale-up of screw configuration from a lab-scale extruder to a larger-diameter production-scale extruder will reproduce the specific mechanical energy input and residence time distribution on the larger TSE.

Power-limited scale-up follows the volumetric relationship where the small and large machines have similar torque density. When the torque density between the two machines is different, the HME process on the larger-scale machine will be different as a result.

Scale-up using heat transfer also results in deviations between the lab-scale and production-scale processes since the volume of TSEs increases with (screw diameter)^3 while the barrel surface area increases with (screw diameter)^2.

The die open area must also be scaled-up according to the volumetric ratio to maintain a constant mass flow per unit time and open area. Accurate scale-up of die open area will then reproduce the melt pressure on the larger-scale extruder.

References

(1) White, J.L. (1990) *Twin-Screw Extrusion*. Hanser Publishers, Munich.
(2) Kohlgrüber, K. (2008) *Co-Rotating Twin-Screw Extruders*. Carl Hanser Publishers, Munich.
(3) Dreiblatt, A. and Eise, K. (1991) Intermeshing corotating twin-screw extruders. In Rauwendaal, C. (ed.) *Mixing in Polymer Processing*. Marcel Dekker Inc., New York, pp. 241–266.

14

Devices and Implant Systems by Hot-melt Extrusion

Andrew Loxley, PhD
Particle Sciences Inc, Bethlehem, PA, USA

14.1 Introduction

Since its introduction in the late 19th century, hot-melt extrusion (HME) has become an essential tool in the food and plastics processing industries. It has only received interest in the last few decades in the pharmaceutical industry, however, where its use has been reviewed [1]. Where it has been adopted in pharmaceutical product development, it has almost exclusively been used to improve solubilization kinetics of poorly water-soluble active pharmaceutical ingredients (APIs) by preparing solid solutions of the APIs in water-soluble thermoplastic polymers. Examples of polymers used in HME processes for oral dosage forms are shown in Table 14.1.

The cooled extrudates are typically hard, brittle materials, and are usually ground into fine powder for pressing into tablets or filling into capsules for the final dosage form. HME is therefore used in the preparation of oral dosage forms almost exclusively as a convenient and cost-effective way to melt the polymer and mix the API into the melt in a homogenous way in a continuous well-controlled process. Twin-screw extruders are most widely used to accomplish this as they provide the most efficient mixing due to the intermeshing self-wiping screws and the ability to achieve the right mixture of dispersive and distributive mixing by proper screw design.

Hot-melt Extrusion: Pharmaceutical Applications, First Edition. Edited by Dennis Douroumis.

Table 14.1 Some polymers used for solid oral dosage forms.

Polymer	T_g (°C)
Polyvinylpyrrolidinone	~125
Soluplus™	70
Eudragit™ L 30 D-55	110
Eudragit™ FS 30D	48
Poly(L-lactic acid)	~55
Poly(L-lactic-co-glycolic)acid	~55

In addition to its usefulness in developing solid oral dosage forms, HME is also an especially useful technique for the preparation of medical devices and implants. This chapter will discuss these products and the role of HME in their development and manufacture.

14.2 HME in Device Development

Devices are made from polymers which often contain well-mixed dispersed materials. In some cases the polymers are melted and mixed with the additives and excipients in batch mixers; however, the advantages of hot-melt extrusion are many:

1. the process is continuous and readily scaled to supply increased demand;
2. materials are usually more homogenously dispersed throughout the extrudate, especially with twin-screw extruders;
3. exposure to heat and shear are minimzed, so reducing polymer or API degradation;
4. thermally sensitive APIs or excipients can be added via ports on the extruder barrel near the output die of the extruder where the polymer is already molten, to further minimize thermal stress;
5. the extrudate can be the final product, or can be converted to the final product with minimal additional processing steps (e.g. catheters or subcutaneous rod implants);
6. product consistency can be well controlled;
7. process analytical technology (PAT) (such as near-infrared or NIR on the extrusion line) is readily applied;
8. the equipment footprint can be smaller;
9. power requirements can be less; and
10. cleaning is usually simpler.

Extruders can be used simply as molten polymer pumps. Alternatively, the extruder can provide the required heat to melt the polymer (from external heaters and the friction of conveying) and shear forces to intimately mix materials with molten polymer, at relatively low T with minimal thermal stress, resulting in reduced thermal degradation of materials.

Furthermore, HME provides the ideal or only means to prepare some devices, e.g. catheter tubing or core-sheath rods and rings for complex devices with very controlled drug-release kinetics.

14.3 Hot-melt Extruder Types

There are essentially two types of extruders used in the development and production of devices: single-screw and twin-screw extruders. Twin-screw extruders are typically more efficient mixers, although single-screw extruders can in some cases also do a sufficiently good job of mixing for the application in hand. Single-screw extruders excel when simple pumping of material is the key requirement, as in the case of for example extrusion of catheter tubing. They are also considerably cheaper than twin-screw units. Since the output of single-screw extruders is more uniform than that of twin screws, they also have the advantage of not requiring additional gear pumps (melt pumps) at the outlet die when used for example as part of a co-extrusion line for producing core-sheath reservoir devices.

The choice of extruder for a particular product depends to some extent on the stage of the product's development or commercialization. In the early developmental stage of a new product when the drug(s) or specialty polymer(s) are scarce, small-scale extruders with minimal waste and capability to prepare small sample volumes are often used to prepare materials for formulation evaluation and material compatibility. Small-scale units of note are the Haake Minilab™ and Haake Minilab II Micro Compounder (Thermo Scientific) and Xcelera™ microextruder (DSM Explore), which both use conical twin-screws and can process very small volumes (~5 mL). Since these units are simpler than their larger cousins, and have conical twin screws instead of parallel twin screws, they do not scale directly to pilot or production scale parallel twin-screw extruders. Manufacturers of parallel twin-screw extruders have added very small-scale units to their products lines, with screw diameters as small as 3 mm (Extrusion Technologies), 5 mm (Three-tec), 12 mm (C W Brabender, Steer America) and 16 mm (American Leistritz and Thermo Scientific). They differ in drive torque, barrel length/diameter ratio, screw design, barrel design, rpm, price, materials of construction, etc., and the buyer should weigh up all factors before buying a small unit.

Once materials are selected, are available in larger quantities and a pilot process is to be developed that could yield clinical trial devices, an 18 mm twin screw is an appropriate-sized unit. Several manufacturers produce them in good manufacturing practices (GMP) versions such as Leistritz's pharma micro 18 mm extruder shown in Figure 14.1. For production, extruders start at 27 mm screw diameters and go up to very large units (for example Coperion's 125 mm ZSK unit), depending on projected supply needs.

As in the case of the non-pharma plastics processing industry using HME, the extruded polymer containing additives and APIs can be the final product (as in the case of for example Implanon®), but is often used in a further step to convert it to the finished product. One particular process, that in fact utilizes another hot-melt extrusion step, is injection molding. Injection molders comprise a single-screw extruder for pumping molten polymer (in this case the molten extrudate of the prior hot-melt compounding run) in the same way as the single-screw extruders discussed above. However, the outlet of the screw is mated to a metal mold whose cavities are the shape of the finished product. This is how some intravaginal devices are prepared from pre-compounded thermoplastics and how all silicone intravaginal rings (such as Estring™ and Femring™) are made. A typical injection-molding unit is shown in Figure 14.2.

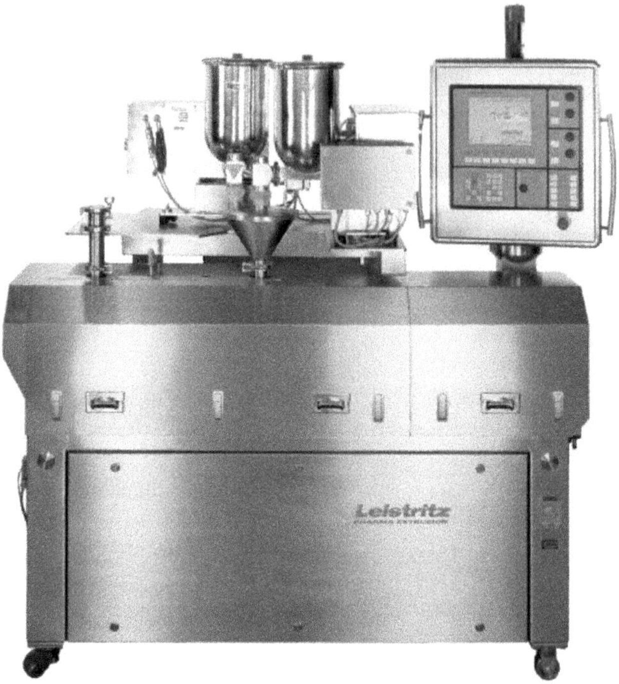

Figure 14.1 *A pharmaceutical-format 18 mm twin-screw extruder. Reproduced with kind permission of the American Leistritz Corp.*

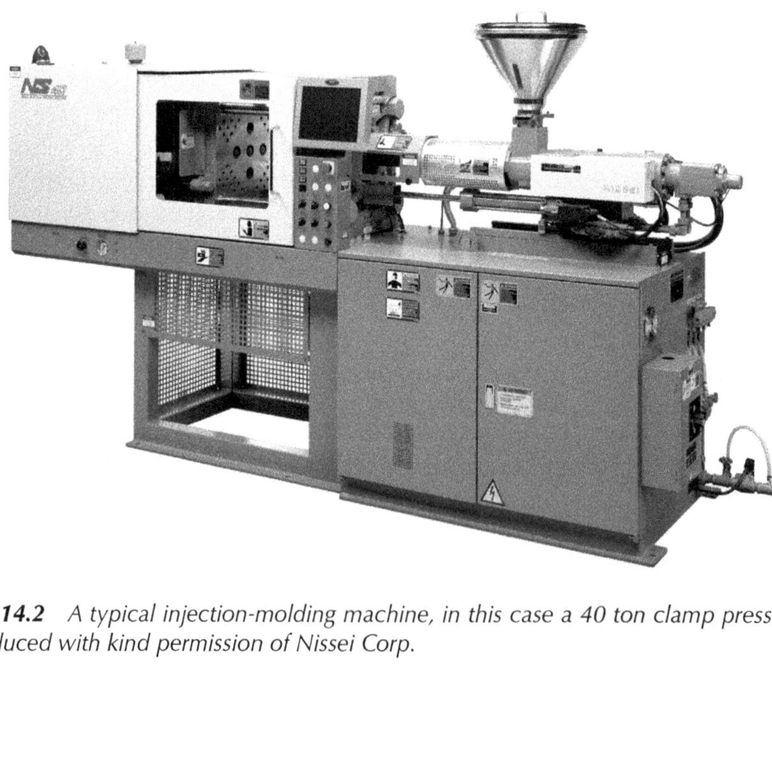

Figure 14.2 *A typical injection-molding machine, in this case a 40 ton clamp pressure unit. Reproduced with kind permission of Nissei Corp.*

14.4 Comparison of HME Devices and Oral Dosage Forms

When developing oral dosage forms, the formulator's intention is typically to maximize the loading of API in the water-soluble polymer to maximize oral bioavailability of the final tablet. Fifty percent drug loading is common, and the drug is ideally in solution or supersaturated solution in the extrudate. If the dose is low, the final tablet can be small; if high, the tablet is large or the patient takes more than one. The tablet size is determined by the drug dose and the polymer is chosen to match the API to obtain the highest concentration of solublilized drug that can remain stable.

With devices, the size of the device is determined by the application not the dose; punctual plugs must be small enough to fit in the ocular puncta and intravaginal rings must be large enough to remain in place once inserted. The concentration of drug in the device is determined by the daily dose and duration of device use. Polymer choice is determined by the required physical properties of the device and on the solubility and diffusion coefficient of the drug as this controls drug release for non-eroding devices. The loading of the polymer is then defined by the daily dose, the duration of the device use and the physical size of the drug-loaded polymer section of the device. In some cases, the drug loading is very low compared to oral products. For example, the Nuvaring® intravaginal contraceptive ring contains 11.7 mg etonogestrel and 2.7 mg of ethinylestradiol in a core 4 mm in cross-sectional diameter, and is used for 21 days. This device weighs around 2 g, so the API loading is just 0.7% w/w. These loadings give the desired daily release of both hormones over the 21-day period. HME is used as a means to mix the API into the molten polymer although, since APIs are soluble both in the molten polymer and the cooled extrudate, twin-screw extruders are not essential for preparing this type of device.

In fact it is often not desired to have all the API in solution in the polymer. For example, Implanon is a contraceptive intra-uterine device (IUD) and the drug-eluting polymer portion has a drug-loaded core with a drug-free release controlling membrane. The core is loaded with approximately 50% API, but the majority of this is in dispersed form. Only the dissolved drug can diffuse from the core to the sheath and out of the device. The API in particulate form maintains the soluble fraction concentration thus maintaining low-rate zero-order drug-release kinetics over the 5-year life of the device. For devices prepared by HME with such high levels of API in the polymer, clearly HME is the mixing method of choice to maximize dispersion homogeneity in the polymer.

Although some biodegradable devices are made from water-erodable polymers, the polymers used for device manufacture are usually not water-soluble and hydrolytically stable. Device polymers are typically hydrophobic non-degradable non-swellable elastomers. Ethylene vinyl acetate (Figure 14.3) is a common choice due to its transparency, flexibility,

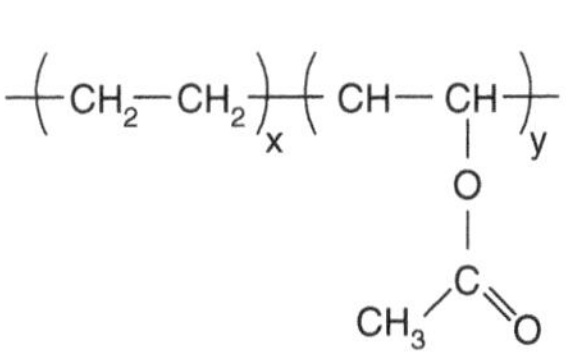

Figure 14.3 *Chemical structure of ethylene vinyl acetate copolymer.*

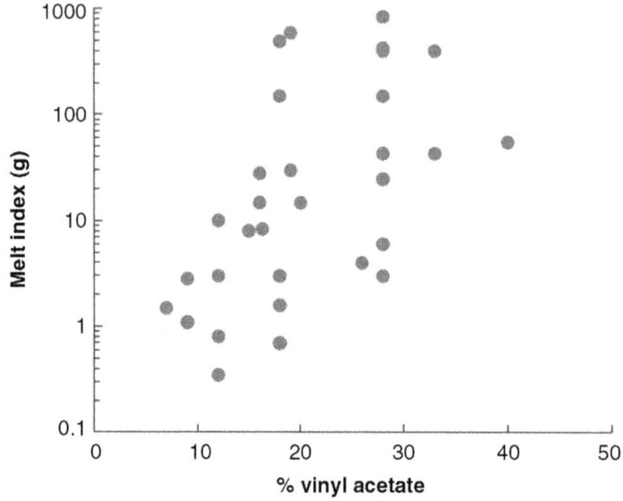

Figure 14.4 Map of melt index and vinyl acetate contents of commercially available EVA.

broad API compatibility, low moisture uptake and long-term stability. It is also available in a broad range of vinyl acetate contents and melt indices (Figure 14.4). Hydrophilic polymers that absorb moisture (such as polyurethanes) need to be dried before processing in the same way as oral product polymers to prevent thermal degradation during HME processing, whereas polymers such as EVA do not require a drying step.

Whereas oral polymers are required to pass toxicity testing, device polymers must pass sets of stringent biocompatibility tests for use in commercial devices. These tests are summarized in ISO10993 (see Figure 14.5).

Finally, the use of HME in device manufacture to melt the polymer and shape the final product is often as important as its use for mixing additives into the polymer.

14.5 HME Processes for Device Fabrication

Since the resins used to make many devices are supplied as pellets rather than powders (as in the case of oral dosage form polymers), blending of the API or powdered additives with the polymer prior to HME is not practical and separate streams of API and polymer may be fed into the extruder by multiple feeders (split feed). When required API loadings in the polymer are low this can present a significant challenge as API feed rates may need to be very low and may reach the limit of standard feeder technology. Figure 14.6 shows the stability of the average feed rate of an antiretroviral powder from two different low-flow-rate feeders, set to target 700 mg/min or 42 g/h. These feeders are rated to about 20 g/h. Figure 14.7 shows the stability of the feed rate of the polymer pellet stream, set to 50 g/min (3 kg/h). This ratio of feed rates corresponds to a drug loading of 1.4% in the extrudate, which had been identified as ideal for the particular drug product. The extruder must be run under 'starved feed' conditions in order to obtain a homogenous extrudate

ISO 10993 BIOCOMPATIBILITY TESTS													
DEVICE CATEGORY			BIOLOGICAL EFFECTS										
Device Type	Body Contact	Contact Duration Limited ☐Less than 24 hours Prolonged ☐ 24 hours to 30 days Permanent ▩ Over 30 days	Cytotoxicity	Sensitization	Irritation or Intracutaneous Reactivity	Systemic Toxicity (acute)	Pyrogenicity	Sub-acute and Sub-chronic Toxicity	Genotoxicity	Implantation	Hemocompatibility	Chronic Toxicity	Carcinogenicity
Surface	Skin	Limited	•	•	•								
		Prolonged	•	•	•								
		Permanent	•	•	•								
	Mucosal Membrane	Limited	•	•	•								
		Prolonged	•	•	•								
		Permanent	•	•	•			•	•				
	Breached or compromised surfaces	Limited	•	•	•								
		Prolonged	•	•	•								
		Permanent	•	•	•			•	•				
External Communicating	Blook Path, Indirect	Limited	•	•	•	•	•				•		
		Prolonged	•	•	•	•	•				•		
		Permanent	•	•		•	•	•	•		•	•	•
	Tissue/Bone/ Dentin	Limited	•	•	•								
		Prolonged	•	•	•	•	•	•	•	•			
		Permanent	•	•	•	•	•	•	•	•		•	•
	Circulating Blood	Limited	•	•	•	•	•				•		
		Prolonged	•	•	•	•	•	•	•	•	•		
		Permanent	•	•	•	•	•	•	•	•	•	•	•
Implant	Tissue/Bone	Limited	•	•	•								
		Prolonged	•	•	•	•	•	•	•	•			
		Permanent	•	•	•	•	•	•	•	•		•	•
	Blood	Limited	•	•	•	•	•	•		•	•		
		Prolonged	•	•	•	•	•	•	•	•	•		
		Permanent	•	•	•	•	•	•	•	•	•	•	•
* Additional tests may be required to satisfy FDA requirements													

Figure 14.5 *ISO10993 biocompatibility tests for device polymers. Reproduced with kind permission of Particle Sciences Inc.*

in this case. Lower API feed rates in such a split-feed process can become unstable over long campaigns, and subject to air currents and static build-up or inhomogenous powder (lumps) in the hopper. In such cases it may be beneficial to feed the API or additive into the polymer at a much higher loading (feed) rate to create a master batch; this is then pelletized and re-fed along with virgin polymer in a second HME step to dilute the master batch to the required device concentration. If the resin is available as a powder (and even elastomer suppliers are beginning to offer their polymers as powders), then the API and

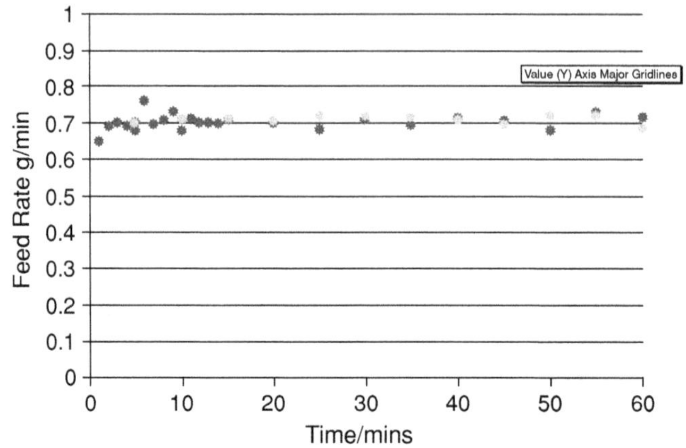

Figure 14.6 *Feed rate stability of two low-rate powder feeders.*

other additives can be preblended and fed as a single feed to the extruder at whatever rate is practical for the size of the extruder.

Downstream processes from the extrudater may include: cooling (air and underwater cooling are used), pelletizing, drawing the hot extrudate into fibers, spooling/pulling (with laser micrometer to maintain tension and therefore extrudate dimensions), cutting, induction welding cut pieces to form rings and injection molding of pellets to form finished parts.

14.5.1 Issues with HME in preparing Drug-eluting Devices

HME is a heat- and shear-intensive process and, although processes should be designed to minimize residence times, barrel temperatures and total shear input, some APIs and polymers may still degrade under the conditions required to melt and pump the materials within

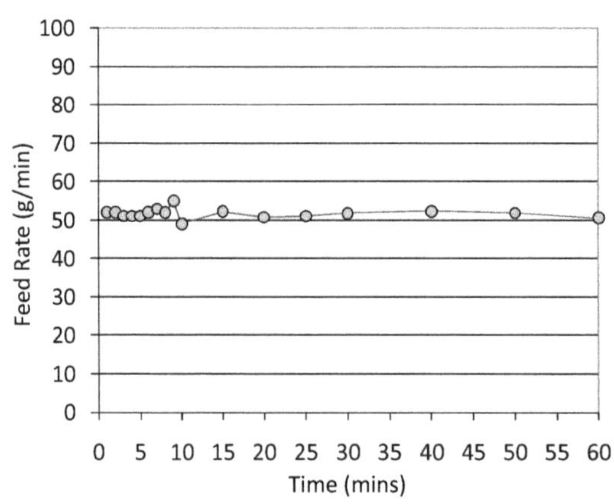

Figure 14.7 *Feed rate stability of an auger-type polymer feeder.*

Figure 14.8 Appearance of IVRs made from various polymers. For a better understanding of the figure, please refer to the color section.

the boundaries of the process design space. Figure 14.8 shows a series of developmental intravaginal rings (IVRs) made from a range of polymers at modest processing temperatures, showing that polymer discoloration can be quite severe. Even when the polymers are stable, the API may not be. Figure 14.9 shows high-performance liquid chromatography (HPLC) chromatograms of solvent extracts of IVRs made with or without antioxidants. The API in the IVR made with the lower level of antioxidant has degraded, as evidenced by the high

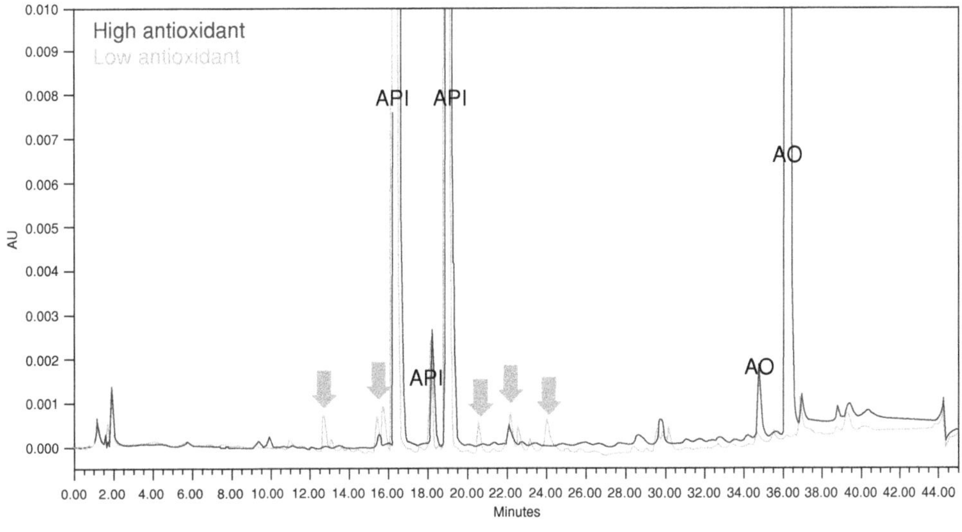

Figure 14.9 HPLC data from extracts of devices made with high and low levels of antioxidant. All arrowed peaks are API-related.

number of related substances observed. The API in the IVR made with a higher level of antioxidant shows much less API degradation (fewer API-related substance peaks).

14.6 Devices and Implants

14.6.1 Anatomical Device Locations

Devices made by HME are used in many locations in the body, as illustrated in Figure 14.10. Some are used topically (transdermal patches or corneal devices such as Occusert^{TM}), some penetrate the skin (venous catheters), some are implanted subcutaneously (e.g. Implanon®), some are implanted internally (e.g. cranial shunts and central venous catheters) and some are 'implanted' mucosally without surgery or tissue damage (intravaginal rings and films, urinary catheters). Figure 14.11 shows an X-ray of a patient with an intravaginal ring in place.

14.6.2 Simple Devices

The simplest medical devices are objects used in the packaging or administration of drug products, but whose contact with human tissue is minimal or non-existent. These devices may be containers such as vials and bags, seals, valves, caps, inserts and closures and administration devices such as syringes and dry powder or metered dose inhalers. The surfaces of the devices may come into contact with the drug product itself, and are therefore able to impact drug product quality and stability. These devices are regulated and, in order to ensure drug product quality, the materials that the devices are made from must pass tests for extractables and leachables. The devices and sub-components must be made in a current GMP (cGMP) environment under proper quality department oversight. Examples of the polymers used for 'simple' medical devices are listed in Table 14.2.

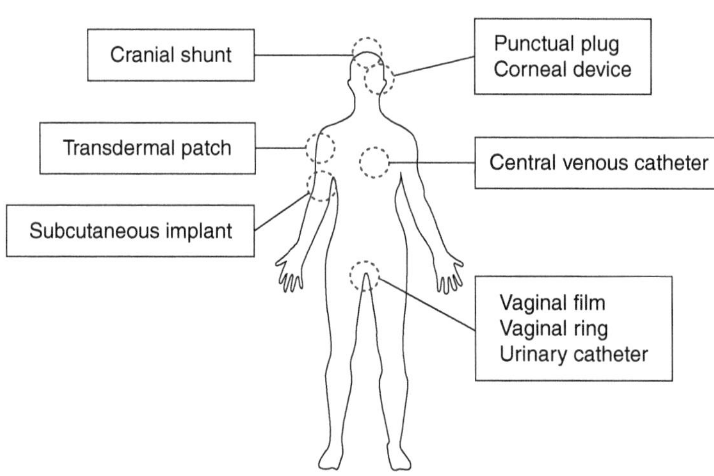

Figure 14.10 Various anatomical locations where devices made using HME are used.

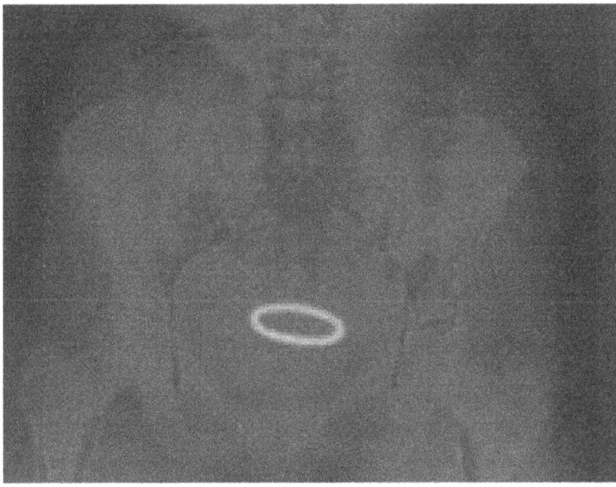

Figure 14.11 *X-ray showing an IVR in position. Reproduced with kind permission of Karl Malcolm, Queens University Belfast. For a better understanding of the figure, please refer to the color section.*

For simple device production, as with solid oral dosage forms, HME is mostly used as a convenient and cost-effective way of mixing additives into plastics. The additives include pigments to impart color to the device, antioxidants and UV protectants to protect the polymer from degradation in processing or device storage and use, nanoparticles and fibers or plasticizers to modify the physical properties of the plastics and fillers to reduce final part cost. Often the additives themselves are products of HME processes. For example, colored pigments are often added as previously mixed 'master batch pellets' (pigment powder in resin) to the device polymer feedstock to simplify weighing and avoid handling powdered pigments in a cGMP clean room. The HME process melts these master batch pellets along with the device resin, and mixes both components sufficiently well to distribute the colored pigment particles uniformly through the bulk. The mixed material is typically extruded cooled and pelletized, and sold as a raw material to the device makers who use them, for example, in injection molding or slot die coating processes to make the final products.

Table 14.2 *Polymers for simple medical devices made using HME.*

Polymer	T_g (°C)	Example devices
LDPE	−78	Bags, frames
PP	~0	Lids, closures, bottles
PS	100	Containers
PMMA	100	Containers
EVA	~−55	IV fluid bags
PTFE	~160–240	Seals, inserts, valve parts, caps
Polycarbonates	Varies	Containers

Table 14.3 Polymers for non-medicated tissue contact devices made using HME.

Polymer	Tg (°C)	Example devices
PLGA and PLA	~55	Dissolving surgical sutures
Polyurethane elastomers:		
aliphatic polycarbonate-based	various, <0	Catheters
aliphatic polyether-based	various, <0	Catheters
aromatic polyether-based	various, <0	Catheters

14.6.3 Non-medicated Prolonged Tissue Contact Devices

A non-medicated implantable device is usually designed to provide some form of physical support or channel. There are many non-medicated implantable devices on the market and in development such as coronary stents, pacemaker pockets, hernia supports and orthopedic implants, and many do not involve a HME step (some are not even made of plastics). Catheters and sutures are examples of products which do utilize HME in production; Table 14.3 lists example polymers used in these cases.

PLGA is a thermoplastic polyester made from copolymerization of lactic and glycolic acids. The chemical structure of PLGA is shown in Figure 14.12. PLGA is used in making surgical sutures as it is strong and holds the wound closed during the healing period. It degrades in vivo due to hydrolysis of the ester linkages, so the stitches dissolve over a few days or weeks and do not need to be removed by a healthcare professional.

Hot-melt extrusion is used in the preparation of sutures for a completely different reason to that of its use in oral dosage forms and most simple devices. Rather than being used for mixing, it is used for the shaping of the molten polymer into the final product form. PLGA is melted in the extruder and the extrudate is pulled under tension, while still hot, to reduce its diameter from the exit die (which might be of the order millimeters) to the final fiber (usually hundreds of microns), then wound on take-up spools. The PLGA fiber is unwound and cut, the needle attached and the device is packaged and sterilized; these various operations are typically conducted in different facilities. This process is similar to that for producing fibers from other thermoplastics such as rayon, polyester and nylon.

Catheters are flexible hollow tubes used for providing a means for fluid flow in the body. They may be inserted into the urinary tract to aid in bladder emptying or may be inserted into major blood vessels in order to introduce drugs into the blood stream and take blood samples. Catheters are made by extruding the molten polymer through a specially designed annular die and, in order to maintain dimensional stability, the catheter is often extruded vertically down as opposed to horizontally as most other extruders. A laser micrometer is used to measure the diameter of the extrudate and this measurement is fed back to the

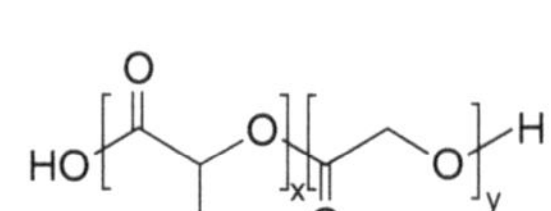

Figure 14.12 Chemical structure of PLGA.

uptake spooler so that the appropriate tension on the extrudate can be maintained to ensure constant diameter. HME is also used in the preparation of catheters simply for melting and pumping the polymer in order to shape it.

Since HME is used in the production of sutures and catheters simply to melt and pump polymers (and not to mix any other materials into them), single-screw extruders are often used rather than twin screws as the additional complexity and cost of twin-screw extruders are unnecessary.

14.6.4 Medicated (Drug-eluting) Prolonged Tissue Contact Devices

A rapidly growing area in the field of drug delivery is drug-eluting implantable devices. A drug-eluting device (DED) is a medical device, usually polymeric, that may be implanted in the body surgically or non-surgically by a healthcare provider (or non-surgically by the patient) and delivers controlled doses of drugs for days, months or even years. For example, Nuvaring™ is a self-administered contraceptive intravaginal ring (IVR) delivering the synthetic progestin etonogestrel and the semisynthetic estrogen derivative ethinylestradiol (120 μg/d and 15 μg/d respectively) for 3 weeks before device removal to allow menses. Implanon® is a surgically implanted subdermal contraceptive device delivering 30–70 μg/d of etonogestrel for up to 5 years. Mirena™ is a contraceptive intra-uterine device (IUD) which is implanted in the uterus during a non-surgical procedure performed by a healthcare professional; it releases 20 μg/d levonorgestrel for up to 5 years.

Typical polymers used for drug-eluting devices are shown in Table 14.4. More extensive lists of commercial (Table 14.5) and developmental (Table 14.6) drug-eluting devices that involve a HME step are also provided below.

Table 14.4 *Polymers for drug-eluting devices made using HME.*

Polymer	T_g (°C)	Example devices
PLGA and PLA	~55	Biodegradable implants
EVA	~−55	Intravaginal rings, transdermal patches, corneal devices, punctual plugs, dental implants
Polystyrene-b-PIB (Kraton™)	-90, 100	Transdermal patches
Poly(ether-b-amide) (Pebax®)	-65	Transdermal patches
Acrylic elastomers	-85 to −20	Transdermal patches
Polyurethane elastomers:		
aliphatic polycarbonate-based	various, <0	Intravaginal rings, drug-eluting catheters
aliphatic polyether-based	various, <0	Intravaginal rings, drug-eluting catheters
aromatic polyether-based	various, <0	Intravaginal rings, drug-eluting catheters
Silicone	~−125	Intravaginal rings (made by reactive injection molding or RIM)

Table 14.5 *Examples of commercialized (on the market or discontinued) drug-eluting devices made using HME.*

Device	Type	API
Implanon®	Contraceptive subcutaneous implant	Etonogestrel
Nuvaring™ [2]	Contraceptive intravaginal ring	Etonogestrel/ethinylestradiol
Actisite™	Periodontal antiobiotc fiber	Tetracycline
Occusert™	Ophthalmic anti-glaucoma device	Pilocarpine
Ortho Evra™	Contraceptive transdermal patch	Ethinylestradiol/norelgestromin
Urinary catheters	Catheter	Chlorhexidine, gentian violet
FemRing™	Contraceptive intravaginal ring (RIM)	Estradiol acetate
Estring™	Contraceptive intravaginal ring (RIM)	Estradiol

14.6.4.1 Form Factors

Drug-eluting devices come in a wide variety of forms: they may be filaments (such as sutures), fibers (such as Actisite™ dental fiber), rods (such as Implanon® subdermal contraceptive), tubes (such as catheters), torroids (such as contraceptive or antiretorovial intravaginal rings and Occusert), free-standing films (such as dissolving strips) and liner-backed films (such as transdermal patches), as shown schematically in Figure 14.13.

Each type of device is made by mixing API (and perhaps other additives) into a molten polymer using a hot-melt extruder (usually a twin-screw to ensure homogenous mixing), then converting the extruded compound into the final dosage form by extruding, pulling, cutting, coating onto a substrate, etc. In some cases, additional downstream processing of the extrudate is required to prepare the final dosage form, such as injection molding or fiber pulling.

14.6.4.2 API Distribution and Device Architecture

If the drug is soluble in the molten polymer, then there are two mixing regimes:

- miscibility regime: $T_{\text{process}} > T_{\text{melt}}$; and
- solubilization regime: $T_{\text{process}} < T_{\text{melt}}$.

Table 14.6 *Examples of developmental drug-eluting devices made using HME [3–6].*

Device type	Polymer	API
Antiretroviral intravaginal ring	EVA	Dapivirine, maraviroc, levonogestrel, UC781, MIV150
	Polyurethanes	Dapivirine, tenofovir
Bioadhesive films [7]	Hydroxypropylcellulose	To be determined
Transdermal patches	PS-b-PIB	Analgesics
	Pebax	Nicotine
Contraceptive implants for dogs	EVA	Levonogestrel
Punctal plug	EVA	Latanoprost

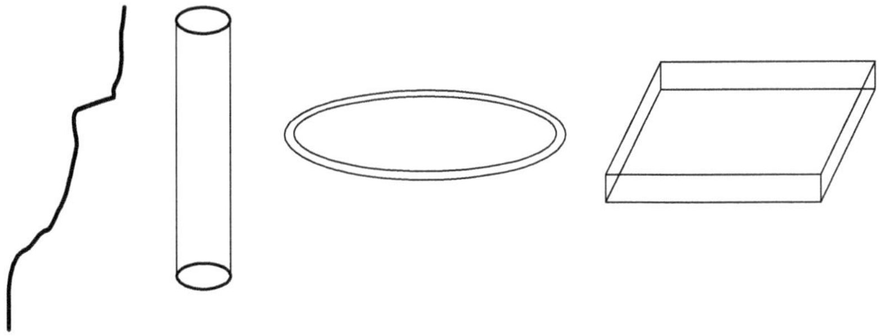

Figure 14.13 *Form factors of devices prepared using HME. From left: threads, cylinders (and tubes), torroids and sheets (and films/strips).*

In either case, if the API is also soluble in the polymer at room temperature, the resulting extrudate is usually a transparent/translucent material that remains this way on storage. Devices made directly from such extrudates are called monolithic solution devices. The molten extrudate is quite viscous, and the viscosity quickly rises further on cooling after it leaves the extruder die. Under these conditions, API crystallization may be hindered as molecular diffusion is too slow to allow crystal growth. The extrudate may cool in the supersaturated state, which appears as a transparent or translucent material. Over time however, and especially at elevated temperature (annealing) and when the API concentration is significantly above the saturation concentration, the API may recrystallize. This instability may not become evident for years, but can be a significant hurdle to device development; this was studied for the etonogestrel/ethinyl estradiol combination as part of the development of Nuvaring [8].

If the API is soluble in the molten polymer at the process temperature but recrystallizes as the extrudate cools, or the API has low solubility in the molten polymer, then the extrudate is usually opaque as a result of light scattering from the API crystals or particles in the polymer. Devices made directly from such extrudates are called monolithic dispersion devices. Both monolithic solution and monolithic dispersion devices are referred to collectively as matrix devices. Examples of these designs are shown in Figure 14.14.

On the other hand, reservoir devices are more complex designs and may have a localized bolus of drug within a pure polymer body, may have several distributed boluses (discontinuous reservoir devices) or may take the form of a continuous core-sheath (such as an electrical cable) where the drug-loaded core is coated with a drug-free polymer layer that controls the release of drug from the core into the patient (continuous reservoir). Such devices typically require co-extrusion equipment. Reservoir devices are preferred when the API release kinetics from matrix devices is not sufficiently constant *in vivo* for the application or indication in mind. A novel extension of the reservoir design, designed to meet the same objective of controlling drug-release kinetics, is Particle Sciences' patent-pending Microreservoir design. In this technology, the API is first encapsulated into polymer microparticles to form drug-loaded microcapsules; these are then mixed with the polymer in a melt-extrusion step to produce an extrudate that has uniformly dispersed microcapsules or 'microreservoirs'. These various architectures are shown schematically in Figure 14.15.

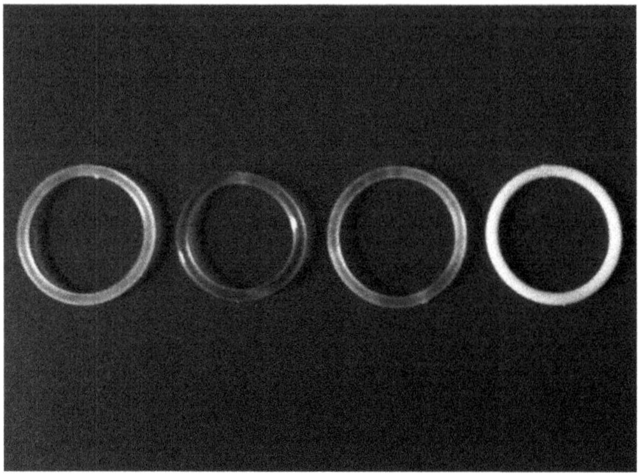

Figure 14.14 Various matrix IVRs, from left: placebo EVA, placebo polyurethane, EVA loaded with API below solubility limit (monolithic solution device) and EVA loaded with API above solubility limit (monolithic dispersion device).

14.6.4.3 APIs Used in Drug-eluting Devices

A wide variety of APIs types are compounded into thermoplastics to prepare drug-eluting devices. These include analgesics, antibiotics, antiretrovirals, hormones (birth control, HRT, etc.), stimulants and glaucoma drugs. Some examples of API molecules that can be found in developmental and commercial devices, broken down by class, are shown in Figure 14.16–14.19. More than one API is used for some devices (such as Nuvaring® and several antiretroviral IVRs). We have found that each API is generally released from the device independently of the other regardless of the ratios of APIs in the device, for example, with the pair levonogestrel and UC781 [4].

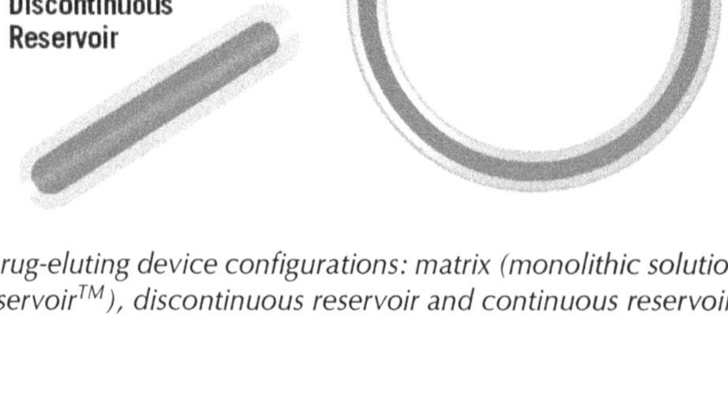

Figure 14.15 Drug-eluting device configurations: matrix (monolithic solution, monolithic dispersion, Microreservoir™), discontinuous reservoir and continuous reservoir.

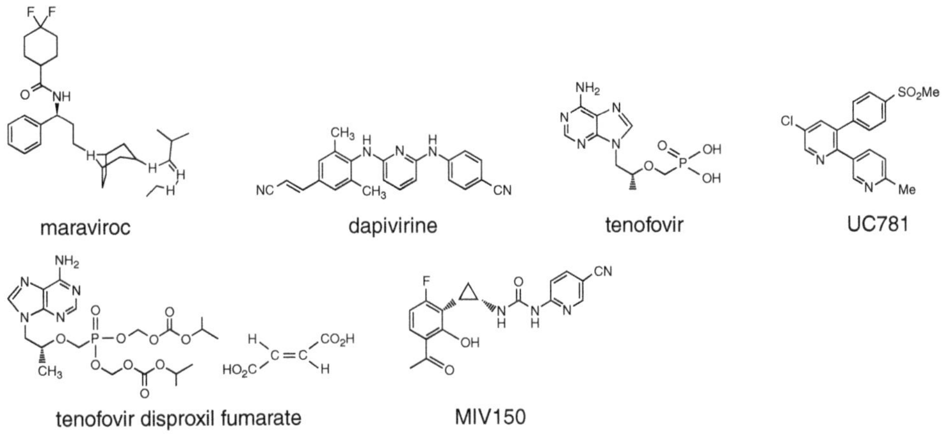

Figure 14.16 *Some antiretroviral APIs used in experimental anti-HIV drug-eluting devices.*

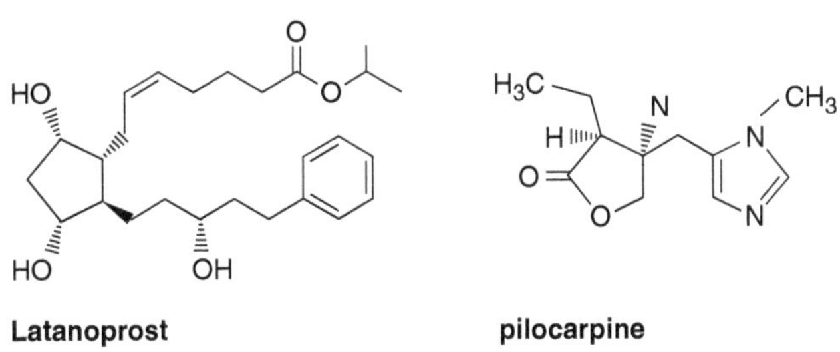

Figure 14.17 *Opthalmic (glaucoma) drugs used in ophthalmic drug-eluting devices.*

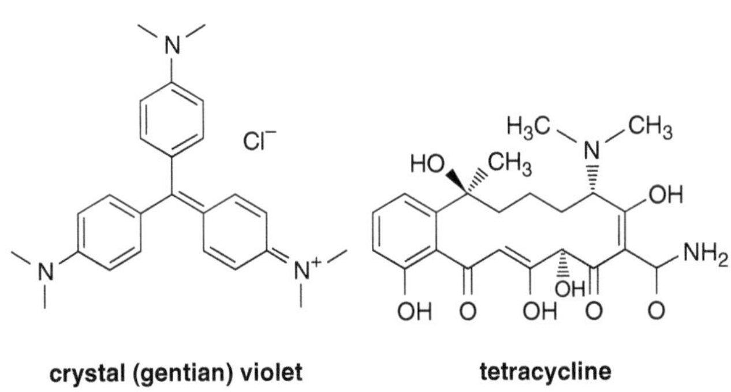

Figure 14.18 *Antibiotics used in catheter drug-eluting devices.*

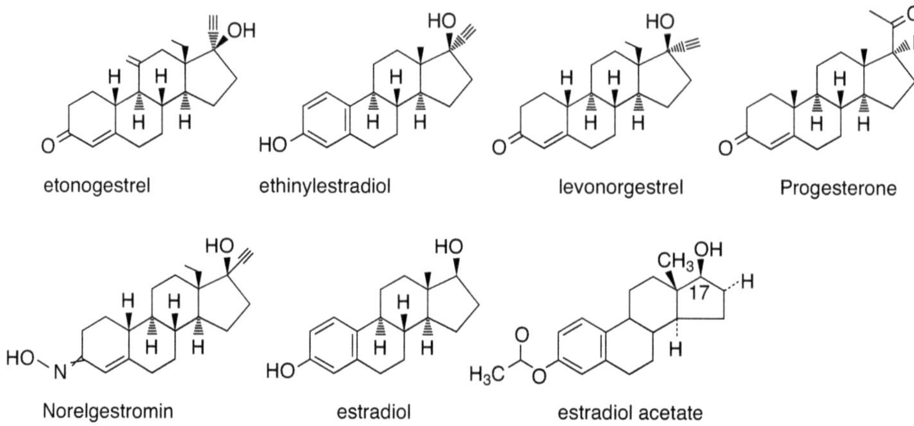

Figure 14.19 Hormones used in contraceptive and HRT drug-eluting devices.

14.7 Release Kinetics

Drug-eluting devices can help to control the release of a drug whose concentration can vary significantly over time after taking a tablet. Release from devices can be zero-order, first-order, Higuchi-type, and so on. The pulsatile drug plasma concentrations that can be associated with oral forms is shown in Figure 14.20, where the first-order release from an example matrix device and the zero-order release of a reservoir device are illustrated.

14.7.1 Mechanisms of API Release

There are three major modes of API release from polymeric drug-eluting devices:

1. bulk erosion of biodegradable polymers;
2. erosion of API from channels in non-erodable polymers; and
3. molecular diffusion of API through the device polymer.

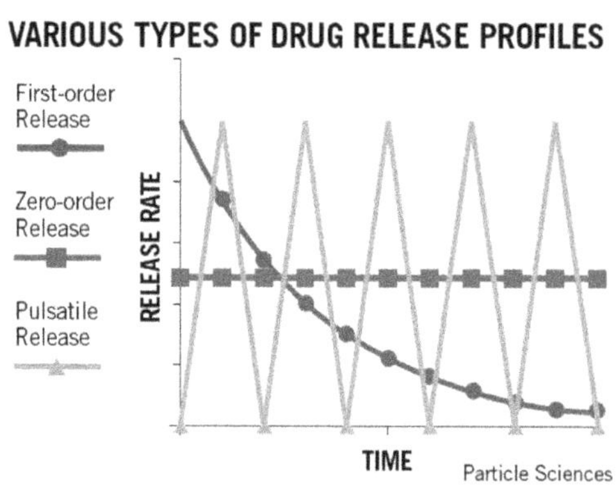

Figure 14.20 Various types of drug-release kinetic profiles. Reproduced with kind permission of Particle Sciences Inc.

Most devices are made from non-biodegradable polymers, and erosion of APIs from channels in non-erodable polymers is limited to water-soluble APIs [9, 10]. The majority of drug-eluting devices therefore release their drugs by solid-state molecular diffusion of the API in the polymer matrix. Solubility of the API in the polymer and the diffusion coefficient of the API in the polymer determine the release kinetics of the API from matrix devices of this type. If the API loading is below the solubility limit in the polymer, then all of the API is in solution. The release from a slab of thickness l of such a monolithic solution device containing M_0 of API at time $t = 0$ and M_t at time t is described by the equations:

$$\frac{M_t}{M_0} = 4\sqrt{\frac{Dt}{\pi l^2}}$$

for < 60% drug release and

$$\frac{M_t}{M_0} = 1 - \frac{8}{\pi^2} \exp\left(\frac{-\pi^2 Dt}{l^2}\right)$$

for > 40% drug release, where D is the drug diffusion coefficient in the polymer.

When the API loading is above the solubility limit, much of the API is dispersed as drug particles within the devices. Drug release from such monolithic dispersion devices is initially from the API in solution in the polymer near the device surface. As this is depleted, drug particles near the surface dissolve to maintain the soluble drug concentration. For devices with up to approximately 5% API, release kinetics are described by Higuchi's equation [11, 12]:

$$\frac{dM_t}{dM_0} = \frac{A}{2}\left[\frac{DC_{s(m)}\left(2C_0 - C_{s(m)}\right)}{t}\right]^{1/2}$$

where C_0 and $C_{s(m)}$ are the total API concentration in the device (dissolved and dispersed) at time $t = 0$ and the API solubility in the polymer, respectively. A is the total surface area of the device slab. Plots of release rate versus $t^{-0.5}$ or cumulative release versus $t^{0.5}$ are straight lines.

When the device is a reservoir type, the sheath modifies the release kinetics which then depend on the relative solubility of API in the core and sheath polymers and the thickness of the sheath layer. In this case, the drug release rate can be essentially zero-order over extended periods.

14.7.2 Example In Vitro Drug Elution Profiles

Figure 14.21 shows data for *in vitro* release of cyclosporine and etonogestrel from matrix EVA devices under sink conditions using an aqueous buffer solution. The release profiles appear to be approximately first-order with a high day-1 release rate, falling over time for each API.

Figure 14.22 shows the in vitro release of dapivirine from EVA intravaginal rings with various loadings of dapivirine, developed as anti-HIV devices for the International Partnership for Microbicides [4]. The results are determined under sink conditions using IPA:water mixture as release medium. The data clearly show the dependence of release kinetics on API loading in the device.

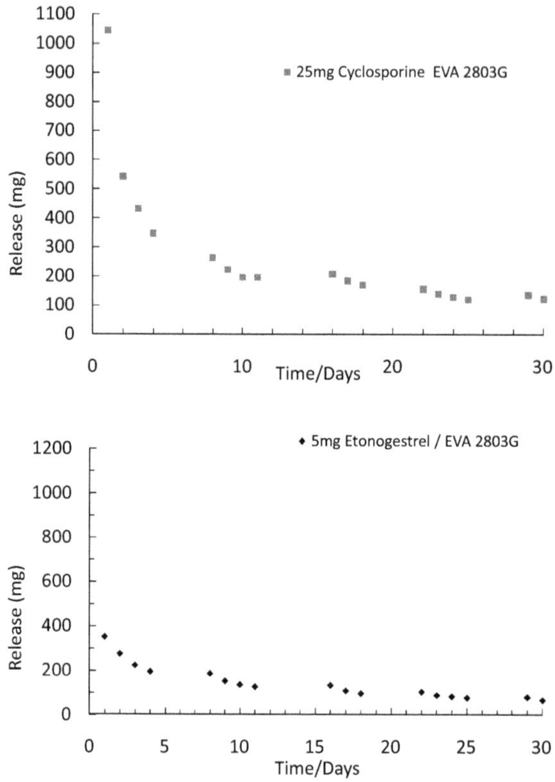

Figure 14.21 *Release of cyclosporine (25 mg, top) and etonogestrel (5 mg, bottom) from a device made from 28% VA EVA, using an aqueous surfactant solution as elution medium. Reproduced with kind permission of Celanese.*

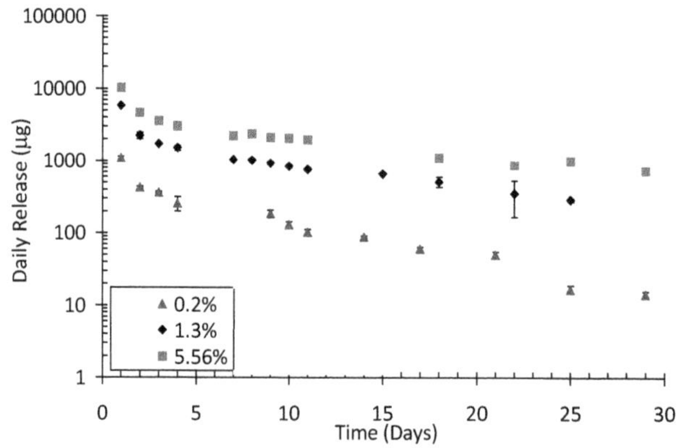

Figure 14.22 *Release of dapivirine from a device made from 28% VA EVA containing various loading of dapivirine, using 1:1 IPA:water as elution medium. Reproduced with kind permission of International Partnership for Microbicides.*

14.8 Conclusions

As in all plastics processing industries, hot-melt processing is useful for mixing, conveying and pumping molten polymers and their additives. A wide variety of device form factors have been made and a range of additives have been incorporated using HME, including APIs. HME is used as a simple mixer or pump, but can also be used to shape the extrudate into the final part shape; some shapes depend on extrusion as their manufacture otherwise would be most complex (e.g. tubes and coextruded rods). Drug-release kinetics of a range of drugs in a variety of polymers has been studied and has led to a number of successfully marketed products.

References

(1) Chebre-Sellaissie, I. and Martin, C. (eds) (2003) *Pharmaceutical Extrusion Technology*. Marcel Dekker, New York.
(2) van Laarhoven, A.H., Kruft, M.A.B. and Vromans H. (2002) *In vitro* release properties of etonogestrel and ethinyl estradiol from a contraceptive vaginal ring. *International Journal of Pharmaceutics*, **232**, 163–173.
(3) Loxley, A., Ghokale, A., Kim, Y., McConnell, J. and Mitchnick, M. (2008) Ethylene-vinylacetate intravaginal rings for zero-order release of an antiretroviral drug. Poster Presentation, Controlled Release Society Annual Meeting, New York.
(4) Loxley, A., Mitchnick, M., Okoh, O., McConnell, J., Goldman, L., Morgan, C., Clark, M. and Friend, D.R. (2011) Ethylene vinyl acetate intravaginal rings for the simultaneous delivery of the antiretroviral UC781 and con*traceptive levonorgestrel. Drug Delivery and Translational Research*, **1**(3), 247–255.
(5) Johnson, T.J., Gupta, K.M., Fabian, J., Albright, T.H. and Kiser, P.F. (2010) Segmented polyurethane intravaginal rings for the sustained combined delivery of antiretroviral agents dapivirine and tenofovir. *European Journal of Pharmaceutical Sciences*, **39**(4), 203–212.
(6) US Patent 5676969, Transdermal patch incorporating a polymer film incorporated with an active agent.
(7) Repkaa, M.A. and McGinity, J.W. (2001) Bioadhesive properties of hydroxypropylcellulose topical films produced by hot-melt extrusion. *Journal of Controlled Release*, **70**, 341–351.
(8) van Laarhovena, J.A.H., Krufta, M.A.B. and Vromansa, V. (2002) Effect of supersaturation and crystallization phenomena on the release properties of a controlled release device based on EVA copolymer. *Journal of Controlled Release*, **82**, 309–317.
(9) US Patent 5470582, Controlled delivery of pharmaceuticals from preformed porous polymeric microparticles.
(10) Saltzman, W.L. and Langer, R. (1989) Transport rates of proteins in porous materials with known microgeometry. *Biophysical Journal*, **55**(1), 163–171.
(11) Higuchi, T. (1963) Mechanism of sustained action medication: theoretical analysis of rate of release of solid drugs dispersed in solid matrices. *Journal of Pharmaceutical Sciences*, **52**, 1145–1149.
(12) Higuchi, T. (1961) Rate of release of medicaments from ointment bases containing drugs in suspension. *Journal of Pharmaceutical Sciences*, **50**, 874–875.

15

Hot-melt Extrusion: An FDA Perspective on Product and Process Understanding

Abhay Gupta and Mansoor A. Khan
Division of Product Quality Research, Office of Pharmaceutical Science, Food and Drug Administration

Disclaimer: The findings and conclusions in this article have not been formally disseminated by the Food and Drug Administration and should not be construed to represent any Agency determination or policy.

15.1 Introduction

Extrusion is a process of converting a material into an object of desired shape and properties by forcing it through a die of desired size under controlled conditions. It may be run continuously, thus offering relatively high throughput potential. Although the extrusion process may be done with the material either hot or cold, the pharmaceutical applications typically involve processing a polymeric material above its glass transition temperature to achieve effective mixing of the drug and excipients/polymer with the application of thermal and/or mechanical energy. The process is therefore referred to as hot-melt extrusion (HME). If the drug is dissolved at the molecular level and forms a one-phase system with the polymer, it is referred to as a solid solution. However, if the drug and polymer forms a two-phase microcrystalline dispersion where the drug is suspended in the amorphous

Hot-melt Extrusion: Pharmaceutical Applications, First Edition. Edited by Dennis Douroumis.

Figure 15.1 *Schematic representation of three modes of incorporation of the drug in a solid dispersion.*

polymer, in amorphous or crystalline state, it is generally referred to as a solid dispersion (Figure 15.1). The polymer used for preparing the HME should ideally be completely miscible with the drug molecule to ensure maximum interaction with the drug molecules at the molecular level in glassy state. The process has been used to improve the dissolution rate and bioavailability of the drug with poor solubility, to control or modify drug release, to mask the bitter taste of drugs and to create implants, depots and topical delivery systems.

HME combines multiple processes such as blending and granulation into a single unit operation and, in many cases, also removes the need for drying the material. The machine used for preparing HME consists of two basic parts: a conveyer system that transports the material (and, in some cases, mixes the material) and a die system that forms the materials into the required shape.The conveyer system may consist of a single screw positioned inside a stationary barrel, whereas more advanced systems may involve multiple-screw systems with screws in co-rotating or counter-rotating configuration. The process offers a number of advantages including low operating temperature, short exposure (< 2 min) of the drug–polymer mix to elevated temperatures, absence of organic and aqueous solvents, minimal product wastage and ease of scaling up.

The most important prerequisite for a material for use in a HME process is its ability to easily deform inside the extruder and solidify upon its exit from the extruder. Most HME dosage forms contain drug along with a complex mix of functional excipients including polymer for matrix formation, plasticizers, release-modifying agents, oxidizing agents, etc. In addition to being safe and pure, these excipients must be thermally stable, although the short exposure to elevated temperature allows for the use of some thermolabile materials in certain cases. Thus, materials that have been used in the preparation of other pharmaceutical dosage forms such as tablets and pellets are routinely used in the pharmaceutical HME process.

Polymer selection is based on the drug–polymer miscibility and drug–polymer stability assessment. Polymers with a high solubilization capacity are particularly suitable because they can dissolve large quantities of drug. Polymers with lower glass transition temperature offer the advantage of not exposing the drug to high processing temperature. Features such as lipophilicity, hydrogen-bonding acceptors, amide groups, etc. are basic prerequisites for a high solubilization capacity, thereby making polyvinylpyrrolidone and its copolymers and acrylic acid copolymer among the most commonly used polymers for the HME process. Plasticizers play an important role in the HME process by occupying sites along the polymer chain thereby preventing chain–chain interactions. They reduce the frictional force between the polymer chains, improving mobility and processability of the polymer and result in lowering the processing temperatures necessary for the HME process. Use of release-modifying agents may be needed to obtain the desired dissolution rate from the HME material, while antioxidants are used with drugs that show oxidative or free radical degradation during processing and/or storage.

The selection of extrusion conditions such as operating temperature, feed rate, screw speed, screw configuration, etc. also play a critical role in ensuring consistent product quality. The operating temperature must be selected by considering both the melting point and thermal degradation of the drug as well as the glass transition temperature of the polymer. Screw configuration may be governed by the viscosity of the material, for example, intermeshing screw design should be avoided when dealing with a high-viscosity system due to the potential for high torque build-up. Screw speed should be selected to allow sufficient time for the material to melt inside the barrel and to allow for uniform mixing of the component.

15.2 Quality by Design

FDA considers high-quality drug products to be those that consistently and reliably deliver the clinical performance and other characteristics stated in the label, are free of contaminations and are available [1]. Quality is defined in ICH Q9 as the degree to which a set of inherent properties of a product, system or process fulfills requirements [2]. A comprehensive understanding of the influence of the formulation components, environmental condition and manufacturing process on the product quality over the life of the product is therefore needed to ensure that the product and the process consistently achieve the desired quality [3]. This can be achieved if the product development encompasses an overall understanding of interaction in between the formulation components and their interaction with manufacturing processes. This will deliver products meeting predefined product quality objectives and not merely empirically derived product performance, which is typically confirmed by end product testing of the batches [4].

The FDA's Pharmaceutical CGMPs for the 21st Century: A Risk-Based Approach (the Pharmaceutical cGMP initiative) was launched in August 2002 with these objectives. The aims of this initiative were to apply scientific and engineering principles in regulatory decision-making, establish specifications based on product and process understanding and evaluate manufacturing processes, thereby improving the efficiency and effectiveness of both manufacturing and regulatory decision-making [5]. The initiative encourages voluntary development and implementation of innovative approaches in pharmaceutical development, manufacturing and quality assurance. It also encourages implementation of risk-based approaches that focuses attention on critical areas.

Quality by design (QbD) is an important element of this initiative. QbD is defined by the FDA and the ICH as a systematic approach to development that begins with predefined objectives and emphasizes product and process understanding and process control based on sound science and quality risk management principles [6]. It provides a sound framework for the transfer of product knowledge and process understanding from drug development to the commercial manufacturing processes and for post-development changes and optimization.

If adequately presented in the regulatory submissions, it has the potential to allow for a more flexible regulatory approach towards post-approval change management of the application and to reduce the role of end product testing.

A good QbD approach to product development includes a systematic evaluation of product attributes, understanding and refining of the formulation and manufacturing processes, application of quality risk management principles to establish appropriate control strategies and use of knowledge management and formal design of experiments to generate and refine the design space throughout the lifecycle of the product. Design space is an important element of QbD. It is the multidimensional combination and interaction of input variables (e.g., material attributes) and process parameters that have been demonstrated to provide assurance of quality [6]. It provides the linkage between the process inputs (input variables and process parameters) and the critical quality attributes (CQAs). A CQA is a physical, chemical, biological or microbiological property or characteristic that should be within an appropriate limit, range or distribution to ensure the desired product quality. An input variable or process parameter need not be included in the design space if it does not influence CQAs when varied over the full potential range of operation. If possible, all variables that affect the process and product quality should be included in the design space. Additionally, explanation should be provided to describe what variables were considered, how they affect the process and product quality and which parameters were included or excluded in the design space.

Design space may be based on an analysis of historical data or it may be defined in terms of ranges of input variables or parameters. It may also be defined in terms of simple or complex mathematical relationships between the CQAs and the input variables or parameters. Individual design spaces, which are often simpler to develop, may be established for individual unit operations. Conversely, a single design space may be established spanning multiple operations based on a comprehensive evaluation, thus providing more operational flexibility. For the design space to be applicable to multiple operational scales, it should be described in terms of relevant scale-independent parameters. Use of scale-independent parameters also provides greater regulatory flexibility during technology transfer and/or site change. The FDA does not consider movement within the design space as a change. Movement out of the design space is however considered to be a change, and would normally initiate a regulatory post-approval change process. Various quality risk management tools and an effective pharmaceutical quality system may be used to deliver consistent product quality as well as to create a basis for operational flexibility [2, 7].

Quality risk management is a systematic process to assess, control, communicate and review risks to the quality of the drug product across its lifecycle (Figure 15.2). It is typically initiated by an interdisciplinary team consisting of experts from appropriate areas (e.g. quality unit, engineering, regulatory affairs, production operations, legal, statistics, etc.) in addition to people with knowledge about the quality risk management process. The risk assessment process begins with a well-defined problem description or risk question and consists of identification of hazards (i.e. the potential source of harm) through systematic use of information. The question concerning what might go wrong is answered and possible consequences when things do go wrong are identified.

Quality risk management also provides a basis for risk analysis and risk evaluation. Risk analysis is the process of estimation of the risk associated with the identified hazards though a qualitative or quantitative process of linking the likelihood of occurrence and severity of

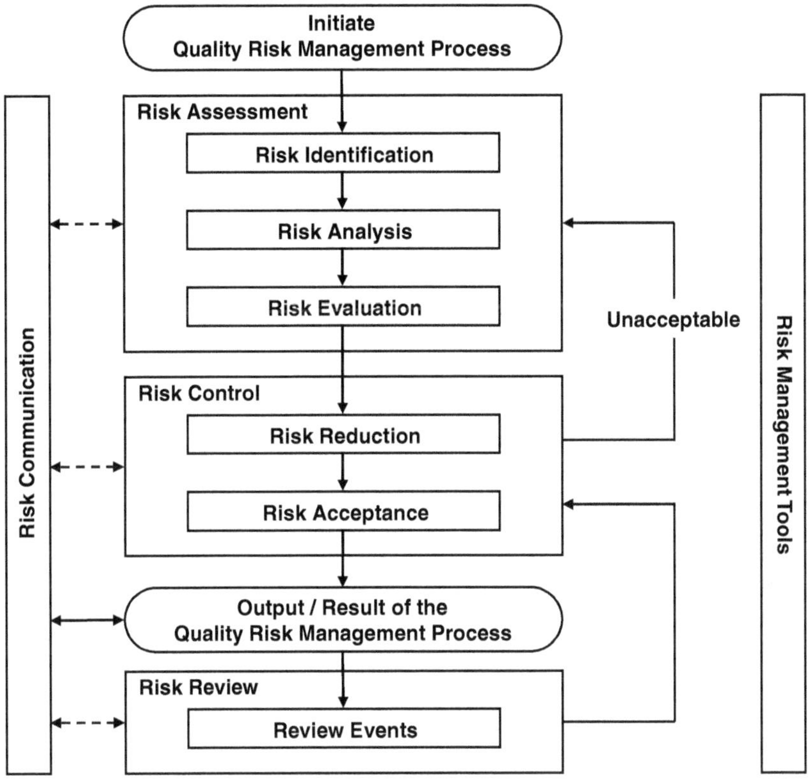

Figure 15.2 *An overview of a typical quality risk management process (from ICH Q9).*

harms. Risk evaluation compares the identified and analyzed risk to given risk criteria and determines the significance of the risk.

The process of decision making to reduce and/or accept risk at an acceptable level is called risk control. Risk reduction focuses on mitigation or avoidance of quality risk when it exceeds a specified (acceptable) level, while risk acceptance is a decision to accept risk at a specific (acceptable) level.

Risk communication involves the sharing of information about risk and risk management between the decision makers and others and should be appropriately documented to allow for periodic review of the risk management process.

Use of quality risk management supports a scientific and practical approach to decision making by providing documented, transparent and reproducible methods for managing risk based on current knowledge about assessing the probability, severity and, sometimes, detectability of the risk. An effective quality risk management approach ensures drug product quality by offering a proactive means of identifying and controlling potential quality issues during drug product development and manufacturing processes.

Pharmaceutical quality system is a management system of directing and controlling a pharmaceutical company by establishing, implementing and maintaining a set of processes that consistently deliver quality products. It includes developing and using effective monitoring and control systems for process performance and product quality, and identifying and implementing appropriate product quality improvements, process improvements, variability reduction, innovations and pharmaceutical quality system enhancements. It involves managing knowledge, from development through the commercial life of the product up to and including product discontinuation, through use of quality risk management tools to identify and control potential risks to quality throughout the product lifecycle. Use of quality risk management and pharmaceutical quality system can improve the decision making if a quality problem arises, facilitate better and more informed decisions, provide regulators with greater assurance of a company's ability to deal with potential risks and affect the extent and level of direct regulatory oversight.

15.3 Utilizing QbD for HME Process Understanding

A QbD approach to design and develop a formulation using HME processes begins with the sponsor defining the quality target product profile (QTPP). QTPPs are the desired quality attributes that are needed to ensure the desired product quality, taking into account safety and efficacy of the drug product. Typical QTPPs for a HME process include good dissolution and bioavailability for BCS class II and IV drugs and presence of drug in amorphous stable state with low potential for crystalline reversion. This may be achieved by utilizing prior knowledge, initial experimental data and/or risk assessment tools, such as Ishikawa (fishbone) diagram (Figure 15.3) or Failure Mode Effects Analysis (FMEA).

Application of these tools may allow the sponsor to identify all potential variables that may be deemed critical to the HME process (Figure 15.4). For drug substance and excipients,

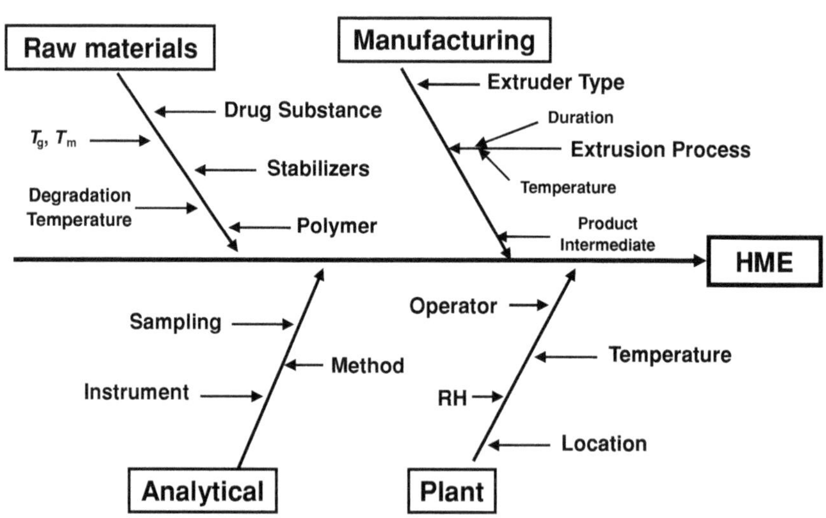

Figure 15.3 Ishikawa diagram.

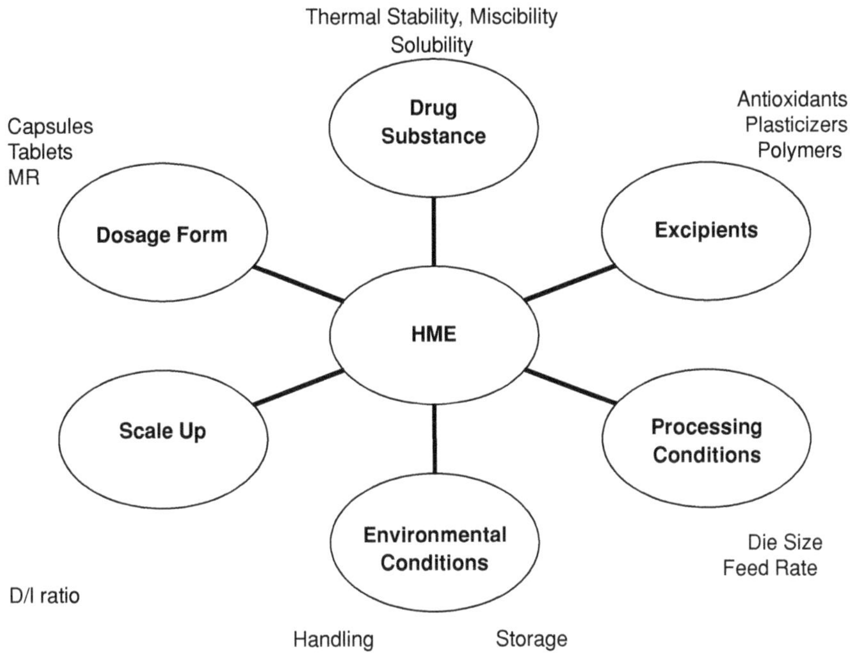

Figure 15.4 *Variables influencing the HME process.*

these may include melting point, glass transition temperature, solubility, thermal stability, drug–polymer ratio etc. The type of dosage form, i.e., immediate release, modified release, etc., may influence the selection of the drug salt and excipient composition.

Critical process and environmental conditions may include screw configuration and speed, screw diameter to length ratio, extrusion temperature, environmental temperature and relative humidity. Scale-up should be based on established scaling laws, dimensionless analysis and/or geometric considerations. Screening design may be used to exclude variables deemed not to be critical to the QTPP.

Formal design of experiments or other experimental approaches may be used to evaluate the impact of the remaining variables on the QTPP to gain greater understanding of the manufacturing process and to develop appropriate control strategies. For HME process, the initial screening may, for example, identify the dissolution rate and the stability of the extruded material as CQAs and the particle size of the drug, operating temperature and screw speed during the HME process as the critical input variables. A full factorial design or a response surface design may then be used to study these critical variables and to identify the potential interactions and individual impact on the CQAs. The information gained could be used to generate the design space, providing the sponsor with the ability to adjust the processing conditions (e.g. screw speed) in response to a change in the properties of incoming raw material (e.g. particle size of the drug substance). Appropriate scale-up studies will then demonstrate that the data is consistently reproducible across different manufacturing scales. Conventional and novel characterization methods, such as differential scanning calorimetry (DSC), powder x-ray diffraction (PXRD) and vibration spectroscopic techniques coupled with appropriate monitoring tools may be used to monitor specific HME process parameters such as temperature, screw speed, etc. in order to ensure consistent product quality.

HME typically results in drug being trapped inside the polymer matrix in an amorphous state (a thermodynamically unstable state). Most materials in their amorphous state exhibit significantly higher solubility when compared to their crystalline state. However, the inherent instability of the amorphous drug in products prepared by the HME process may result in recrystallization of the drug substance during storage, leading to significant changes in dissolution and/or stability of the drug product. The mobility of the amorphous drug–polymer molecules may be stabilized by choosing excipients with high drug–polymer interaction. It is important to characterize and study changes, if any, in the crystalline/amorphous state of the material prepared by the HME process during early stages of product development. DSC and PXRD are the most common analytical techniques used for this purpose. DSC detects temperatures at which thermal events occur. Thermal events may be a glass to rubber transition, (re)crystallization, melting or degradation. PXRD detects material with long range order, with crystalline material giving sharper diffraction peaks. Absence of peaks in the DSC and PXRD patterns of material prepared by HME process may be used to confirm the presence of drug in the amorphous state and absence of crystalline material (Figure 15.5 and 15.6). Scanning electron microscopy may be used to study the homogeneous distribution of drug and polymer, while infrared and/or Raman spectroscopy may be used to confirm drug–polymer interactions at a molecular level. Near infrared chemical imaging may also be used for non-destructive qualitative and/or quantitative determination of the drug in the extruded material.

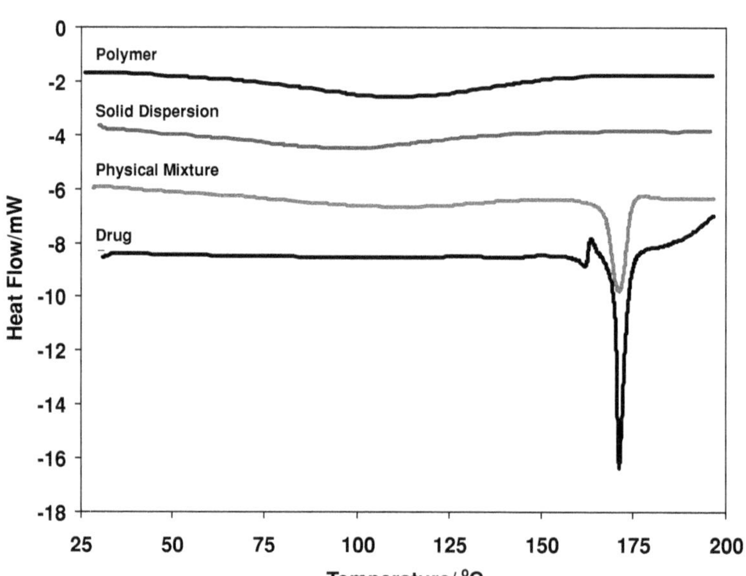

Figure 15.5 *DSC profiles of pure drug, polymer, physical mixture and solid dispersion.*

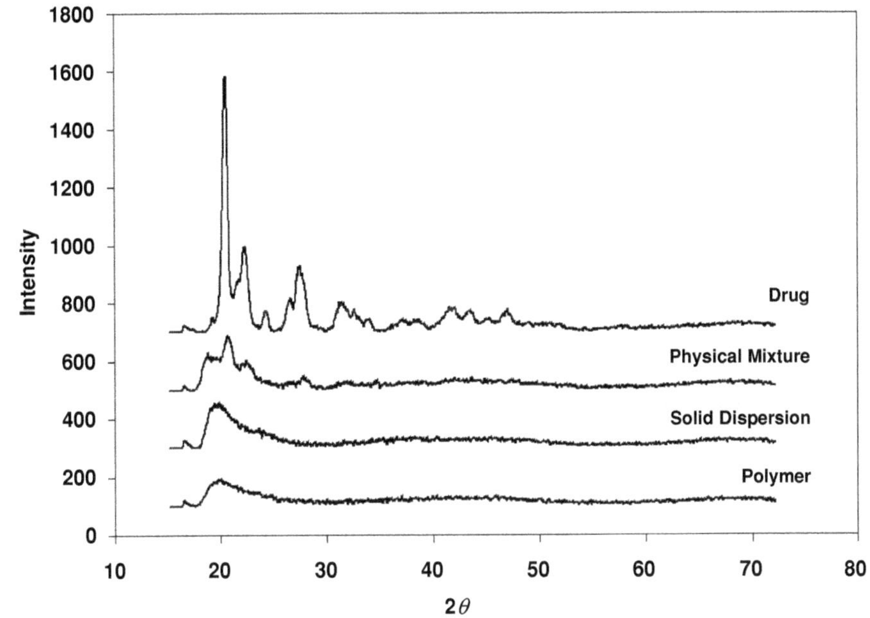

Figure 15.6 *Powder x-ray diffraction patterns of pure drug, polymer, physical mixture and solid dispersion.*

References

(1) Woodcock, J. (2004) The concept of pharmaceutical quality. *American Pharmaceutical Review*, **7**, 10–15.

(2) ICH Q9 (2006) Quality Risk Management. ICH Harmonised Tripartite Guideline. http://www.ich.org.

(3) Nasr, M. (2007) FDA's quality initiatives: An update. http://www.gmp-compliance.com/daten/download/FDAs_Quality_Initiative.pdf (accessed 08/02/2011). APIC/CEFIC European Conference on Active Pharmaceutical Ingredients, 24–26 October, Warsaw, Poland.

(4) Yu, L. (2008) Pharmaceutical quality by design: product and process development, understanding, and control. *Pharmaceutical Research*, **25**, 781–791.

(5) FDA (2004) Pharmaceutical cGMPs for the 21st Century: A Risk-Based Approach. Final Report, Fall 2004. http://www.fda.gov/cder/gmp/gmp2004/GMP_finalreport2004.htm.

(6) ICH Q8(R2) (2009) Pharmaceutical Development. ICH Harmonised Tripartite Guideline. http://www.ich.org

(7) ICH Q10 (2009) Pharmaceutical Quality System. ICH Harmonised Tripartite Guideline. http://www.ich.org.

16

Improved Process Understanding and Control of a Hot-melt Extrusion Process with Near-Infrared Spectroscopy

Chris Heil
Chief Scientist, Product Specialist, Thermo Fisher Scientific, Madison, USA
Jeffrey Hirsch
Chief Scientist, Thermo Fisher Scientific, Madison, USA

16.1 Vibrational Spectroscopy Introduction

This chapter describes the use of near-infrared spectroscopy for understanding and controlling the hot-melt extrusion process in the pharmaceutical industry. Vibrational spectroscopic theory is reviewed as well as the basic layout of near-infrared instrumentation. In addition, basic application studies showing the use of a near-infrared analyzer with pharmaceutical samples are detailed as well as more online applications which include a fiber-optic probe integrated into a hot-melt extruder.

Spectroscopy is the interaction of light with matter to gain information that informs a conclusion. The type of information gained depends on the frequency of light used. For example, visible light (400–750 nm) can provide information about electronic states in molecules while near-infrared (NIR, 750–2500 nm) and mid-infrared (IR, 2500–25,000 nm) light provide information about vibrational states. Vibrational energy states are typically plotted as a spectrum in two dimensions showing the frequency of light on the x axis versus the amount of light absorbed on the y axis allowing for comparison between molecules and,

Hot-melt Extrusion: Pharmaceutical Applications, First Edition. Edited by Dennis Douroumis.

ultimately, identification. The specificity of vibrational spectroscopic information enables a hazardous materials (HazMat) team, for example, to definitively identify an unknown sample on the street, providing the unambiguous chemical information necessary for safe removal. In quality analysis and quality control (QA/QC), vibrational spectroscopy provides proof that an incoming raw material matches its label for safe entry into production. Online applications that include NIR, as for hot-melt extrusion, measure the integrity of the product by determining the quality or relative amount of component materials permitting real-time adjustment to maximize yield of a suitable quality.

Knowing the precise energy of each vibration in a molecule is critical to the utility of infrared spectroscopy. As this energy is, by definition, equal to the energy of the light that excites it, we will look briefly at some basic equations on the nature of light. Electromagnetic radiation has two defining characteristics: wavelength λ, the distance between successive peaks on the wave, and frequency ν, the number of peaks per unit distance or time (measured in units of cm^{-1} or Hz, respectively). These two properties are linked by Equation (16.1) below where n is the index of refraction of the medium and c is the speed of light in a vacuum, 2.99×10^8 m/s:

$$\frac{c}{n} = \lambda\nu \tag{16.1}$$

Equation (16.2) describes the energy of light E where h is Planck's constant with a value of 6.6×10^{-34} Joule second.

$$E = \mathrm{h}\nu = \frac{\mathrm{h}c}{n\lambda} = \frac{\mathrm{h}c\,\bar{\nu}}{n} \tag{16.2}$$

Substituting Equation (16.1) into Equation (16.2), it becomes apparent that the energy of light is directly proportional to its frequency and inversely proportional to its wavelength. This relationship will be very useful in understanding spectroscopic measurement of samples, especially regarding the proportionality of energy to frequency in wavenumber or cm^{-1} ($\bar{\nu}$).

One of the reasons vibrational spectroscopy is so useful is its ability to map out the large number of energy states in which molecules or parts of molecules may vibrate, allowing the unambiguous identification of complex materials or mixtures of materials with a single technique. Predicting the energy states vibrations is, therefore, critical.

The energy of a molecular vibration can be approximated by analogy with the classical case for the potential energy (PE) of a spring. This is referred to as Hooke's Law and takes the form:

$$\mathrm{PE} = \frac{1}{2}kx^2 \tag{16.3}$$

where k is the spring force constant and x is the displacement of the spring from its equilibrium resting position. As the spring compresses or stretches away from its natural resting point, the potential energy of the system will increase because there will now be a restoring force trying to bring the spring back to its equilibrium position. To further the analogy between molecular vibration and classical mechanics, we include balls attached to the ends of the spring representing atoms. This new model system is referred to as the simple harmonic oscillator (SHO) and has a vibrational frequency ν described by:

$$\nu_{\mathrm{SHO}} = \frac{1}{2\pi}\sqrt{\frac{k}{\mu}} \tag{16.4}$$

where k is the spring constant from Hooke's law and μ is the reduced mass of the two balls or atoms including terms for their respective masses, m_1 and m_2. The reduced mass is defined as:

$$\mu = \frac{m_1 m_2}{m_1 + m_2} \tag{16.5}$$

The simple harmonic oscillator is a purely classical law and does not account for the known quantization of energy levels in molecular vibrations. Equation (16.6) is a solution to the Schrödinger wave equation for a simple harmonic oscillator that accounts for this difference and predicts the energy E_υ of a quantized harmonic vibrational state using the integer-term upsilon υ to represent the quantum level. The term for frequency is identical to that in Equation (16.4), illustrating the pivotal role that reduced mass plays in the energy of quantized molecular vibrations. For a detailed mathematical derivation of the energy of a quantum vibration seen below, the reader is directed to texts by Hanna [1] and Harris and Bertolucci [2].

$$E_\upsilon = \left(\upsilon + \frac{1}{2}\right)\frac{\mathrm{h}}{2\pi}\sqrt{\frac{k}{\mu}} = \left(\upsilon + \frac{1}{2}\right)\mathrm{h}\nu \tag{16.6}$$

Several things are clear from Equation (16.6): (1) only certain discrete vibrational energy levels are permitted ($\upsilon = 1, 2, 3$, etc.); (2) there is a residual, finite energy of the system at $\upsilon = 0$ called the zero point energy that is innate to any vibrational system even at absolute zero and is equal to $\mathrm{h}\nu/2$ (note that vibrational levels here are denoted by upsilon υ and vibrational frequency by nu ν); (3) the energy level spacing between adjacent vibrational levels is the same; and (4) the energy of the vibration is directly proportional to its frequency in wavenumbers and inversely proportional to its reduced mass.

Classically, this latter observation is intuitive. We can imagine that the natural vibrational frequency of a golf ball suspended from the ceiling on a given spring will be higher (i.e. will vibrate faster) than a bowling ball attached to the same spring. This effect is readily demonstrated when comparing infrared vibrational frequencies of molecules such as HX (where X=F, Br). The molecule containing the lighter of the two halogens, F, will have a peak in the infrared spectrum at a higher frequency (3962 cm^{-1} for HF versus 2558 cm^{-1} for HBr), denoting a higher energy vibration [3].

The harmonic oscillator approximation for molecular vibration has several critical drawbacks when applied to well-known properties of molecular vibration. The wavefunction for a harmonic oscillator defines the permissible energy transitions as $\Delta\upsilon = \pm 1$. Restated, this constraint of the harmonic model says that only transitions between adjacent vibrational energy levels are allowed. This model cannot explain observed resonances in the NIR portion of the spectrum called overtones where $|\Delta\upsilon| \geq 2$ or greater (e.g. a transition from $\upsilon = 0$ to $\upsilon = 2$). In addition, a harmonic wavefunction does not consider what happens to a real chemical bond at the extremes of vibration. When a bond is stretched so the potential energy exceeds its bond dissociation energy (D_e), it will break. When a bond is compressed it will reach a point where internuclear repulsion (positively charged nucleus against like-charged nucleus) will prevent the two from coming together, making the potential energy curve extremely steep at very short internuclear distances.

Shortcomings of the harmonic oscillator approximation are remedied by using an anharmonic function such as that proposed by Morse [4]. The Morse potential, as it is commonly known, describes the potential energy (V) of a diatomic molecule and includes a term for the molecular dissociation energy, D_e:

$$V = D_e\left(1 - e^{-\beta q}\right)^2 \tag{16.7}$$

where q is the displacement of the atoms from equilibrium and β is a term containing molecular information (as for the already familiar reduced mass, μ). It is clear from the Morse potential that as the atoms are pulled further and further apart from each other (larger values of q), the second term in the parenthesis will tend to zero. As this happens, the potential energy will reach a maximum at the bond dissociation energy (D_e) as intended. Applying an anharmonic potential to this system also changes the vibrational wavefunctions themselves. This leads to new selection rules for vibrational transitions, so resonances other than $\Delta\upsilon = \pm 1$ are now allowed. In addition, anharmonicity results in successively smaller energy gaps between adjacent levels with increasing values of υ. This behavior is in stark contrast to the SHO model which predicts all energy levels are evenly spaced (Figure 16.1).

There are several criteria that need to be satisfied to see a peak in an infrared spectrum: the energy of the incident photon must exactly match the energy of the transition, the dipole moment of the molecule must also change during the course of the vibration and its vector must be in the same direction as the electric field vector of the exciting infrared radiation. Vector alignment is typically not a concern for most samples as, at a molecular level, there will be enough molecules in the correct orientation to provide adequate signal. The requirement of a dipole moment change upon vibration is, however, critical to infrared activity and necessitates a discussion of degrees of freedom and molecular symmetry.

One atom in space can move in any one of the three Cartesian dimensions, so we say that the atom has 3 degrees of freedom. If we then try and describe the possible ways two particles can move, we have 3+3 or 6 degrees of freedom and so on. For a molecule with N

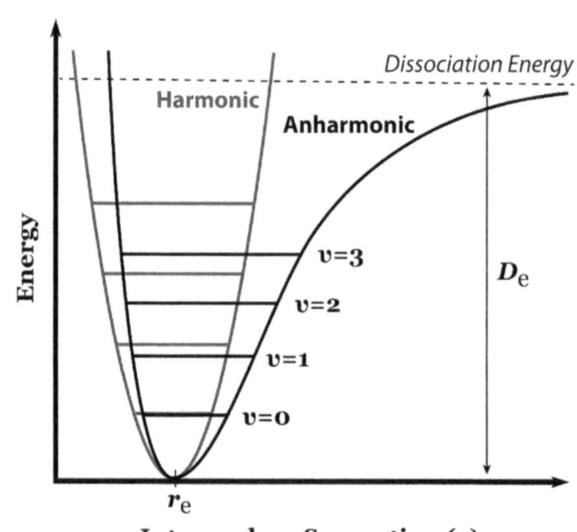

Figure 16.1 Morse potential.

atoms, it follows that it has $3N$ degrees of freedom. Motions arising from a rotation (around the x, y and z axes for a total of 3) or translation of the molecule (in the x, y and z directions for, again, a total of 3) do not result in any vibrational motion so we can assert that any non-linear molecule will have $3N-6$ degrees of vibrational freedom. Linear molecules will have $3N-5$ degrees of vibrational freedom (rotation around the bond axis in a linear molecule produces no net molecular displacement so it is not counted as one of the possible degrees of vibrational freedom).

Nitrogen gas, $N_{2(g)}$,is a homonuclear diatomic molecule. Using the degrees of freedom calculation for a linear molecule, we see that it has only one possible vibrational mode (i.e. $(3 \times 2)-5 = 1$). The same is true of the heteronuclear diatomic molecule carbon monoxide $CO_{(g)}$. Even though these two molecules possess the same number of degrees of freedom, their vibrational spectroscopic behavior is different based on the nature of their constituent atoms. In summary, degrees of freedom analysis provides the maximum number of total vibrations available to any molecule based simply on the number of constituent atoms. The task of identifying the exact appearance and spectroscopic activity of each vibration is left to the application of symmetry and group theory.

The symmetry classification of any object, from the macroscopic to the molecular, is derived from the combination of its known symmetry elements. It is this precise combination of symmetry elements for any given molecule that explains the presence or absence of spectral peaks in infrared due to the required dipole moment change upon vibration or Raman due to the required polarizability change upon vibration. The three major categories of symmetry elements are rotation, inversion and reflection with element names such as C, an axis of rotation σ, a mirror plane i, a center of inversion S, an improper rotation axis (rotation followed by inversion) and the identity element E where the molecule is unchanged in space.

Take for example the case of a soccer ball, a highly symmetric collection of pentagonal and hexagonal units. The ball will look exactly the same if you turn it one-fifth of a circle (72°) on an axis going straight through the middle of one pentagon. The ball therefore possesses one of the known symmetry elements (in this case, an axis of rotation referred to as C_5 denoting a one-fifth of a circle rotation). The molecular equivalent of the soccer ball, C_{60} or buckminsterfullerene, has 120 symmetry elements. Using only symmetry considerations we can predict that C_{60} should have 4 peaks in the IR (all F_{1u} vibrations) and 10 peaks in the Raman (2 A_g and 8 H_g) which, indeed, it does [3]. The complete list of one molecule's symmetry elements combine in specific characteristic ways, forming what is referred to as a point group (icosahedral or I_h for C_{60}). The point groups and their associated character tables allow the rapid identification of Raman- and infrared-active vibrations.

The two main branches of vibrational absorption spectroscopy are mid-infrared spectroscopy and near-infrared spectroscopy (IR and NIR, respectively). IR looks primarily, although not exclusively, at the fundamental vibrations of an IR-active molecule ($\Delta \upsilon_{0\to1}$) via absorption of light. NIR works similarly and looks at overtones ($|\Delta\upsilon| \geq 2$ or greater) and combination bands. Both IR and NIR have the ability to ascribe peaks in a spectrum to specific molecules or parts of molecules based on the identity of the atoms and their symmetry properties as the discussion above has shown. This is in contrast to techniques such as chromatography, for example, that look at molecular retention factors. These can be specific for certain functional groups in a molecule, but are not based on reduced mass or molecular symmetry.

Both IR and NIR have innate advantages that arise from the wavelength of light used and the method of acquisition. IR has a high signal-to-noise ratio allowing for rapid, unambiguous identification of any IR active material using automated database searching. Historically, IR analysis of solids in transmission suffered from a heavy burden of sample preparation which required samples be mixed with KBr than pressed into pellets. The use of attenuated total reflection (ATR) where samples are compressed against a window of high refractive index by a simple slip-clutch mechanism has all but eliminated the difficulty of pressing pellets in IR spectroscopy. Near-infrared, in contrast, requires no sample preparation and can analyze through bags for raw material ID or through fiber-optic probes for online QA/QC. NIR achieves the ability to quantify materials in both synthetic and natural matrices through the use of calibrations.

NIR light can travel via fiber-optic cable, much as it does in the telecommunications industry, without significant degradation of the signal because the absorbance of the light is low. This also provides a greater interrogation of the sample as the light can travel further without being fully absorbed by the material. In contrast, IR light will typically only penetrate a few microns into a sample while NIR can penetrate several millimeters. For these reasons, NIR spectroscopy has become widely accepted in the pharmaceutical industry as an accurate, robust spectroscopic technique for the analysis of materials for QA and QC. Whereas IR typically dominates the research lab as sharper peaks allow for higher resolution studies and better signal-to-noise, NIR is prevalent in online and QA/QC applications due to its ability to travel through fibers and interrogate more of the sample such as in transmission analysis of a tablet.

There are several major types of commercially available NIR instrument: filter, dispersive and Fourier transform. Although many NIR spectrometers use a common light source (a tungsten-filament quartz-halogen bulb) and detector (InGaAs), they differ significantly in the method of wavelength selection.

Filter instruments are the simplest design, using a series of bandpass filters on a wheel to selectively allow specific wavelengths of light to pass. That light then impinges on the sample where the frequencies corresponding to the infrared-allowed vibrational energies can be absorbed. The remaining light travels to the detector creating a response that is recorded. Once all the filters have passed in front of the source bulb, the detector response is then stitched together to create a spectrum.

Dispersive systems fall into one of two categories based on their detection scheme: single or array. In both layouts, the source light is guided to a grating where it produces a diffraction pattern composed, in part, of light that has been dispersed into its component frequencies. If the system only uses a single detector, the dispersed light is then guided through a slit frequency by frequency by turning the grating (i.e. a monochromator). In contrast, array technology makes use of multiple detectors lined up next to each other allowing the entire dispersed spectrum to fall across the detector array simultaneously without the need to move the grating (i.e. a polychromator or spectrograph).

The final configuration for NIR spectrometers is Fourier-transform near-infrared (FT-NIR). A simplified diagram of a FT-NIR spectrometer is shown in Figure 16.2. FT-NIR uses an interferometer to modulate the light analogous to the modulation of radio signals. The component frequencies are all 'packaged' into an interferogram that interacts with the sample without the need for a scanning grating or filters. Any specific frequency absorbed by the sample changes the interferogram. Since the interferogram cannot be interpreted

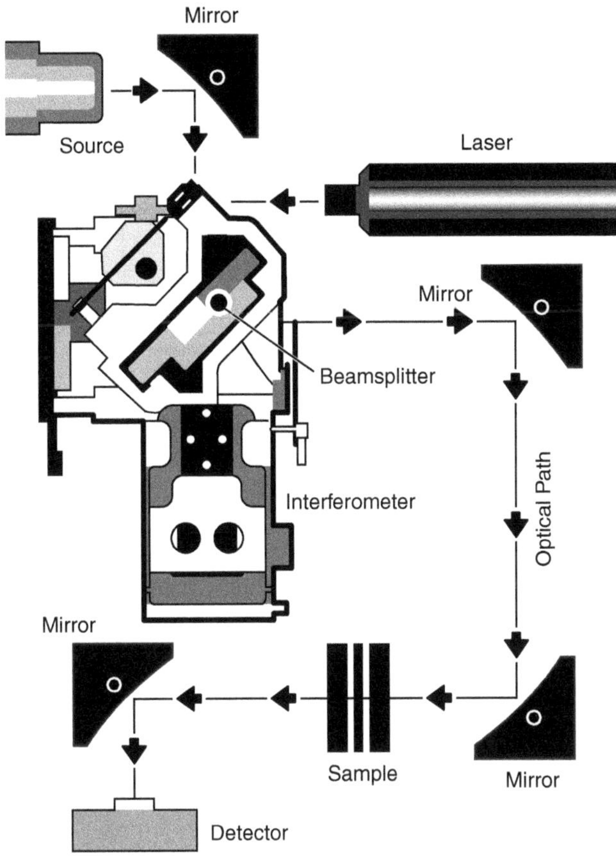

Figure 16.2 *Optical layout of an FT-NIR spectrometer.*

directly, a Fourier transform is performed by a PC which presents the user with a spectrum showing sample absorbance versus wavenumber (frequency). A spectrum taken with the sample absent, a background, is ratioed to one with the sample present to provide a ratioed absorbance spectrum showing the spectral contributions of the sample.

16.2 Near-infrared Method Development

NIR spectra have a unique appearance relative to IR spectra (Figures 16.3a and b, respectively) due to the nature of the vibrations. Fundamental vibrations typically have narrow peaks with good separation between neighboring resonances. Combination bands and overtones are significantly smaller on the absorbance scale, highly overlapped and much broader. The combination band region in NIR, roughly 4000–6000 cm^{-1}, has peaks that are smaller than those found in the IR and are slightly broadened. In the overtone region (6000–10,000 cm^{-1}), the peaks continue to get smaller and broader and begin to overlap

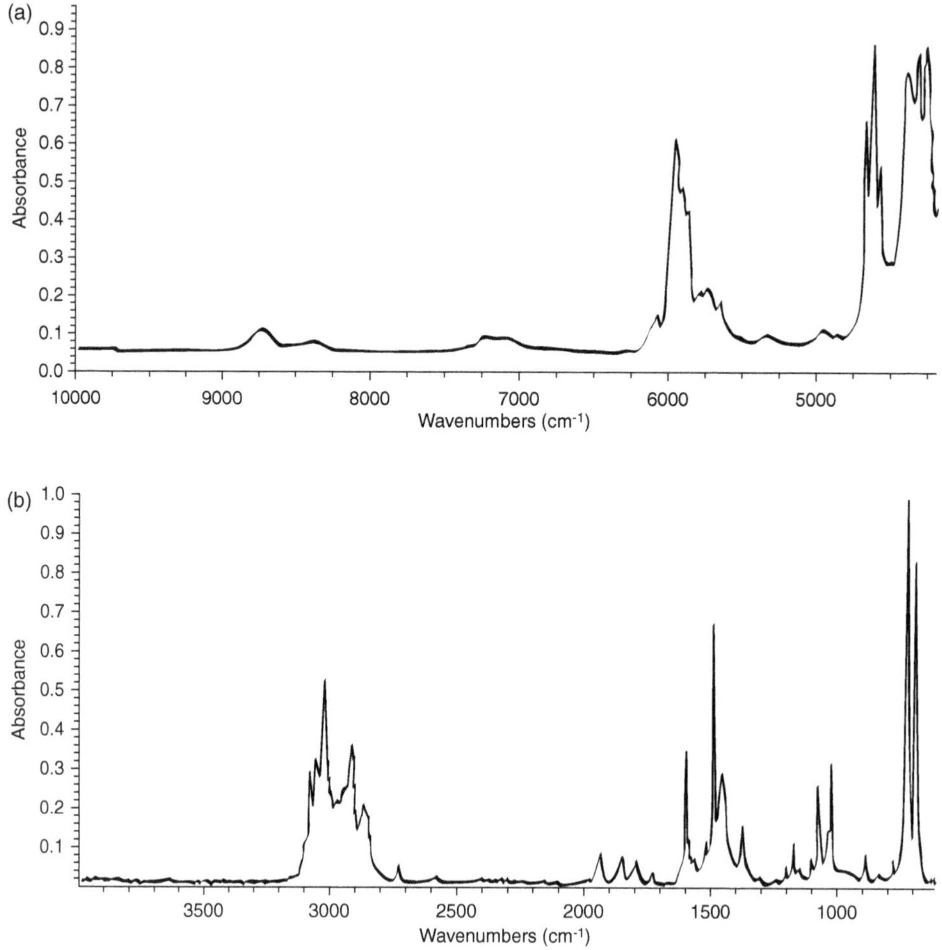

Figure 16.3 *(a) NIR and (b) IR spectra of toluene.*

each other substantially. This type of spectral character necessitates the use of statistical techniques such as principal component analysis (PCA), principal component regression (PCR) or partial least squares (PLS) for tasks such as multi-component quantification, material identification and qualification.

The relationship between spectral data and known properties of the samples must be established from standard spectra either by collecting light reflected off a sample (reflection) or light transmitted through a sample (transmission). Quantification requires a calibration model to be developed with the accuracy of the model tied to the accuracy of a primary quantitative analytical method (e.g. HPLC). For qualification analysis, spectra of standards with known classes are used to develop a qualification model. For identification analysis, spectra of known chemical identification are used as standards for identification model development. For all three analysis types in the pharmaceutical industry, the calibration model must be validated prior to it being put into use for QA/QC testing.

Quantitative analysis for multiple components is most commonly performed by a PLS regression. The PLS model is a statistical analysis technique that develops a prediction model based on the correlation between absorbance in the sample that is related to the changes in the component concentration. There are multiple classification algorithms that can be used for identification or qualification calibration model development with advantages and disadvantages based on the specific application. Classification models are sensitive to spectral changes caused by chemical differences in the different compounds. The classification model determines the degree of similarity or difference or performs a spectral match to standards and compounds used to develop the calibration model.

During NIR calibration model development, it is important to include all known and expected chemical and physical variations from the sample and process such as temperature and humidity as these will have effects on the spectra. It is also important to use appropriate spectral regions that show absorbance changes related to changes in component concentration. Appropriate spectral pre-processing such as smoothing or derivitization can, for example, remove unwanted spectral variation such as shifting or sloping spectral baselines. By using the correct standards, regions and spectral treatments in the calibration model, the model will have the necessary robustness to produce accurate predicted results on 'unknown' samples.

NIR calibration models must also be properly validated to verify the model prediction accuracy prior to the model being put into use. Standards with component concentrations that cover the entire range of the quantification calibration model should be used for validation. For identification and qualification calibration models, the validation standards should contain both positive and negative challenge samples of known chemical composition. The negative challenge samples should give a fail result, indicating that the compound could not be identified as a compound in the calibration model. The positive challenge samples should be conclusively identified or qualified as a compound or class in the calibration model and given a pass result.

The pharmaceutical industry has taken a major role in adopting and implementing FT-NIR spectroscopy for several key reasons: NIR is chemically specific, non-destructive, has a high penetration depth for experiments such as tablet transmission, is easily validated and can go through fiber optics for online analyses. Common applications for NIR analysis in pharmaceutical lab and line operations include moisture analysis in fluid bed dryers, blend homogeneity, raw material identification (RMID) and active pharmaceutical ingredient (API) quantification in hot-melt extrusion. The United States [5] and European Pharmacopoeias [6] have written chapters on NIR analysis, <1119> and 2.2.40, respectively, making it generally accepted for pharmaceutical analysis.

Content uniformity in tablets is one of the most common applications for NIR and will serve as an instructive example for the use of NIR spectroscopy in pharmaceutical analytics. Tablets are highly compacted mixtures of powders that can measure 10 mm thickness or more, making their interrogation by optical techniques challenging. Due to the low absorbances of light in the NIR as discussed previously, transmission analysis can provide a quantitative value for API in tablets with more accuracy than a reflection measurement.

In building a calibration curve for NIR content uniformity analysis, the goal is to predict the API concentration in unknown tablets accurately and precisely. Calibration curves are made with spectra from tablets with known concentrations spanning the desired prediction range (Figure 16.4). The challenge for content uniformity predictions is that %API in

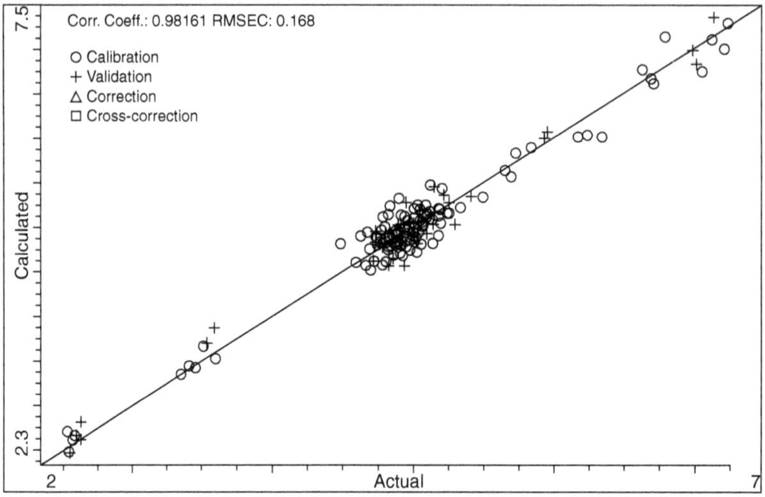

Figure 16.4 *Tablet calibration curve.*

commercial product, by definition, should always be equal to approximately 100% label claim. In order to build a calibration for this type of analysis, multiple API dosage strengths must be created synthetically to extend the calibration curve.

To further this point, if one is using USP Chapter 905 on content uniformity, the calibration curve would have to include values allowing for quantification at levels between 75% and 125% label claim. In practice, this range should be extended further by adding standards with API concentrations that are 10% higher than the known analytical range as well as standards that are 10% lower than the desired range, resulting in a calibrated range that exceeds the desired analytical range. This is good calibration practice, minimizing error in the calibration especially at the extremes of the desired analytical range. This means that synthetic samples ranging from approximately 68% to 137% label claim (75–125% ±10%) are required for NIR analysis adhering to USP <905>. In the current experiment, low-dose tablets (< 2.0% w/w API, ~5 mg) were analyzed using transmission analysis in the NIR. Synthetic calibration samples were created to span the range from 50% to 150% label claim and production samples were culled to be included in the validation set. The primary method in this case was an already-validated HPLC determination.

The NIR calibration included a set of 149 calibration standards, some synthetic and some from production batches. Also included were 49 validation standards spanning the same concentration range as the calibration standards. The API ranged from ~50% to ~150% label claim (from approximately 2.5 mg active to 7.2 mg active). Label claim in this case was approximately 5.0 mg active, making this a low-dose tablet at approximately 2% w/w API for a 250 mg uncoated tablet. The most common multivariate algorithm for calibrating NIR spectral data is PLS as it has been shown to provide accurate, precise results with spectral peaks that are naturally broad and overlapped. The data were pretreated as a second derivative, a common procedure employed to sharpen broad NIR peaks without altering their innate characteristics. A Norris smoothing filter was applied after the derivative was introduced. The spectral region was approximately 8650–8880 cm^{-1}.

There are multiple ways to diagnose regression accuracy. The most critical gauge of performance is the calibration curve, a plot showing predicted concentrations using NIR spectra and regression on the *y* axis against a set of accepted API concentration values on the *x* axis. The NIR spectra in this case are calibrated against data from high-pressure liquid chromatography (HPLC), a well-established separations technique in pharmaceutical science. Once the regression is completed, API concentrations of unknown material can be rapidly predicted (~10 seconds per measurement) simply by taking a NIR spectrum. As discussed above, this makes NIR a so-called secondary technique as its capacity for calibration is pegged to values determined by another technique. The residual, a relative of the calibration curve, allows the user to understand the percent deviation (positive or negative) of the predicted values from the known values (Figure 16.5). Residuals are meant to provide a quick percent error snapshot for a linear regression by taking the NIR-predicted API values and dividing them by the HPLC value at every data point. Multiplying by 100 gives the percentage accuracy for each data point. Typical residuals start with high percent residuals closer to zero concentration and become smaller as the concentrations become larger due to increasing signal versus a finite, constant spectral noise floor.

Calibrations and residuals, along with an independent validation set, provide an unvarnished look at the performance of a NIR method. Other performance measures in regression analyses include: root mean square error of calibration (RMSEC, to see how close all the calibration standards are to the theoretical regression line), root mean square error of prediction (RMSEP, how close the validation standards are to the same line), root mean square error of cross-validation (RMSECV, where one calibration standard is removed from the calibration and then predicted against the regression that remains, a common method for

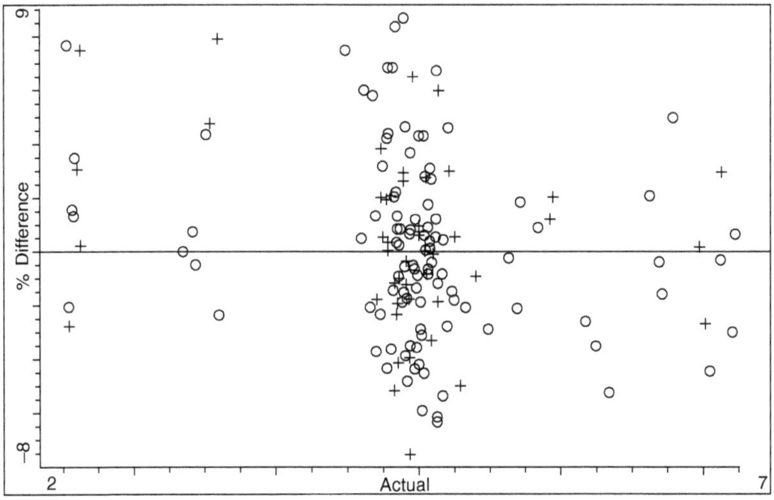

Figure 16.5 *Calibration residual.*

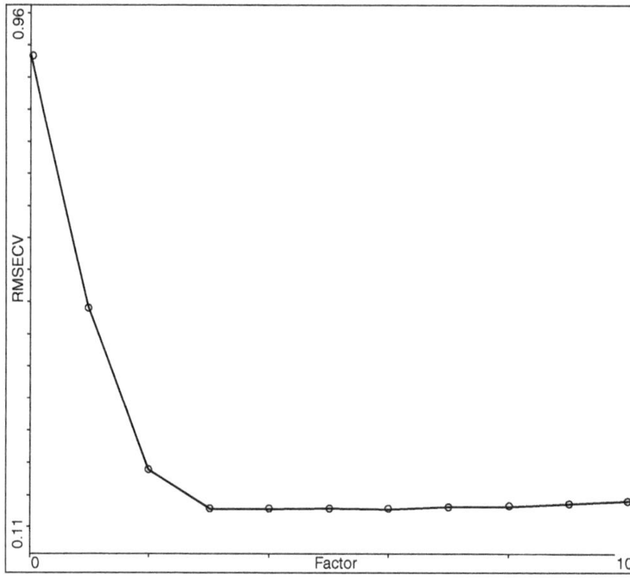

Figure 16.6 PRESS plot.

determining robustness) and predicted error sum of squares (PRESS). The RMSECV plotted versus the number of principal component factors should show a clear global minimum, allowing the user to select the appropriate number of factors for the analysis (Figure 16.6).

The correlation coefficient for the tablet transmission example application is 0.9816, which shows strong correlation between the HPLC measurements and the NIR predictions. In addition, a RMSEC and RMSEP of 0.168 and 0.149, respectively, also support a high degree of correlation in this method. For this method the RMSEC, expressed as a percentage of label claim, is 3.3% and the RMSEP is 2.9%. The shape of the PRESS plot (Figure 16.6) in this case is indicative of a reasonable correlation with the HPLC data. The number of factors chosen in this case was 4 as there is no noise factored into the principal component spectra up to the 4th factor. There were also no spectral outliers found in this calculation.

Method repeatability was determined in this experiment by analyzing the same tablet six times without changing its presentation to the instrument. Altering tablet presentation, especially with the presence of embossing or stamping, can alter the NIR predictions. This issue must be addressed either via sampling (present the sample to the instrument the same way every time) or by factoring this variability into the method by including calibration standards for all different positions. The former will result in a more robust, precise method whereas the latter allows operators to orient the tablet in any fashion. The repeatability test was run on five separate tablets and resulted in an RSD of $<1.0\%$.

16.3 Near-infrared Probes and Fiber Optics

NIR is a unique analytical technique since the analysis can be performed with the instrument remote from the sample point. This remote analysis capability is made possible through the use of fiber-optic cables to connect the probe, installed in the process, to the NIR instrument. Probes combined with fiber optics are commonly used for analyzing materials directly in tanks, pipes, blenders, dryers and extruders. Near-infrared probe design allows for flexibility in sampling for diffuse reflection, transmission or transflectance analysis. Probes can have different lengths, diameters and attachment mechanisms to ensure that they can be properly inserted and removed from the process. The probe design must meet requirements for spectroscopic performance and robustness for use in a production environment. To meet these requirements, a large percentage of instrument source energy must be transmitted from the instrument to the sample and back to the instrument detector. This ensures that a high-quality spectrum with good signal-to-noise is attained for the in-process measurement.

The tip of the probe is the location where the source energy interacts with the sample to produce the NIR spectrum. For analysis directly on a hot-melt extruder, both reflection and transmission probes can be used. The reflection probes are used to analyze extrudates that are opaque or reflective in nature, while the transmission probes analyze extrudates that are transparent. For an accurate analysis, the probe must consistently collect representative spectra off the material in the process by making contact with the material.

The design and materials of construction used for fiber-optic probes must be capable of withstanding production process conditions including highly corrosive environments, high temperatures and pressures. It is not uncommon for a process probe to be rated to 300°C and 5000 psi. Proper location and installation of the probe is critical for NIR analysis of the process. Materials commonly used in pharmaceutical, chemical and food production facilities are often used for fiber optic and probe construction. These materials include stainless steel, sapphire, silicone, Kalrez®, Viton® or Teflon®. Extruder probes will have a sensing window made of sapphire, due to its strength and ability to resist high temperature, pressure and abrasion. The extruder probe body is most often constructed of stainless steel for corrosion resistance and ease of cleaning.

Probes that are used for continuous process control and monitoring are often installed in the production process for long periods of time. The probe attachment mechanism must be designed to allow for removal of the probe for cleaning and inspection. The most common attachment mechanisms use Swagelok® fittings, tri-clamps and bolt-on flanges. Tri-clamp and bolt-on flanges require a flange piece to be welded onto the shaft of the probe that mates up to a flange permanently connected to the process sampling point.

Extruder probes are typically attached by threading the probe tip into a port on the extruder with physical dimensions that exactly match that of the probe tip. The extruder probe tip is typically the standard 1/2-20 UNF Dynisco thread design commonly used for extruder temperature and pressure sensors. For proper connection to the extruder and interaction of the probe with the sample, the probe must have the proper tip diameter, pitch, thread type and depth of insertion. Figure 16.7 shows an example of a diffuse reflection 1/2-20 UNF Dynisco extruder probe.

The extruder probe must make contact with the moving extrudate in order to collect the most representative and highest-quality spectra. The design of the probe and extruder port must ensure that the probe window is being continuously cleaned by the moving extrudate. For transmission analysis, a pair of probes is installed on the extruder with the probes located 180° directly across from one another. The NIR source light travels from one probe through the transparent extrudate material to a collecting probe located directly across from the source probe. The design of the extruder port must guarantee that

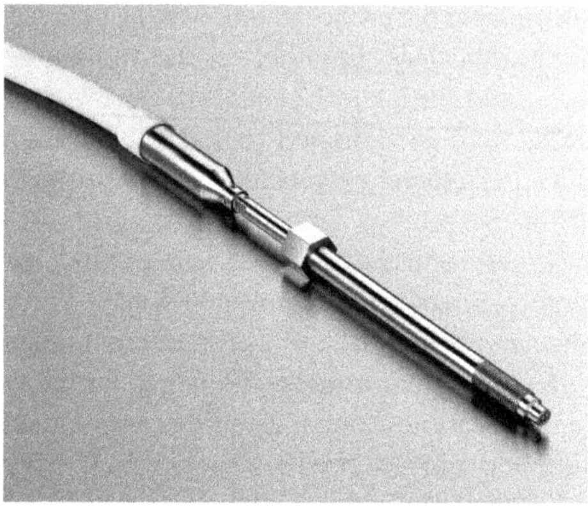

Figure 16.7 Diffuse reflection 1/2-20 UNF Dynisco extruder probe (courtesy of Precision Sensing Devices).

the pair of probes is exactly 180° across from one another and that they are separated by exactly the same distance each time they are installed. The light travelling through the extrudate will be absorbed according to the amount of material present. Changes in the thickness of the extrudate between the pair of probes will change the magnitude of the peaks in the absorbance spectra. For this reason, it is critical to maintain consistent thickness or pathlength between the probe tips in order to develop and implement accurate NIR calibration models for transmission analysis. For reflection analysis, a probe threaded into an extruder port analyzes opaque or semi-opaque extrudate. The source light interacts with the extrudate and the light that is diffusely reflected by the extrudate is collected by the probe. It is critical that the probe is fully threaded into the port, makes consistent contact with the extrudate and that the material is sufficiently reflective for accurate NIR analysis. The probe(s) can be located on any zone of the extruder with a suitable threaded port. The most important port is the final port prior to the extrudate exiting the extruder. This port will give spectra that best represent the physical and chemical make-up of the extrudate, since all the heating and mixing will have occurred prior to this point.

The specially designed extruder probe shown in Figure 16.7 has an external metal shaft that can twist independent of the internal fiber-optic probe shaft. This design allows removal of the probe with the fiber bundle still attached to the probe, thus eliminating the issue of twisting and torque forces being applied to the fiber bundle during removal. Removing the probe while the extruder material is hot and liquid will aid in the cleaning of any excess material that has become attached to the probe threads or body. Removal of the probe from a cold extruder can be difficult and result in damage to the probe.

The optical fibers used in NIR fiber-optic cables are made of silica glass. These fibers serve to transmit NIR energy to the sampling point and back after the source energy has interacted with the sample. Standard fiber optics used for NIR analysis are referred to as 'ultra-low OH' because they have a hydroxyl (OH) content less than 1 ppm. The ultra-low OH fibers transmit light across the entire NIR wavelength range from $4000\,cm^{-1}$ to $12{,}000\,cm^{-1}$. Optical fibers come in multiple diameters, with the NIR application dictating which diameter of fiber to use. The fiber optics contain a silica glass-fiber core that is commonly 200 μm in diameter for reflection analysis and 600 μm in diameter for transmission analysis. For reflection analysis, multiple smaller-diameter fibers are arranged into bundles giving a larger sampling area compared to a single fiber. Larger-diameter fibers used for transmission analysis allow light to be transmitted for hundreds of meters when analysis remote from the NIR spectrometer is required.

The fiber-optic cables for both single and bundled fibers will have jacketing to give the fiber-optic assembly both chemical and physical protection. Silica fibers are very fragile and susceptible to being damaged by twisting, bending or crushing forces. For this reason, the jacket is constructed of protective materials such as PVC/Kevlar furcation and stainless steel sheath or mesh with polymer cover.

The most common connections of fiber-optic cables to the NIR spectrometer use SMA (SubMiniature version A) type connectors. For certain probe types, the connection of the fiber-optic cables to the probe is also by SMA connectors. SMA connectors are screw-type connectors that allow for quick and easy connection of fibers to optical hardware such as a NIR spectrometer. The SMA connectors mechanically couple and align the fiber core(s) to the NIR spectrometer in order to minimize the light loss at the connection.

Considerations must be taken for proper routing of fiber-optic cables between the NIR spectrometer and the sampling port to avoid damage to the fiber-optic cables. It is common in production environments to have fiber-optic cables installed in metal conduit for permanent protection of the fibers. The fiber-optic cables are also susceptible to damage at the connection interfaces to the NIR spectrometer and fiber-optic probe. For this reason, strain relief mechanisms are often installed at the fiber-optic connection interface to the NIR spectrometer.

NIR is an ideal technique for remote process analysis since fiber-optic cables can carry the NIR energy between the NIR spectrometer and probe installed on the extruder. This allows the NIR spectrometer to be located in a separate room or area away from the extruder. Data collection and control software for the NIR spectrometer operates on a computer which can be near or remote to the NIR spectrometer. Depending on the environment where the extruder is installed, the probe and fiber-optic cables might be the only equipment in the production environment. In that case, the NIR spectrometer and controlling computer would be installed in a separate 'safe' and often climate-controlled area. In an R&D or pilot plant installation, the NIR spectrometer and controlling computer are often located near the extruder allowing the user easy access for making software and hardware changes.

The NIR controlling software often will have external data communication protocols for sending data and information to and from external process automation hardware and software. These data can be used for process monitoring and trending and control strategies including real-time feedback control of the hot-melt extruder. Commonly used process communication protocols include OLE (object linking and embedding) for process control (OPC), bus technologies, analog and digital I/O, as well as options to report data to text files to be accessed by Laboratory Information Management System (LIMS) or automation software packages.

16.4 NIR for Monitoring the Start-up of a HME Process

Changes in NIR spectra over time are directly linked to changes in the chemical and sometimes physical make-up of the extrudate. Historically, extruder processing time has been used to determine when the extruder process has reached a steady state. The extrudate

Figure 16.8 1/2-20 UNF Dynisco diffuse reflection probe installed on the last port of a HME. For a better understanding of the figure, please refer to the color section.

at the beginning of a production run would be discarded until a certain time had elapsed. Often extrudate of good quality was discarded because the time at which consistent product was exiting the extruder was unknown.

An application study was performed on the Thermo Scientific Pharma 16 HME using a mixture of theophylline, polyethylene oxide and lactose to demonstrate how NIR can determine when a hot-melt extruder has reached a consistent operating state. The Thermo Scientific Antaris FT-NIR with a 1/2-20 UNF Dynisco diffuse reflection extruder probe (Figure 16.8), installed at the last port on the extruder, was used to collect the standard spectra. Spectra were collected with varying amounts of theophylline in the mixture to develop a calibration model. Theophylline was used as the model drug API in 0%, 5%, 10%, 15% and 20% concentrations. The polyethylene oxide was used as the polymeric matrix. The lactose was used as an excipient in the mixture. The mixtures were fed using a single-screw feeder into the Pharma 16 HME at a feed rate of 500 g/h. The material was extruded at a screw speed of 100 rpm with a constant temperature of 120°C at the output die. The extruder screw profile contained two kneading sections for melting and mixing of the mixtures.

A PLS calibration model for theophylline % was developed from several spectra collected at each model drug API%. The calibration model results can be seen in the predicted NIR versus actual theophylline % plot (Figure 16.9). With only 2 PLS factors, a calibration model was developed with a high correlation (correlation coefficient 0.994) between the predicted and actual theophylline %.

The calibration model for theophylline % was deployed in the Thermo Scientific RESULT software package for real-time prediction and trending of the start-up of the HME process. The RESULT™ software package was used to acquire spectra, predict theophylline % and

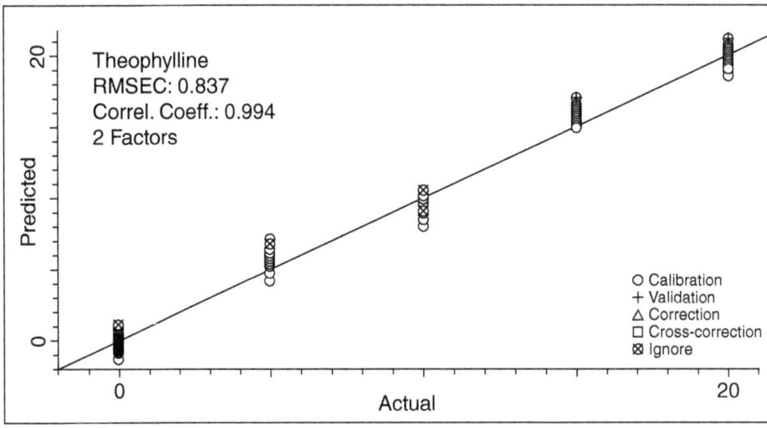

Figure 16.9 *Calibration plot of theophylline % in extrudate.*

automatically export the results for trend charting. The trend of the FT-NIR predicted values for a mixture of 15% theophylline is shown in Figure 16.10. This trend chart demonstrates how FT-NIR can determine when the extrusion process reaches a consistent operating state. As shown in Figure 16.10, the extrusion process stabilizes at the FT-NIR reading number 14. It can also be seen that the predictions are all well within the error of the calibration model, as shown by the relatively small spread in theophylline predicted values compared to the magnitude of the RMSEC of 0.837.

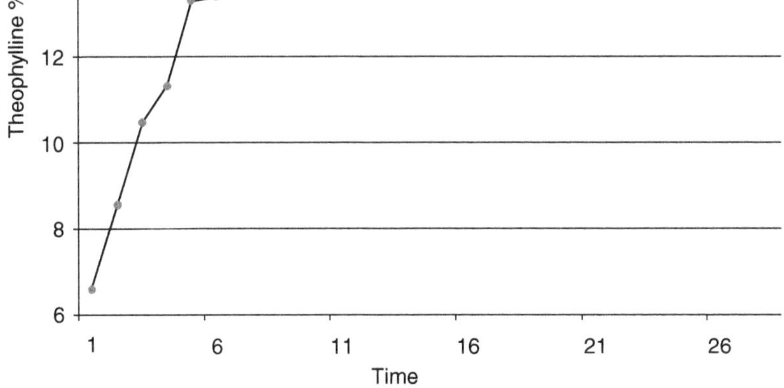

Figure 16.10 *FT-NIR-determined stabilization of a hot-melt extrusion process after start-up.*

16.5 NIR for Improved Process Understanding and Control

Traditionally, the optimization of the hot-melt extrusion process for a specific drug formulation required experimentation and validation on multiple extruder production runs. Chemical analysis of the extrudate required post production and off-line reference methods to verify the extruder was generating product of consistent quality. The optimization process led to wasted product and a significant investment in resources. An application study was performed on the Pharma 16 HME to demonstrate how NIR can be used to optimize and gain understanding of the hot-melt extrusion process. This study showed how NIR can determine and quantify the influence of processing parameters, such as extruder screw speed and feed rate, on the blend uniformity of the extrudate. Other extruder processing parameters that can be studied by NIR include extruder barrel temperature and screw configuration. NIR is commonly used as a process analytical technology (PAT) tool for monitoring blend uniformity of solid dosage forms in the pharmaceutical industry. The ability of NIR to monitor blend uniformity of extrudate in-line and in real-time demonstrates how NIR can be used as a PAT tool for both development of new processes and control of existing processes.

The drug formulation used for this study was a mixture of ibuprofen, Kollidon® and lactose. The ibuprofen served as the drug load API. The Kollidon® was the polymeric melt matrix. Lactose was in the formulation as an excipient. A reflection fiber-optic probe, installed near the exit of the extruder, was used to collect spectra on the Antaris™ FT-NIR. The extruder was operated at a consistent flow rate of 1 kg/h and screw speed of 200 rpm for the entire study.

To learn how the changes in concentration and process conditions are represented as changes in the NIR spectra, PCA was used. Principal components (PC) are orthogonal vectors that explain the spectral variation in a set of sample spectra. This spectral variation can be due to both chemical and physical differences in the sample spectra. The first PC explains the most spectral variation and each additional PC explains the residual spectral variation. The PCA of a sample set is displayed as principal component scores plots. It is common to display a PC scores plot as an x–y plot of the score values for two of the principal components. The PC scores plot is an integral tool for developing and optimizing near-infrared calibration models. The PC scores plot can be used to detect outlier samples, choose calibration and validation samples and, most importantly, detect similarities, differences or patterns in a sample set. PC scores plots are used to gain insight into what is causing the spectral variation in a set of sample spectra. This spectral variation is often difficult to detect in the raw unprocessed absorbance spectra. By studying the PC scores plots, it is often possible to determine that a PC is highly correlated to changes in a chemical component or a change in a physical parameter such as temperature or density.

A PC scores plot was developed from spectra collected while the extruder operated with constant process parameters and only the ibuprofen concentration was varied (Figure 16.11). The PC scores plot clearly shows that the first PC describes spectral variation caused by the changes in ibuprofen concentration. This is a clear indication that the set of NIR spectra are suitable for developing a quantitative calibration model to predict ibuprofen %. Ellipses were drawn around points in the PC scores plot, corresponding to specific production runs of fixed ibuprofen %. It can be seen that the size of the ellipses are very similar, demonstrating that there is very little variation not attributed to the ibuprofen concentration differences. If

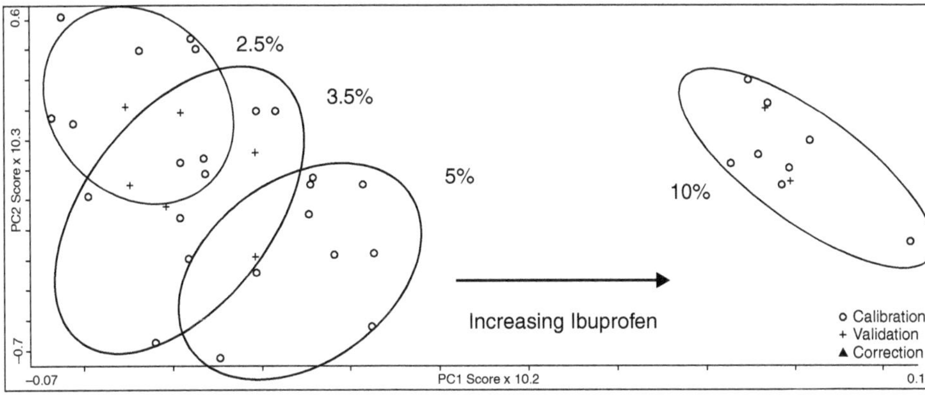

Figure 16.11 *Principal component scores plot showing the ability of FT-NIR to detect changes in ibuprofen concentration.*

spectral variation sources other than changes in ibuprofen % were being explained by PC1 and PC2 shown in Figure 16.11, the ellipses would expand in size.

A partial least squares (PLS) calibration model was developed and optimized for ibuprofen in the range 2.5–10%. The calibration model was used to predict ibuprofen % in the extrudate for production runs with differing process parameters. The changes in the predicted ibuprofen % for these production runs demonstrated the effect of changing process parameters on the blend uniformity.

Prior to developing the calibration model, the spectra were processed with a second derivative to enhance the peak shape and spectral separation of the calibration standards. Even though no unique peak was found for the ibuprofen, three spectral regions were identified that showed spectral variation caused by changes in the ibuprofen concentration. The calibration model obtained for ibuprofen showed a very good correlation (correlation coefficient 0.992) and a calibration error of 0.4%.

A design of experiments was constructed to evaluate the influence of changing the screw speed and the feed rate on the blend uniformity of the extrudate. The design of experiments consisted of a 3^2 factorial design where screw speed and feed rate were varied. The hot-melt extruder was operated at a consistent output die temperature of 160°C during the study. The screw speed varied between 200, 400 and 600 rpm. The extruder hopper feed rate varied between 1, 1.5 and 2 kg/hr. A consistent drug formulation of 5% ibuprofen and 95% Kollidon®/lactose (95/5 g ratio) was used for the experiments. The Antaris FT-NIR was used to collect 2 spectra, containing 32 co-added scans, every minute. Approximately 30 NIR predicted values for ibuprofen were produced during each experiment. For each individual experiment, approximately 15 minutes of extrudate was produced with the first 10 spectral readings being discarded to account for start-up instability of the extruder. Once the predictions settled to a stable value, data were compiled and used to calculate a relative standard deviation (RSD) value for each experimental condition.

Principal Component Analysis (PCA) was used to investigate whether NIR could sense changes in extruder screw speed and hopper feed rate. The PC scores plot generated from the spectra collected under various process conditions is shown in Figure 16.12. The major

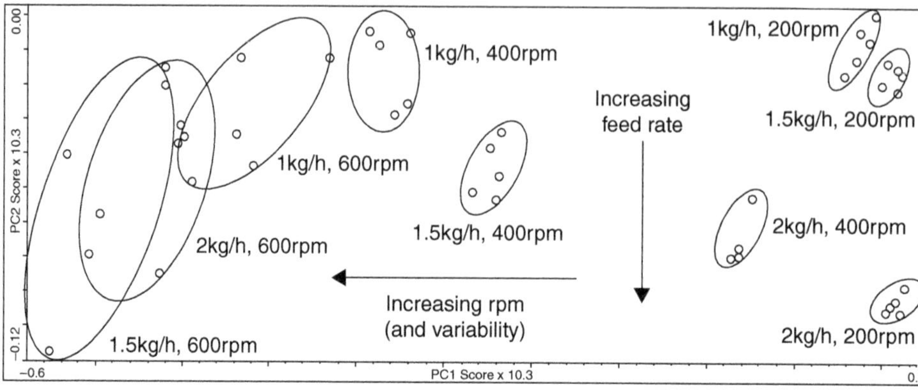

Figure 16.12 *Principal component scores plot showing the ability of FT-NIR to detect changes in screw speed and feed rate.*

source of variation in the NIR spectra was the change in screw speed (rpm) as shown along PC1. In addition, the ellipses are smaller for lower rpm sample sets indicating that higher screw speeds introduce more variability into the extrudate formulation. The second PC score shows there is smaller variation in the spectra due to feed rate than screw speed since the second PC explains less variation than the first PC. The principal component data clearly show that NIR can detect changes in hot-melt extrusion processing conditions.

As stated earlier, the RSD was calculated once the NIR-predicted values for ibuprofen stabilized for each experiment. A plot of the RSD values with changes in screw speed is shown in Figure 16.13. The RSD data showed that the 200 rpm and 1.5 kg/h experiments produced the least variability in predicted ibuprofen compared to the other screw speeds

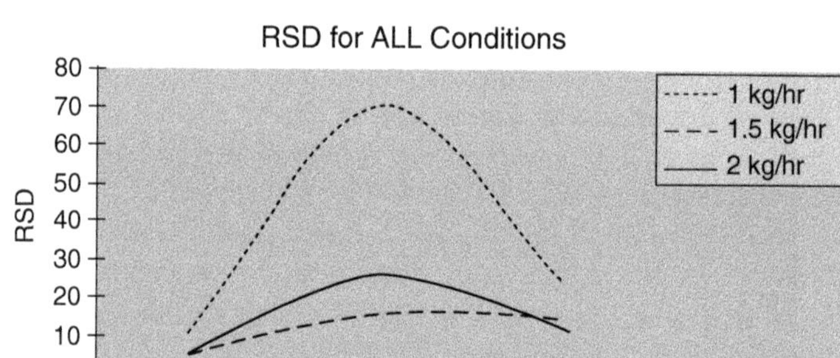

	200 rpm	400 rpm	600 rpm	
...... 1 kg/hr	10.9	69.7	22.5	
– – – 1.5 kg/hr	5	15.7	14.3	
—— 2 kg/hr	5.3	25.9	11.6	

RPM

Figure 16.13 *Relative standard deviation for predicted ibuprofen % in extrudate for different screw speeds and feed rates.*

and feed rates. A purely quadratic relationship can be seen for the influence of screw speed on the blend uniformity. In general, the 200 rpm and 1.5 kg/h series showed the best repeatability, thus demonstrating the best conditions for processing this particular formulation. Looking back at the PC scores plot in Figure 16.12, the conditions that produced small RSDs generally have smaller ellipses. Smaller ellipses in the PC scores plot correspond to smaller NIR spectral variation in the sample set. The correlation between small ellipses and small RSDs reinforces the previous observation that NIR can determine the conditions that produce the least variability for a particular formulation.

The results from this design of experiments demonstrate that the near-infrared technique can be used for optimizing extruder processing conditions without referring to reference methods. By using NIR to replace the reference method, significant time and expense can be saved during optimization of the extruder processing conditions. For example, processing conditions could be chosen with regard to the least sample-to-sample spectral variability. This is an example of how NIR can be used to design quality into the extruder process in the quality by design (QbD) framework to produce consistent extruder product. These studies have shown how NIR can be used as a process analytical technology (PAT) tool for real-time in-process analysis for monitoring and controlling the hot-melt extrusion process.

References

(1) Hanna, M.W. (1981) *Quantum Mechanics in Chemistry*, 3rd ed. Benjamin/Cummings, Menlo Park.
(2) Harris, D.C. and Bertolucci, M.D. (1978) *Symmetry and Spectroscopy: An Introduction to Vibrational and Electronic Spectroscopy*. Dover, New York.
(3) Nakamoto, K. (2009) *Infrared and Raman Spectra of Inorganic and Coordination Compounds Part A: Theory and Applications in Inorganic Chemistry*, 6th ed. Wiley, Hoboken.
(4) Morse, P.M. (1929) Diatomic molecules according to the wave mechanics. II. Vibrational levels. *Physical Review*, **34**, 57–64.
(5) United States Pharmacopeia (2011) USP 34. Vol 2. Rockville, MD: Unitetates Pharmacopeia Convention, 649–654.
(6) European Pharmacopoeia (2005) EMEA 2.2.40. 5 Ed. Strasbourg, Council of Europe, 59–63.

Index

Hot-melt Extrusion: Pharmaceutical Applications, First Edition. Edited by Dennis Douroumis.

www.ingramcontent.com/pod-product-compliance
Ingram Content Group UK Ltd.
Pitfield, Milton Keynes, MK11 3LW, UK
UKHW051035091026
15624UKWH00011B/435

9 780470 71118